Science, Technology and Medicine in Modern Histo

General Editor: **John V. Pickstone**, Centre f
Medicine, University of Manchester, UK (www

One purpose of historical writing is to illumina
nium, science, technology and medicine are e
is little studied.

The reasons for this failure are as obvious as they are regrettable. Education in many countries, not least in Britain, draws deep divisions between the sciences and the humanities. Men and women who have been trained in science have too often been trained away from history, or from any sustained reflection on how societies work. Those educated in historical or social studies have usually learned so little of science that they remain thereafter suspicious, overawed, or both.

Such a diagnosis is by no means novel, nor is it particularly original to suggest that good historical studies of science may be peculiarly important for understanding our present. Indeed this series could be seen as extending research undertaken over the last half-century. But much of that work has treated science, technology and medicine separately; this series aims to draw them together, partly because the three activities have become ever-more intertwined. This breadth of focus and the stress on the relationships of knowledge and practice are particularly appropriate in a series concentrates on modern history and on industrial societies. Furthermore, while much of the existing historical scholarship is on American topics, this series aims to be international, encouraging studies on European material. The intention is to present science, technology and medicine as aspects of modern culture, analysing their economic, social and political aspects, but not neglecting the expert content which tends to distance them from other aspects of history. The books will investigate the uses and consequences of technical knowledge, and how it was shaped within particular economic, social and political structures.

Such analyses should contribute to discussions of present dilemmas and to assessments of policy. 'Science' no longer appears to us as a triumphant agent of Enlightenment, breaking the shackles of tradition, enabling command over nature. But neither is it to be seen as merely oppressive and dangerous. Judgement requires information and careful analysis, just as intelligent policy-making requires a community of discourse between men and women trained in technical specialities and those who are not.

This series is intended to supply analysis and to stimulate debate. Opinions will vary between authors; we claim only that the books are based on searching historical study of topics which are important, not least because they cut across conventional academic boundaries. They should appeal not just to historians, nor just to scientists, engineers and doctors, but to all who share the view that science, technology and medicine are far too important to be left out of history.

Titles include:

Julie Anderson, Francis Neary and John V. Pickstone
SURGEONS, MANUFACTURERS AND PATIENTS
A Transatlantic History of Total Hip Replacement

Roberta E. Bivins
ACUPUNCTURE, EXPERTISE AND CROSS-CULTURAL MEDICINE

Linda Bryder
WOMEN'S BODIES AND MEDICAL SCIENCE
An Inquiry into Cervical Cancer

Roger Cooter
SURGERY AND SOCIETY IN PEACE AND WAR
Orthopaedics and the Organization of Modern Medicine, 1880–1948

Catherine Cox, Hilary Marland
MIGRATION, HEALTH AND ETHNICITY IN THE MODERN WORLD

Jean-Paul Gaudillière and Ilana Löwy (*editors*)
THE INVISIBLE INDUSTRIALIST
Manufacture and the Construction of Scientific Knowledge

Jean-Paul Gaudillière and Volker Hess (*editors*)
WAYS OF REGULATING DRUGS IN THE 19TH AND 20TH CENTURIES

Christoph Gradmann and Jonathan Simon (*editors*)
EVALUATING AND STANDARDIZING THERAPEUTIC AGENTS, 1890–1950

Aya Homei and Michael Worboys
FUNGAL DISEASE IN BRITAIN AND THE US 1850–2000
Mycoses and Modernity

Sarah G. Mars
THE POLITICS OF ADDICTION
Medical Conflict and Drug Dependence in England since the 1960s

Alex Mold and Virginia Berridge
VOLUNTARY ACTION AND ILLEGAL DRUGS
Health and Society in Britain since the 1960s

Ayesha Nathoo
HEARTS EXPOSED
Transplants and the Media in 1960s Britain

Neil Pemberton and Michael Worboys
MAD DOGS AND ENGLISHMEN (hardback 2007)
Rabies in Britain, 1830–2000

Neil Pemberton and Michael Worboys
RABIES IN BRITAIN (paperback 2012)
Dogs, Disease and Culture, 1830–1900

Cay-Rüdiger Prüll, Andreas-Holger Maehle and Robert Francis Halliwell
A SHORT HISTORY OF THE DRUG RECEPTOR CONCEPT

Thomas Schlich
SURGERY, SCIENCE AND INDUSTRY
A Revolution in Fracture Care, 1950s–1990s

Eve Seguin (editor)
INFECTIOUS PROCESSES
Knowledge, Discourse and the Politics of Prions

Crosbie Smith and Jon Agar (editors)
MAKING SPACE FOR SCIENCE
Territorial Themes in the Shaping of Knowledge

Stephanie J. Snow
OPERATIONS WITHOUT PAIN
The Practice and Science of Anaesthesia in Victorian Britain

Carsten Timmermann
A HISTORY OF LUNG CANCER
The Recalcitrant Disease

Carsten Timmermann and Julie Anderson (editors)
DEVICES AND DESIGNS
Medical Technologies in Historical Perspective

Carsten Timmermann and Elizabeth Toon (editors)
CANCER PATIENTS, CANCER PATHWAYS
Historical and Sociological Perspectives

Jonathan Toms
MENTAL HYGIENE AND PSYCHIATRY IN MODERN BRITAIN

Duncan Wilson
TISSUE CULTURE IN SCIENCE AND SOCIETY
The Public Life of a Biological Technique in Twentieth Century Britain

---

Science, Technology and Medicine in Modern History
Series Standing Order ISBN 978–0–333–71492–8   hardcover
Series Standing Order ISBN 978–0–333–80340–0   paperback
(outside North America only)

You can receive future titles in this series as they are published by placing a standing order. Please contact your bookseller or, in case of difficulty, write to us at the address below with your name and address, the title of the series and one of the ISBNs quoted above.

Customer Services Department, Macmillan Distribution Ltd, Houndmills, Basingstoke, Hampshire RG21 6XS, England

---

Eugenics and Nation in Early 20th Century Hungary

Marius Turda
Reader, Oxford Brookes University, UK

To Bradley
with best wishes.
Marius
Oxford
17th Oct 2016

palgrave
macmillan

First published 2014 by
PALGRAVE MACMILLAN

Palgrave Macmillan in the UK is an imprint of Macmillan Publishers Limited,
registered in England, company number 785998, of Houndmills, Basingstoke,
Hampshire RG21 6XS.

Palgrave Macmillan in the US is a division of St Martin's Press LLC,
175 Fifth Avenue, New York, NY 10010.

Palgrave Macmillan is the global academic imprint of the above companies
and has companies and representatives throughout the world.

Palgrave® and Macmillan® are registered trademarks in the United States,
the United Kingdom, Europe and other countries.

ISBN 978-1-349-45121-0   ISBN 978-1-137-29353-4 (eBook)
DOI 10.1057/9781137293534

This book is printed on paper suitable for recycling and made from fully
managed and sustained forest sources. Logging, pulping and manufacturing
processes are expected to conform to the environmental regulations of the
country of origin.

A catalogue record for this book is available from the British Library.

A catalog record for this book is available from the Library of Congress.

Typeset by MPS Limited, Chennai, India.

For Ariadne

Contents

List of Figures and Tables

Figures

Tables

Acknowledgements

Now that this book is finally completed, it is a pleasure to thank two institutions that supported this research from the very beginning: the Wellcome Trust in London and Oxford Brookes University. I have received generous funding through the Wellcome Trust's Strategic Award (Grant no. 082808), for which I am grateful. For the past ten years, I was fortunate to work in the Centre for Health, Medicine and Society at Oxford Brookes University. I want to express my gratitude to former and current members of the Centre for their collegial support and for providing a stimulating academic environment.

Many long conversations with Tudor Georgescu have helped to clarify my arguments, and he also deserves my recognition for his loyal friendship. As I became more and more enmeshed in my own line of reasoning, I needed a fresh and objective perspective on many obscure arguments that I imperfectly thought were comprehensible to the reader. Matthew Feldman stepped in and offered to help. I benefitted greatly from his remarkably attentive suggestions and comments. Anne Digby, Maria Sophia Quine and Stephen Byrne deserve special thanks for calling to my attention some matters of style and substance. Donal Lowry and Elizabeth Hurren graciously read chapters of this manuscript in its last stages and offered valuable advice both in terms of content and style. I am grateful to both of them.

I am also fortunate to have wonderful and supportive friends, whose generosity was always unsurpassed. I want to sincerely thank some of them: Răzvan Pârâianu, Paul J. Weindling, Daniela Sechel, Stefano Bottoni, Gábor Palló, Eric Weaver, Francesco Cassata, Benedek Varga, Christian Promitzer, Herwig Czech, Margit Berner, Maria Teschler-Nicola, Tibor Szász, Gábor Komjáthy, Chris Davis, Thomas A. Loughlin and Bernadette Baumgartner. Barnabás Kalina graciously read every chapter of the book and offered ways to improve it. When it became impossible for me to travel to Budapest, Barnabás generously offered to retrieve missing articles and books for me. I thank him wholeheartedly.

Very special thanks are due to Tania Mühlberger and Eric Christianson for reading through the entire manuscript so carefully. They pushed towards greater clarity and this is certainly a better book as a result of their comments and suggestions.

Many archivists and librarians have assisted me in my research. In particular, I want to show my appreciation and gratitude to the staff at the Semmelweis Museum, Library and Archives of the History of Medicine, the Library of Parliament, Széchényi National Library, the Hungarian National Museum and the Library of Hungarian Central Statistical Office, Budapest; the National Archives, Kew Gardens; the Wellcome Library, the British Library and the Archives of the London School of Economics, London; the Bodleian Library, Oxford; the Austrian National Library and the Museum of Natural History, Vienna; the Rockefeller Archive Center, Sleepy Hollow and the Library of the Academy of Medicine, New York.

I also want to express my deepest gratitude to László András Magyar, the director of the Semmelweis Medical History Library in Budapest. His knowledge of Hungarian medical history is impressive. I am also grateful to him and to Benedek Varga for facilitating the permission to reproduce the photos of the Hungarian eugenicists discussed in this book. I also extend special thanks to the late Elaine M. Doak and Amanda Langendoerfer at Truman State University in Kirksville, Missouri for facilitating access to Harry H. Laughlin's papers; to Mary Ann Quinn at the Rockefeller Archive Center for granting me permission to reproduce the photo of the Child Welfare Centre of the Stefánia Association in Gödölő; to Dirk Ullmann at the Archive of the Max-Planck Institute in Berlin for providing access to Adolf von Harnack's personal papers; to Katalin Czellár, who graciously provided me with photos of her maternal uncle, Lajos Dienes; to Ákos Lencsés for supplying the photo of the Central Statistical Office; and, last but not least, to Pál Macskásy (Hoffmann), for occasioning one of the greatest moments during the research for this book, photos of Géza Hoffmann, Mr Macskásy's grandfather. None of the above-mentioned, however, should be held responsible for any of the opinions expressed in this book, which are entirely those of the author.

Finally, it is to my wife Aliki Georgakopoulou that I owe my ultimate gratitude. She gave me the most beautiful gift I have ever received – our daughter Ariadne, to whom this book is dedicated.

Note on place names. As this book focuses on the period between 1900 and 1919, all places are given first the Hungarian name in use at the time followed by its usage in German, Romanian and so on.

Prologue

In one of the many letters to his niece Milly, Francis Galton – the English scientist who pioneered modern research on heredity and eugenics – muses about a green woodpecker visiting the garden and the "uncommonly attractive" neighbours' daughter before making an important, if brief, remark about a request he had received a day before, on 1 December 1907.[1] It was made by "a man with a much more horrid name, which I can't venture to reproduce from memory", Galton remarked unsympathetically. The correspondent was modestly requesting Galton's "permission to translate my recent 'Herbert Spencer Lecture' into *Hungarian*, for his *Sociological Review*, of which he enclosed a prospectus".[2]

The gentleman whose name Galton could not remember was none other than Oszkár Jászi, the progressive Hungarian sociologist. Jászi had written to Galton on 26 November 1907, praising his "endeavours for propagating the new science [of eugenics]". Galton was, Jászi assured him, "already well-known to the public of this review and your powerful essay will surely awake a still greater interest in the Hungarian readers".[3] Galton agreed to have his lecture translated and published in Hungarian. "They do these things well in Buda-Pest", he conceded to Milly.

Jászi's letter to Galton serves to introduce the subject of this book: the history of eugenics in early twentieth-century Hungary. Effectively, it illustrates not only the remarkable level of communication between scientists across borders, cultures and languages at the time, but more importantly the widespread circulation of Galton's ideas of eugenics during the first decade of the twentieth century. The vision of social and biological improvement associated with eugenics became central to various programmes of social reform and national progress elaborated by Hungarian intellectuals, scientists and politicians after 1900.

For too long, however, eugenics has been ignored by historical scholarship on Hungary, paralleling another historiographic neglect, that of Hungary in the scholarship on international eugenics. This book should hence be read as a contribution to both historiographic traditions, generating – it is to be hoped – a meaningful dialogue between modern European history and the history of eugenics.

Introduction

In 1900, Hungary appeared to be a confident country. Four years earlier, the Millennium celebrations had paraded Hungarian achievements to the outside world, in fields as diverse as music, the fine arts, ethnography, justice, forestry and public health.[1] The director of the Royal Statistical Office, József Jekelfalussy, enthusiastically described the exhibition organized for the occasion as "a summary of the results of the development of a [nation over a] thousand years".[2] Hungarian officials finally seemed to have succeeded in establishing the myth of national homogeneity, glossing over important differences in language, religion and regional traditions within what was broadly defined as the Kingdom of St Stephen. These officials may have been, according to Lee Congdon, "intoxicated with the heady wine of nationalism",[3] but their confidence was nonetheless largely justified.

During the last decade of the nineteenth century, and the first decade of the twentieth, Hungary and Budapest, in particular, underwent a spectacular transformation, fostered by extensive urbanization and continuous industrialization. This transformation of Hungary's capital, reflecting the country's broader changes, fashioned a new physical environment, one to which the city's social and health reformers actively responded.[4] Increasingly, state authorities – rather than private initiatives – began to mediate the pursuit of national welfare.[5] This social and economic climate also promoted new and vibrant intellectual activity, highlighting Budapest's importance as one of the important centres of modernist Central European culture.[6]

There was also another side to Hungary's increasing modernity, one that historians have neglected so far. In his lecture to the National Association of Public Health (Országos Közegészségi Egyesület) delivered on 19 December 1900 in Budapest, the physician Mór Kende warned of the widespread "degeneration of the human race". According to Kende, the health of the individual and that of the national community were under threat due to a wide range of social and medical problems, as well as forms of physical and mental degeneration.[7] He was not the only Hungarian physician to link degeneration to modernity, especially industrialization and urbanization. An entire section (VII) focused on those with "physical and mental defects" at the 8th International Congress on Hygiene and Demography held in Budapest in 1894.[8] The topic was also discussed at the International Congress on Child Protection, organized in Budapest in 1899.[9] Moreover, the degeneration of the human body was given extensive treatment in the clinical literature on neurology and psychiatry. The prominent physiologist Ernő Jendrassik, for example, proposed the theory of "heredodegeneration" to explain the hereditarian nature of various nervous and muscular diseases, as well as the interrelation between degeneration, gender and biological inheritance.[10]

Artists, social reformers, intellectuals and progressive politicians in Europe and elsewhere increasingly adopted this new vision of degeneration provided by medicine, biology and anthropology. In Hungary, moreover, degeneration – whether social, cultural or biological – was simultaneously viewed as the emblem of modernity and an impious transgressor of traditional national values. In this context, the future of the Hungarian state depended on the protection of the Hungarian nation and race. It is precisely these different responses to the alleged social and biological degeneration brought about by modernity that must be stressed when analysing the emergence of eugenic theories in early twentieth-century Hungary.

To be sure, the sociopolitical emphasis on national regeneration was an important element in forming various ideologies of culture across Europe and the USA at the beginning of the twentieth century.[11] Yet this was also a period when, according to Michel Foucault, "the medical – but also political – project for organizing a state management of marriages, births, and life expectancies" received widespread support.[12] In this view, modernity fundamentally challenged existing interpretations of the human body, bringing them into contact with new cultural, political and epistemological arrangements of state power. Modernity also connected social control with the biological ideals promoted by the nation state.

Research into eugenic ideas of social, biological and national renewal, therefore, consistently reinforced the general significance of modernity as a central site of national identity formation, while at the same time offering a way of overcoming ideological volatility and asymmetrical cultural and political practices. In engaging with these issues, this book endeavours to redress the neglect of various eugenic narratives of national improvement, which have been largely marginalized in historical accounts of Hungarian culture. More detailed work into Hungary's individual and collective eugenic narratives is required in order to appreciate what was ultimately a much wider, indeed European project after 1900: the biological transformation of the modern state. Eugenics was predicated upon the idea of fusing scientific research on biological improvement with social and cultural critiques of modernity. Correspondingly, eugenicists addressed not only abstruse scientific topics related to mechanisms of heredity and evolution, but more general problems perceived to characterize Hungarian society as well.[13] This nascent eugenic ideology thus aimed to offer a totalizing, progressive and rational social vision both on, and for, modern Hungarian society.

The period under examination here is of particular significance in Hungarian history. In 1900 Hungary was a regional power in Europe with imperial pretensions; by 1919 it was reduced to the status of a small Central European country, crippled by profound territorial, social and national transformations. Yet, in the span of these two decades, Hungary experienced unrivalled cultural dominance in Central Europe, with Budapest becoming the impressive metropolis that we know today. Eugenics was an integral part of this dynamic historical transformation, serving as a vehicle for transmitting social and biological messages that transcended the differences between political parties and opposing ideological worldviews. Hungarian eugenicists not only engaged in the same speculative debates concerning heredity and evolution as their counterparts did elsewhere in Europe and the USA, they also conjured up a national interpretation of the application of eugenics to society, one which aimed at solving long-standing social, economic and medical problems specific to Hungarian society.[14]

Methodology

Recovering Hungary's eugenic past is a complicated task. First, one must excavate a large mosaic of hitherto unknown eugenic texts. Second, these texts and their meaning must be understood both historically and conceptually. At the beginning of the twentieth century, eugenics was a collection of disparate social, medical and biological arguments concerning human improvement, which gradually grew into an articulated system of ideas defined in opposition to rival cultural, social and political movements. Consequently, it is imperative to examine how and where eugenics intersected with culture and politics in order to acquire a better understanding of the extensive attraction that eugenics had for Hungarian intellectuals of various sociopolitical and professional orientations. Conceptual frontiers and ideological barriers were often much more fluid than has previously been assumed. With these considerations in mind, this book evaluates and describes the emergence of Hungarian eugenics from a comparative perspective, while simultaneously highlighting its specific national character.

The present approach draws sustenance from comparative and intellectual history as well as from the history of science and the social history of medicine. Eugenicists were part of coexisting cultural environments, public as well as professional, and attention must be given to their points of intersection. By occupying central positions in the scientific community, moreover, eugenicists were able to provide the reading public with the necessary concepts, references and symbols to define their collective attempt at creating a eugenic culture in Hungary. In broader terms, tracing the development of eugenics during the first two decades of the twentieth century illuminates many overlooked moments, which historians have repeatedly edited out of Hungary's national past. In attempting to restore these suppressed historical nuances and conceptual idiosyncrasies, focus will be placed upon different individual trajectories; the general sense of innovation and excitement; as well as the uncertainty, an absence of coherence and personal rivalries that characterized the eugenic movement in early twentieth-century Hungary.

After 1900, eugenics gradually became a dominant scientific language in which health experts, reform-oriented politicians and intellectuals expressed their duties and responsibilities towards the nation and state. Articulating a prospective eugenic programme, Francis Galton summarized it thus in 1904:

> firstly, [eugenics] must be made familiar as an academic question, until its exact importance has been understood and accepted as a fact; secondly, it must be recognised as a subject whose practical development deserves serious consideration; and, thirdly, it must be introduced into the national conscience, like a new religion.[15]

This was a promising attempt at clarification and practical systematization, and one which eugenicists the world over were to embrace enthusiastically in subsequent decades. Significantly, in 1904, the first professorial chair in eugenics was inaugurated at University College

London, followed by the formation of the Eugenics Education Society in 1907. For their part, American eugenicists had established the Station for Experimental Evolution at Cold Spring Harbor in 1904, followed by the Eugenics Record Office in 1910. By then, the Society for Racial Hygiene (Gesellschaft für Rassenhygiene), established by Alfred Ploetz in 1905, was leading the way in disseminating eugenic ideas to the general public, both at home and abroad, as illustrated by the Hygiene Exhibition held in Dresden in 1911. These eugenic societies heralded the search for new forms of social engineering and biological propaganda that eugenicists everywhere were soon to undertake. There was a growing appreciation among cultural and political elites at this time that the nation's physical existence was wedded to its biological future; and eugenicists were the experts to supervise it.

By the time the First International Eugenics Congress convened in London during July 1912, Galton's first commandment – the popularization of eugenics "as an academic question" – had been embraced by more than 400 participants. In fact, as American eugenicists proudly praised their domestic achievements, it appeared that some were already experimenting with practical eugenics. Indiana introduced the first sterilization law in 1907, one targeting "undesirable" individuals, especially those with physical disabilities, the mentally ill and criminals. When the First National Conference on Race Betterment met in Battle Creek, Michigan, in 1914, the consensus among eugenicists held that members of "undesirable groups" should be prohibited from reproducing. Yet the early twentieth-century debate on eugenics was not only about the biological management of the population. Eugenicists also campaigned for the improvement of living conditions, gender equality, social progress and public health reforms; in short, the creation of a modern society and state.

With the turn of the twentieth century, eugenicists everywhere had become increasingly concerned with the social and biological implications of accelerated urbanization – served by large-scale internal migrations from rural to urban areas – and industrialization, which resulted in a deteriorating standard of living and worsening hygienic conditions in working-class social environments. The nation's health was thus viewed and interpreted through the lens of eugenics. Within this context, the individual's alleged biological deterioration became conterminous with a perceived collective degeneration, an imbalance that had to be remedied through appropriate eugenic, social and medical interventions.

There were, however, two other essential features of this emerging eugenic vocabulary. The first encouraged a hereditarily defined sociobiological hierarchy, while the second relied on interventionist state policies to maintain this hierarchy, including social segregation and even sterilization. These two directions led scholars to catalogue eugenic activities as positive and negative, respectively. Galton himself referred to this dual function when he offered his oft-cited definition of eugenics as "the study of agencies under social control that may improve or impair the racial qualities of future generations either physically or mentally".[16] His compatriot Caleb W. Saleeby agreed, defining "positive eugenics, as the encouragement of parenthood on the part of the worthy", and of "negative eugenics as the discouragement of parenthood on the part of the unworthy".[17] Eugenics, meanwhile, was expanding its purview in other ways and in other countries as well. Germany gave priority to ideas of racial improvement very early on, given that eugenicists like Alfred Ploetz and Wilhelm Schallmayer were much admired and emulated at the beginning of the twentieth century. Racial hygiene (*Rassenhygiene*), Ploetz's own term for eugenics, was focused towards increasing the number of racially "superior" individuals, while decreasing – through elimination, if possible – those considered racially undesirable.[18] The biological language used to describe and justify these eugenic projects connected the individual with a larger collective, namely the national community.

Correspondingly, this book will focus on eugenics as understood by its Hungarian supporters. This recourse to the original language and terminology is especially useful when trying to understand how eugenics emerged as a movement concerned concurrently with improving social conditions (education, better living standards, public assistance) and the population's health more generally (alcoholism, infectious and sexually transmitted diseases, differential fertility). Eugenics was in this context a complex constellation of ideas that linked social and health reform to scientific communities and state institutions. This was a process primarily focusing on protecting racial qualities deemed to be superior, while simultaneously introducing preventive measures against dysgenic individuals or racial groups perceived to be inferior and thus a threat to the nation. As a result, the nation's body politic was eugenically choreographed, thereby prompting another phenomenon: the biologization of national belonging. These two developments complemented each other. In the broad discourses on eugenics developed after 1900 in Hungary, the biologization of national belonging underpinned both theories concerning social reform and progress and theories about racial improvement.

In demanding that the modern state pursue the social and biological improvement of its national community, eugenicists frequently depicted the nation as a living organism, functioning according to biological laws. Eugenicists, whether situated on the left or the right of the political spectrum, invested the modern state with the specific mission of not only improving the life of the individual but also regenerating the national community. A corollary aim was to direct disparate narratives of historical experience and cultural traditions towards the overarching idea of improving the national community's racial qualities. The nation was seen to function according to biological laws and to embody certain key genetic qualities. These symbols of an innate biological character were transmitted from generation to generation. According to this line of reasoning, eugenics operated through the investigation of biological processes that regulated the sacred trinity connecting the individual to the nation and the nation to the state. Thus it is particularly important to understand how notions of social and racial homogeneity and protectionism informed eugenic conceptions of a healthy Hungarian nation. The eugenic dream of a modern state pointed to the creation of a racially unified society in which social and ethnic distinctions, divisions between the cultural and the political, no less than between the individual and the collective, would be controlled and managed according to scientific norms. This emphasis on science, in turn, empowered the eugenicists, who were heralded as the national community's ultimate defenders.

Debates on national identity endowed eugenics with a cultural significance we have still to appreciate. As in other European countries at the beginning of the twentieth century, Hungarian eugenics embraced this new nationalist ethos, placing it within a scientifically grounded discourse: one whose legitimacy stemmed from the dual claim that it could both improve the population's health and protect the nation's racial qualities. The oft-studied process of Jewish assimilation into Hungarian language and culture deserves to be mentioned here, as there were many active eugenicists of Jewish origin in Hungary. However, no "Jewish eugenics" developed in Hungary along the lines explored for Germany by Veronika Lipphardtand or for Poland by Kamila Uzarczyk[19] – a phenomenon explained largely by the Jews' complete identification with Hungarian nationalism and its claims for a strong and Hungarian-dominated state, especially after 1867. The effects of this entrenched assimilationist ethos are primarily in evidence in the case of the term *race* (*faj*) which, until the end of the First World War, was routinely employed as a synonym for the nation (*nemzet*). As eugenicists (particularly those associated with the political right) began to insist upon separating culture (nurture) from biology (nature) in determining the racial character of the Hungarian nation, they also began to expose the fragility of defining the nation in religious, cultural and linguistic terms, thus offering unexpected, ominous possibilities for the racial appropriation of nationalistic thinking.

Eugenics eventually engaged both prominent and peripheral figures from various professional disciplines, extending to medicine, biology, sociology and anthropology. A comparative approach is therefore required in order to uncover the sheer variety of approaches advocated at the time. Indeed it is essential to consider how the eugenic vision of a modern state in Hungary reflected more extensive European developments, as well as the multiplicity of cultural and political contexts underpinning the transmission of ideas of social and biological improvement. The corresponding objective is to assemble the scattered elements of this neglected history into an integrated narrative that accounts for its various individual components and which, ultimately, may help in shedding light upon its historical meaning.

Historiography

The emergence of eugenics in early twentieth-century Hungary was also essentially linked to a remarkable degree of institutional networking. At the time, British, American and German eugenicists were praised for their commitment to practical schemes of social and biological improvement. More often than not, developments in other national contexts exhibited a similar character. Eugenics in France, Italy, Russia and the Scandinavian countries, for example, emerged both as a response to local conditions and as an emulation of the above-mentioned hegemonic models.[20] This intermingling of internal and external factors dominates, in fact, all national histories of eugenics, and nowhere has this been more pronounced than among the lesser-known eugenic movements in Central Europe. Prior to the First World War, Austrian, Hungarian, Czech and Polish eugenicists sought to imitate European eugenic movements, particularly the German and the British.[21] However, eugenics in Central European countries, as existing scholarship on the interwar period demonstrates, retained distinctive national overtones, differentiated by the region's individual culture and social context.[22] The preoccupation with eugenics may not have been as strongly represented in Central Europe as it was in Western Europe and the USA, but eugenic and racial ideas, like their practices, were nonetheless present to a much greater degree than has generally been acknowledged.

The history of eugenics in early twentieth-century Hungary power-fully illustrates how seemingly universal eugenic ideas concerning social and biological improvement were nationalized through a convoluted process of negotiation, refutation and appropriation. Yet a systematic attempt to understand this transmission of eugenic ideas to, and within, Hungary has never been made. Before 1989, few Hungarian historians of medicine had acknowledged eugenics and, even when the topic was remarked upon, most authors gravitated towards the historiographic cliché that eugenics was inseparable from, if not identical to, National Socialist biomedical racism.[23] To speak of eugenics in Hungary at the time was, perversely, to speak in the vein of endorsing the atrocities committed by the "superior German race" during National Socialism. By and large, this generalized historiographic attitude developed as a result of official dogma which, following the Soviet model, condemned eugenics as "racist" and "fascist". Even scholars from other disciplines, who were generally less inclined to anachronistic generalizations such as these, mentioned eugenics hesitantly.[24]

It was only after the collapse of communism that scholars in Hungary and elsewhere rediscovered the history of eugenics.[25] Collectively, these studies have appropriately viewed Hungarian eugenics from the vantage point of the history of anti-Semitism, nationalism and racism.[26] For example, in his noteworthy survey of Hungarian political culture, Miklós Szabó extended this framework of analysis to examine, albeit succinctly, the relationship between eugenics, nationalism and nascent Hungarian racism prior to 1918.[27] Of late, preoccupations with the controversial politician Pál Teleki prompted Balázs Ablonczy to consider some of the early twentieth-century debates over eugenics and racial hygiene in Hungary.[28]

However none of this literature has attempted a systematic investigation of the eugenic ideas consistently professed by Hungarian intellectuals, not to mention the genesis, evolution and internal contradictions of eugenics in early twentieth-century Hungary, particularly in the light of its relationship to similar eugenic movements in Europe and the USA.[29] Such an omission is surprising, for even a cursory review of early twentieth-century medical, social and political literature reveals the depth of eugenic practices in Hungary. It is time to direct the historiographic gaze towards this neglected eugenic movement, and thus to advance our knowledge of the history of international eugenics through an exploration of the Hungarian case and of how it compares in relation to wider eugenic debates concerning social and biological improvement during the first two decades of the twentieth century across Europe.

Objectives and Organization

This book has three major objectives. First, it aims to identify the most important Hungarian eugenicists and to contextualize their arguments within their corresponding discursive cultures. After 1900 prominent physicians, biologists, social scientists, intellectuals, religious and political leaders in Hungary consistently expressed their support for eugenic ideas. Consequently, many of them advanced projects for protecting the Hungarian nation and race from its alleged biological degeneration, as well as strategies for improving the health of the population and for increasing the number of healthy families. An analysis of what was a highly biologized social and nationalist discourse thus offers a new perspective on the institutionalization and professionalization of social reform in early twentieth-century Hungary.

Second, this book explores and explains the interconnections between eugenics and nationalism, involving sociology, medicine, anthropology, biology and population policies in early twentieth-century Hungary. An integrated approach to these disciplines facilitates, in turn, a more extensive view of how scientific ideas about health and hygiene were couched in eugenic idioms, in addition to the means by which these idioms became embedded in social, political and national agendas. Eugenicists were interested in both biological and social reproduction, and as a result came into conflict with the interests of individuals and families. Eugenics hence serves as an ideal locus through which to illuminate the complexity of formal and informal relations between professionals, the state and its national community.

Third, this book portrays eugenics in Hungary as part of an international movement for social and biological improvement. Rather than merely adding another chapter to the general history of Hungary, or of treating it as an unfamiliar instance of more illustrious developments in other European countries, this history of eugenics in Hungary restores it to its place within a more general European context. This in turn facilitates a more nuanced interpretation of the relationship between science and politics during this period in general. Hungarian eugenicists were striving to save the nation in order to secure a healthy, protected and luminous racial future. Their quest for a Hungarian national state was translated into a quest for an organic, racial community, one completely integrated within its own geographical space. Consequently, the chapters that follow revisit the intellectual origins of eugenics, not only to improve our understanding of the history of Hungary, but more importantly to illuminate the currently popular debates on the relationship between science and politics in the twentieth century as a whole.

Chapter 1 thus opens with a wide-ranging discussion of the main European theories of eugenics and their reception in early twentieth-century Hungary. This is well illustrated by articles published in journals as thematically diverse as *Huszadik Század*, *Athenaeum*, *Egészség*, *Fajegészségügy*, *A Társadalmi Muzeum Értesítője* and *Magyar Társadalomtudományi Szemle*. Eugenics was, from the very beginning, portrayed as a concrete strategy to improve the biological possibilities of the Hungarian nation through modern medicine and technology. For authors like Pál Teleki, Károly Balás, Lajos Hajós, Gyula Donáth, Gyula Kozáry and Péter Buro, eugenic knowledge was to be applied for an exclusively social and national benefit, and eugenicists commented upon, and offered solutions to, a wide range of issues, concentrating on the protection of the family, child welfare, and state-controlled schemes of social hygiene and public health. Early indications of intellectual support for eugenics was shown by an invitation to the Austrian eugenicist Max von Gruber to address the 16th International Congress of Medicine, held in Budapest in 1909, on the topic of heredity and eugenics. In 1910, the Society of Social Sciences (Társadalomtudományi Társaság) organized a series of public lectures on eugenics in Budapest. This was followed by a lively public debate in 1911.

Chapter 2 suggests that the eugenic debate was influenced by wider developments within European evolutionary science. Through eugenics, supporters of social and national improvement – like Lajos Dienes, Zsigmond Fülöp, József Madzsar, István Apáthy, René Berkovits, Leó Liebermann, Vilma Glücklich and others – sought to determine the relative degrees of reciprocity existing between those scientific theories of biological perfection and the evolutionary language utilized by them. Seen in this context, the public debate on eugenics has a double significance: it gave supporters of eugenics in Hungary the necessary opportunity to synthesize their views on social and biological improvement while additionally introducing a new dimension to general discussions on social and political transformation, which characterized the evolution of social reform in Hungary at this time.

Chapter 3 explores the contested location eugenics inhabited at various intellectual crossroads, within and outside Hungary. In the years preceding the outbreak of the First World War, Hungarian eugenicists – like eugenicists everywhere – called for the supervision of the nation's body to be moved from the private into the public sphere. A number of important international exhibitions and congresses also took place during this period when eugenic theories were presented alongside other ideas advocating social and biological reform. This complex process of reconfiguring the individual – so as to match the broader biological canvas of the nation – was part of a much wider set of intellectual concerns that evolved. These included biological arguments for improving the living and working conditions of the peasantry and the urban working class, the nature of human reproduction, the inheritance of physical and mental traits, neo-Malthusian ideas of birth control, the potential of moral and religious education to shape racial character and, crucially, the individual's recognition as a member of the ethnic body. For eugenicists like Géza Hoffmann – the only Hungarian eugenicist to have achieved wide international recognition at the time – defining the nation in biological and eugenic terms also incorporated a greater attempt to find an alternative national experience for Hungary. Hoffmann also turned to American and German eugenics for ideas and practices that he then filtered and adapted to the Hungarian context.

Prior to the beginning of the First World War, Hungarian eugenicists succeeded in establishing their own eugenic organization. As discussed in Chapter 4, the Eugenic Committee of Hungarian Societies (Egyesületközi Fajegészségügyi Bizottság) was created in 1914 and entrusted with both the popularization of eugenics among the general public and the coordination of the dialogue between eugenicists and the state. What was proposed was a new form of cultural and political modernity, one adapted to the unique conditions resulting from the fusion between Hungarian nationalism and eugenic projects of state-building. This process presupposed inclusion and exclusion as well as new racial hierarchies. Essentially, it biologized national belonging.

The First World War marked a period of intensive eugenic activities, and it is not difficult to see why this international conflict was to become the central rite of ideological passage for eugenicists in Hungary and elsewhere. In reality, as revealed in Chapters 5 and 6, the war was the ultimate frontier that eugenic ideas of social and biological improvement had to traverse. If, prior to the war, eugenics had preponderantly advanced social and medical concerns, during the war these concerns were increasingly connected to a nationalist agenda based upon ideas of race-protectionism and national survival. Indeed, the vision of a healthy Hungarian race served as the eugenic programme upon which societies like the Stefánia Association for the Protection of Mothers and Infants (Országos Stefánia Szövetség az Anyák és Csecsemők Védelmére), the Association of National Protection against Venereal Diseases (Nemzetvédő Szövetség a Nemibajok Ellen), the National Military Welfare Office (Országos Hadigondozó Hivatal) and, finally, the Hungarian Society for Racial Hygiene and Population Policy (Magyar Fajegészségtani és Népesedéspolitikai Társaság) were all established between 1915 and 1917. These societies, and in particular the Society for Eugenics and Population Policy, promoted the image of a national eugenic movement in Hungary that had gradually been building since the beginning of the war in 1914.

This process of internal consensus-building among Hungarian eugenicists was to be short-lived, however. The conclusion to the war and the subsequent democratic and communist revolutions of 1918 and 1919 brought an end to the Austro-Hungarian Monarchy. As discussed in the final chapter of this book, the regimes that followed – while redefining the contours of the social and national body in Hungary – nevertheless retained the appeal and primacy of eugenics. Highlighting the importance of national health, the Ministry of Welfare (Népjóléti Minisztérium) was re-created as the Ministry of Labour and Social Welfare (Munkaügyi és Népjóléti Minisztérium), only to be subsequently transformed into the Commissariat for Labour and Social Welfare (Munkaügyi és Népjóléti Népbiztosság) by the National Council of Health (Országos Egészségügyi Tanács). Social hygiene and public health were invoked as possible eugenic strategies that would suitably connect the emerging proletariat to the new Hungarian state. A propitious set of circumstances for the unfolding of the eugenic ideal of a healthy nation and its corollary, the political project of creating a modern Hungarian state, was thus briefly achieved during this revolutionary period.

Yet these turbulent and often violent political changes, combined with the country's military occupation and a hostile international environment, ultimately brought the Hungarian nation to its knees. By the end of 1919, however, it became obvious that neither the revolutionary nor the counter-revolutionary governments were able to avert Hungary's national disaster. The use of eugenic arguments at the peace negotiations, while a clear indication of the importance afforded to eugenics by the Hungarian delegation, was to no avail. With the signing of the Treaty of Trianon in June 1920, Hungary lost two-thirds of its territory and over three million Hungarians, who were now living in the successor states of Austria, Czechoslovakia, Romania and Yugoslavia. The eugenic dream of a healthy and numerically strong Hungarian nation turned into a nightmare. Hungary, as described in this book, was no more. It became an ideal lost country and the central reference within a new mythology of nation and state, one that is still with us today.

1
A New Dawn

In the first decade of the twentieth century, the debate over the nature and content of eugenics intensified. Interpretations differed from country to country, depending on eugenicists' cultural, social and political backgrounds. In Hungary, the ambition of the first generation of eugenicists was, first and foremost, to construct a *social* science which could be used as an instrument to facilitate social reform. This does not imply that the *biological* and *medical* dimensions of eugenics were ignored. On the contrary, next to sociology and anthropology, biology and medicine were seen as two essential disciplines underpinning eugenic claims for social re-engineering and national protection. This convergence – between social and medical dimensions of eugenics – deserves to be highlighted, both historically and theoretically.

In Hungary, the interpretation of eugenics as a social theory was most successfully popularized by the progressive journal *Huszadik Század* (*Twentieth Century*). It featured a wide range of intellectual arguments and controversies, centred on culture and society, right up to the publication of its last issue in 1919. No less a celebrity than the English philosopher and sociologist Herbert Spencer blessed *Huszadik Század*'s first issue with his encouragement. "I rejoice", he wrote to the editors, "to learn that you propose to establish a periodical having for its special purpose the diffusion of rational ideas – that is to say, scientific ideas, – concerning social affairs".[1] But Hungarian intellectuals associated with the journal had hoped for more than just a constructive and creative intellectual disposition. As Oszkár Jászi confessed to Bódog Somló in 1907, "We intend to not simply create well-written monographs but to stir up the intellectual life of this dark, backward country."[2] And they did.

Huszadik Század promoted an intellectual programme based on a mixture of positivism, socialism and Darwinism.[3] From its beginning in 1900, the journal attracted a large number of Hungarian social and natural scientists, including the historian Gusztáv Gratz, the economist Pál Szende, the sociologist Oszkár Jászi, the legal scholar Bódog Somló and two of Hungary's most promising philosophers, Gyula Pikler and Ervin Szabó. Many of these intellectuals also pursued political careers and played important public roles over the following years, fully justifying the journal's editorial credo that politics and science should be part of the same cognitive effort to both grasp social reality and advance scientific progress.

Huszadik Század's intricate history, and the cultural and political movements it generated, have been the subject of much debate in Hungarian historiography.[4] Less so, however, the eugenic texts published in this journal. Among the eugenicists who contributed to *Huszadik Század* are, for example, physicians József Madzsar and René Berkovits, the biologist Lajos Dienes, the natural scientist Zsigmond Fülöp, the geographer and politician Pál Teleki and the diplomat Géza Hoffmann. These authors participated in shaping Hungary's "complete *Weltanschauung*"[5] – the reconfiguration of intellectual traditions that Oszkár Jászi, one of the journal's editors-in-chief, identified, in 1899, as the rationale behind launching the new publication.

Equally important, in this journal more than any other, eugenics was conceptualized as an integral component of the Darwinian revolution and the newly institutionalized social sciences. In 1901, *Huszadik Század* mobilized a number of Hungarian scientists and intellectuals who, in turn, constituted the Society of Social Sciences (Társadalomtudományi Társaság), with the sociologist Ágost Pulszky as its first president.[6] At the time institutionalized sociology was in its infancy across Europe. The Sociological Society of London was only formed in 1903, followed by the Sociological Society (Soziologische Gesellschaft) of Vienna and the German Society for Sociology (Deutsche Gesellschaft für Soziologie) established in 1907 and 1909, respectively.[7]

To grasp the complex conditions that contributed to the development of eugenics in Europe, one must also recapture the emergence of sociology as the "science of society".[8] Not surprisingly – as Philip Abrams and R. J. Halliday have argued – the origins of sociology in Britain can be found in Émile Durkheim and Ferdinand Tönnies's methodologies to much the same extent as in Francis Galton's unified conception of statistics and biology.[9] In France, too, as Terry N. Clark has noted, "much of anthropology, segments of statistics and political science, and sizable elements of history, economics and geography emerged from identical sources".[10] Paul Weindling has identified a similar confluence of interests between eugenics and social sciences in Germany, where "advocates of social hygiene and demographic studies oriented to the European problem of a declining birth rate conflicted with those concerned primarily to establish sociology as an academic discipline".[11] As it was understood at the beginning of the twentieth century, eugenics – like sociology – was concerned with the rational regulation and management of both the individual and society.

However, it was not just the popularization of ideas of social and biological improvement but the fashioning of the entire edifice of the modern state and society that Hungarian eugenicists demanded. As simultaneous products of both Hungary's own particular conditions and its participation in larger intellectual European currents, eugenics and sociology engaged in public debates over how modern Hungarian society ought to be organized, and on which cultural and biological values it should be based. Intellectual and political change was thus recast by means of social and biological diagnoses. Yet the imposition of biological precepts, simultaneously both specific and idealistic, on modern society did not go unchallenged. As soon became clear, eugenicists in Hungary – embedded as they were initially in a dialogue between disciplines that wanted to assert their conceptual distinctiveness – would find it problematic to claim their own intellectual identity. Asserting this identity would ultimately become coterminous with the eugenic vision of a modern Hungarian state.

Crossing Boundaries

The question of whether biology can have a recognized social and moral role in society had preoccupied sociologists and eugenicists alike since the late nineteenth century. "Does a real biological science of the evolution of human societies exist?", pondered the English biometrician and eugenicist Karl Pearson in his 1909 study, *The Groundwork of Eugenics*.[12] Why this question should concern the eugenicist has everything to do with the fact that some of the most powerful critiques of eugenics have been found in the works of sociologists strenuously denying the significance of race as a factor in social improvement. Based on this conceptual framework, eugenics would not only study the biological basis for social evolution, but would investigate the ethics and morality of human improvement as well. What implications, then, does such a claim have for the reading of eugenic texts in Hungary?

Locating the intellectual genealogy of Hungarian eugenics within a broader European intellectual tradition – one in which various academic definitions of natural and social sciences competed for pre-eminence – contributes to an understanding of eugenics as a symbiosis of social and medical discourses geared towards both individual and collective improvement. It is thus worthwhile to reflect upon the specific lines of reasoning within which demands for population management were voiced by the eugenicists. For example, the first issue of the *Archiv für Rassen- und Gesellschaftsbiologie* (*Journal of Racial and Social Biology*), the prestigious German periodical edited by Alfred Ploetz, was published in the same year as Galton's much-quoted 1904 article on the definition and aims of eugenics. With the founding of the Society for Racial Hygiene (Gesellschaft für Rassenhygiene) in Berlin the following year, it seemed that German eugenicists had finally put their differences aside and transformed their irregular networks into a formalized constituency. The new journal's aims were both managerial and conceptual, as Ploetz not only wanted to unite German eugenicists, but to also provide them with a correspondingly attractive theoretical platform. *Rassenhygiene* (racial hygiene), Ploetz's own idiom for eugenics, was exclusively concerned with the hereditary qualities of the race. As such its aims were twofold: to encourage the reproduction of those individuals deemed hereditarily "superior" on the one hand, and on the other to decrease – if elimination was not possible – the number of those considered racially undesirable. The protection of existing hereditary racial qualities was given impetus by Ploetz's eugenic vision of a new racial community to be built on scientific rationality, biological solidarity and control over reproduction. Racial hygiene, as Ploetz conceived it, was ultimately a vast experiment in biological and social engineering.[13]

The *Archiv für Rassen- und Gesellschaftsbiologie* was immediately recognized as providing a much-needed forum for the growing German-speaking eugenic community.[14] Questions of scientific complexity aside, Ploetz enlisted a number of disciplines – including the social and economic sciences more generally – along with history and psychology, in order to complement racial hygiene's provocative demand for scientific recognition. Central to this tendency was his explicit insistence on the primacy of *biology* (nature), as the necessary alternative to *culture* (nurture), that would set in motion the nation's social and political progress. It was a daring objective, and one with which those encountering and reading Ploetz's journal readily engaged.[15]

One of the first critical commentaries to confront the conceptual mosaic into which Ploetz fused social biology, racial hygiene and anthropology came from Hungary and was published in *Huszadik Század*. This testifies to the *Archiv für Rassen- und Gesellschaftsbiologie*'s immediate impact outside Germany. But it also makes visible *Huszadik Század*'s unmistakable pride in presenting the Hungarian public with the latest cultural and scientific debates in Europe. Indeed, the journal's publication of articles on eugenics are best read in this light. Eugenics, ultimately, represented one of the strategies both editors and contributors employed towards achieving two of the journal's main goals: asserting scientific knowledge and charting "Hungary's anatomy and physiology".[16]

The latter goal, in particular, was to prove especially influential in generating wide-ranging eugenic narratives of social and biological improvement. It is also notable that the person offering this assessment of the new German journal was none other than Count Pál Teleki. Born to one of the most esteemed aristocratic Transylvanian-Hungarian families in 1879, Teleki was a comparatively multifaceted, if controversial, individual, who, among other things, served twice as Hungary's prime minister (1920–21 and 1939–41). Prior to the First World War, Teleki was less interested in politics – though he was a member of Parliament, being elected three times between 1905 and 1911 – and more preoccupied with science in general, and geography in particular.[17] Considering Teleki's central role in the institutionalization of eugenics in Hungary a decade later, his 1904 review of the first issue of the *Archiv für Rassen- und Gesellschaftsbiologie* already displays a remarkable familiarity with contemporary debates on racial and social sciences.[18]

From the outset, Teleki expressly indicated his commitment to the social and political significance of eugenics. The ostensible purpose of the new "discipline" was understood to be intellectually indebted to both social and biological sciences. This was not merely an incidental comment: Teleki believed that cross-disciplinary collaboration was essential in avoiding "the vortex of prejudicial development and dogmatism".[19] This intellectual pluralism was matched, however, by biological determinism, for Teleki believed that "all the basic laws of human existence are biological". Within this all-encompassing social and racial biology, eugenics emerged as the guardian profession for individuals and communities alike. Its ultimate goal, Teleki noted, was to "explain the past, present, and future of state and society".[20]

These postulates were accompanied by a strong belief in the regulatory mandate of biology over other social and cultural factors. Yet Teleki was not a racialist in the narrow sense of the term; that is, he did not believe in ideas of racial superiority. His views on the relationship between eugenics and racial essentialism were more nuanced, expressed especially in the next part of the review, where Teleki engaged with Ploetz's definition of race. Embracing Ploetz's preference for the importance of race – namely "the group of morphologically similar individuals who share common ancestry and are capable of producing similar offspring" – over the individual, Teleki was, however, less inclined to accept the former's view that racial biology should become a distinct branch of medical science, divided into "anatomy, physiology, pathology, and therapeutics". Teleki believed that analogies between racial hygiene and general biology were "conceptually confusing". This objection to Ploetz's methodology was, in fact, an indication that Teleki was not comfortable with analogies between biological and social organisms. He thought the practical application of eugenic precepts premature. "The aspirations of racial hygiene", he noted further, "although they can certainly be expected to bring large benefits in time, do not possess significant practical value and will remain sterile until racial biology has developed into a systematic science".[21]

A closer inspection of this argument reveals, however, a broad acceptance of eugenics on Teleki's part. He agreed with Ploetz that the community's racial welfare was closely connected to that of the individual. Societies, both authors agreed, did not exist outside nature, and were essentially constituted as much by biological as by social factors. Ploetz simply took the argument further than Teleki, berating modern societies' harmful features for the quality of the race. Substantive eugenic engineering was required. Consequently, Ploetz proposed to protect the race from damaging social conditions through "excluding unfit and defective individuals from procreation", that is, by controlling social selection to prevent the transmission of undesired hereditary qualities. Teleki deemed both methods untenable due to the inadequate scientific knowledge about heredity and the widespread resistance to the radical encroachment on human liberties such a practice would entail.[22] Teleki's overall assessment of the first issue of the *Archiv für Rassen- und Gesellschaftsbiologie* was, however, constructive. "The future of the journal is", he noted, "in its own hands because the science to which it is devoted is expected to experience an impressive development." While praising the journal and its contributors, Teleki simultaneously expressed his support for biological theories of human development. Biology, he concluded, was "the only real foundation of the social sciences".[23]

Eugenics constituted a modern, scientistic worldview, one that resonated favourably with *Huszadik Század*'s intellectual strategies. By accentuating its scientific claims, eugenics neatly dovetailed with the Society of Social Sciences's rejection of intellectual isolationism and its promotion of the country's modernization.[24] This was to be a necessary synthesis between Hungarian national traditions and modern science. From the outset, *Huszadik Század* was emphatically open to all intellectual trends, as the editors completely rejected parochialism and dogmatism, whether political or cultural.[25] One should not, then, be surprised to find members of the aristocracy like Teleki contributing to a progressive journal like *Huszadik Század*. There can be no denying that Teleki was a political and social conservative, inspired by Christian morality and Hungarian nationalism. Yet in many important respects, his modern and progressive scientific outlook was no different than Jászi's. Both authors shared a broad conception of social reform and cultural enlightenment, constantly combining their individual scholarly preferences with an assortment of political ideas. It is only by working through their ideas that we can begin to engage with the ebb and flow of eugenic thinking and how it influenced the making of a modern state in early twentieth-century Hungary.

Subversive Affinity: Race and Society

The attachment to a biological definition of identity was not only an essential component of eugenics, but also of a racial philosophy of history, prominently on display in various European cultures since the mid-nineteenth century. Seen in this context, eugenics was a form of "biological determinism", one which presupposed that "shared behavioural norms, and the social and economic differences between human groups – primarily races, classes, and sexes – arise from inherited, inborn distinctions and that society, in this sense, is an accurate reflection of biology".[26]

Although aware of the importance of environment and education in shaping individual and collective character, most eugenicists in Europe and the USA were generally hereditarian in their interpretation of social and national improvement. Francis Galton himself phrased the debate in terms of a distinction between "race" and "nurture".[27] In his 1873 study "Hereditary Improvement", Galton accepted that nurture was essential to eugenic betterment. "An improvement in the nurture of the race", he noted, "will eradicate inherited diseases; consequently, it is beyond dispute that if our future population were reared under more favourable conditions than at present, both their health and that of their descendants would be greatly improved".[28] Envisioning a shared role for racial and cultural development, Galton thus focused on the need to create an effective correlation between social environment, education and sanitary welfare, warning that the quality of the race would be damaged if these conditions were not improved. Yet nurture was of secondary importance. "I look upon race", Galton emphasized,

> as far more important than nurture. Race has a double effect, it creates better and more intelligent individuals, and these become more competent than their predecessors to make laws and customs, whose effects shall favourably react on their own health and on the nurture of their children.[29]

Although nurture was not entirely abandoned, it was race that served as the underpinning concept of Galton's eugenics. As phrased in the *Inquiries into Human Faculty and Its Development*, eugenics engaged with "various topics more or less connected with that of the cultivation of the race".[30]

Other eugenicists followed a similar hereditarian, biological understanding of race. In a letter to Galton from 1905, Ploetz described him as "the senior of the practical application of the principles of evolution", expressing the hope that he may find the time, "in an hour of leisure", to read the *Grundlinien einer Rassen-Hygiene* (*Foundations of Racial Hygiene*) and convey his "cool judgement". Referring to the origins of this book, Ploetz candidly remarked that it was "written mostly in a small town, where I practiced as a physician, absent from a good library and therefore without much knowledge of current literature" on eugenics. What is more, Ploetz credited Galton's influence for his preference of the term *Rassenhygiene*: "I started from an English use of the word 'race' and tried to investigate the conditions of preserving and developing a race-hygiene ('Rassen-hygiene')".[31]

Ploetz further explored the racial adaptation of eugenics in the first issue of the *Archiv für Rassen- und Gesellschaftsbiologie* published in 1904, and then again at the first conference of the German Society for Sociology in 1910. By now, there was no shortage of disapproving views on the racial component of eugenics. The list of conference participants included illustrious names in German social sciences, philosophy, theology and history, such as Martin Buber, Hermann Kantorowicz, Werner Sombart, Ernst Troeltsch, Ferdinand Tönnies, Georg Simmel and Max Weber.[32] Alfred Ploetz presented his concepts of race and society – already outlined in his 1904 article – which nevertheless occasioned a heated debate on the meaning of eugenics, social biology and hereditary determinism in the social sciences.[33] Max Weber, in particular, reacted vehemently to Ploetz's assertion that social improvement was closely connected with eugenics and social biology.[34] This was not Weber's only objection to Ploetz. The *Archiv für Rassen- und Gesellschaftsbiologie*, for instance, he described rather unsympathetically as "a sheer arsenal of boundless hypotheses (which to some extent are presented with an enviable fullness of spirit)".[35]

These postulates were accompanied by Weber's strong belief in the role of religious, social and economic factors in shaping the individual and society. Weber was, in fact, deeply sceptical over the extent to which race and eugenics were even commensurate with sociology. "Does there exist even today", he asked, "a single fact that would be relevant for sociology, a single concrete fact which in a truly illuminating, valid, exact and incontestable way traces a definite type of sociological circumstance back to inborn and hereditary qualities which are possessed by one race or another? The answer", he declared, was "definitively no!"[36]

Ploetz may have been unsuccessful in persuading Weber to grant racial hygiene a decisive role in shaping social development, but other European eugenicists were more compelling when engaging with this issue. In 1909, for instance, Pearson spelled out what he called the "bricks for the foundations" of eugenics. The new scientific edifice was based on "three fundamental biological ideas", namely:

(a) That the relative weight of nature and nurture must not *a priori* be assumed but must be scientifically measured; and thus far our experience is that nature dominates nurture, and that inheritance is more vital than environment.

(b) That there exists no demonstrable inheritance of acquired characters. Environment modifies the bodily characters of the existing generation, but does not modify the germ plasm from which the next generation springs. At most environment can provide a selection of which germ plasms among the many provided shall be potential and which shall remain latent.

(c) That all human qualities are inherited in a marked and probably equal degree.[37]

The argument conveyed here synthesizes the social, cultural and political possibilities resulting from a hereditarian vision of the world. As nature was held to be more important than nurture, eugenics and race seemed inseparable. For his part, Ploetz viewed race as the heuristic strategy needed for explaining both biological diversity and notions of collective social and cultural improvement. Biological anxieties over

race had initially emerged, he maintained, precisely because the hereditary potential of a given culture was allowed to diminish and eventually
perish, as was the case supposedly with the ancient Greeks and Romans.
Race, moreover, was not an anthropological category. "When I say",
Ploetz insisted, "'the blooming of the race is a necessary foundation for
the formation of the society', quite naturally I am not referring to an
anthropological variation but to biological constancy".[38]

A new and formal eugenic language, with its own conceptual refine-
ments and vocabulary, was thus gradually emerging during the first
decade of the twentieth century. And through this language, race was
expressed as an important and methodologically formative force: it
conditioned national identity and identification, as well as the eugenic
resources available to renew them. In a world that was widely diagnosed as prone to biological and social degeneration, the idealization
of eugenic control over the population confirmed what Pearson termed
as the "real enlightenment" to follow the "scientific treatment of the
biological factors in race development".[39]

Outlined by Galton, Ploetz and Pearson is an element to be frequently
encountered in Hungarian texts on eugenics: a distinctly hereditarian
understanding of the national body. As the nation's aspiring biological epistemology, eugenics was inseparable from a model of identity
based on physical and racial typologies. Just as the debate over social
and biological improvement became intertwined with the idiom of
eugenics, so too the operative link between race and national community underpinned a new form of construing Hungarian identity in
this period – one described above and elsewhere as the biologization of
national belonging.[40]

Health and Degeneration

At the beginning of the twentieth century, Hungarian physicians preoccupied with improving the health of the population never lost sight of
the relationship between heredity and social problems. The first national
conference on psychiatry organized between 28 and 29 October 1900
in Budapest, for example, made this connection explicit. Attended by
the most important Hungarian psychiatrists, including László Epstein,
Gusztáv Oláh, Károly Décsi, Hugó Lukács, Ernő Moravcsik, Gyula
Donáth, Károly Laufenauer and Kálmán Pándy, this scholarly meeting was a convincing example of how theories of psychiatric illness
intersected with direct medical intervention and scientific research. In
the context of a developing interest in eugenics, the participants also

lost no opportunity to emphasize individual health in terms of the perceived protection of the national community.[41] This was a period when psychiatry, both in Hungary and elsewhere, developed new strategies for limiting the threats that the mentally ill posed to society.[42]

New connections between the individual and society were established, conjuring up a modern medical culture imbued with cultural and biological concerns about the health of the population. Reflecting these concerns, a more assertive connection to race was made by those situated in-between medicine and anthropology. One such author, the psychiatrist Béla Révész from Nagyszeben (Hermannstadt, Sibiu) in Transylvania, focused on the association between disease and racial improvement, offering a biological narrative centred on the medical polarity between the normal and the pathological.[43] Arguments such as these demonstrate not only how medicine encompassed the concrete transformations of society brought about by scientific discoveries in pathology, embryology and immunology, but also the growing speculations of a wide range of scholars on the application of ideas of heredity and evolution to physical and mental illnesses.[44] Medical knowledge helped rationalize health problems in society in order to reflect new theories of heredity, but the much-disputed bearing of race upon eugenics remained a theoretical dilemma for many intellectuals in Hungary.

A summative account of the relationship between existing theories of evolution and race was provided by Bernát Alexander and Mihály Lenhossék's two edited volumes on *Az ember testi és lelki élete, egyéni és faji sajátságai* (*Mankind's Physical and Spiritual Life, Its Individual and Racial Characteristics*) published in 1905 and 1907.[45] It was – according to Lenhossék – the first comprehensive attempt in the Hungarian language to understand "the human body and its spiritual life as well as the past and present evolution of the human race".[46] The editors' combined expertise is illustrative of the volumes' scholarly interdisciplinarity: Alexander was the founder of modern Hungarian aesthetic theory and a distinguished philosopher, while Lenhossék was one of Hungary's foremost medical anthropologists. Contributors included the biologist and zoologist Sándor Gorka, a disciple of the German embryologist Ernst Haeckel, the physician Mihály Pekár, the neurologist and psychologist Pál Ranschburg, the Director of the Ethnographic Section of the Hungarian National Museum in Budapest, Vilibáld Semayer, and the physical therapist and pioneer of Hungarian sports medicine, Zoltán Dalmady. Both editors and contributors agreed that the relationship between race and environment was integral to the changing cultural, social and biological understandings of human societies.

Health, Dalmady argued, was the most important constituent of modern life, shaping not only the life of the individual, but also the social fabric; it was thus a normative category, operating in both medical and social environments.[47] With great assiduity, Dalmady then built a typological construction of health, whereby the connection between one's genetic history and disease was paramount. In doing so, Dalmady wielded the modern vocabulary of Mendelism to strengthen his medicalized discourse on health and society. He believed that health could evolve into a commanding medical philosophy, attentive to social, cultural and racial conditions in specific national contexts. Hereditary knowledge about social problems and diseases was coupled with the language of scientific hygiene and modern conceptions of health. The eugenic ideal of a healthy nation, Dalmady suggested, ought to provoke a rethinking of the accepted cultural boundaries separating the individual from society.[48]

What emerges from this argument is a fluctuating and permeable definition of race and its relationship to health. In his eclectic treatise on social demography and population policies published in 1905, the economist Károly Balás further contributed to this discussion, arguing that demographic growth, social dynamics and reproductive patterns could not be isolated from racial theories of national health. It was obvious to him that a dynamic theory of population needed to include not only social, economic and cultural factors, but racial determinants as well.[49] Like other Hungarian social scientists at the time, Balás was fascinated by the topic of racial character, both in the individual and the collective, and he engaged extensively with contrasting methodologies and theorists concerned with biological variability, including Malthus, Spencer, Darwin, Galton and Ploetz. But in reviewing the popular racial theories offered by Arthur de Gobineau, Houston Stewart Chamberlain and others, Balás rejected the omnipotence of race in the shaping of human history. In the case of the Hungarians, in his view, it was language, social solidarity and patriotism that most determined the narrative of national identity.

Balás also noted the importance of biological theories of human improvement to the development of Hungarian demography.[50] Population policies and eugenics thus emerge in his study as complementary answers to a single problem: the nation's racial potential to prevent degeneration. Like elsewhere in Europe at the time, Hungarian eugenicists were worried about the nation's degeneration, which they imputed to precarious social conditions, poor hygiene and lack of medical assistance – all factors contributing to the proliferation of individuals of supposedly mediocre racial quality. According to the paediatrician Sándor Szana, for instance, this imbalance had to be remedied by sanitary improvements, appropriate child and family policies, as well as by encouraging modern welfare programmes.[51] He also highlighted the eugenic importance of children and recommended "social hygiene" as a means to protect their quantity and the biological quality.[52]

Indicative of this association between societal health and biological degeneration, Hungarian medical and legal literature from the early twentieth century displayed great awareness of the nation's declining health. In a lengthy article published in four instalments in *Huszadik Század* between 1900 and 1901, for example, the psychiatrist Lajos Hajós located health, hygiene and eugenics within a matrix of biological renewal and degeneration. If the former was based on a scientific knowledge of society that subjected it to the rigours of health regulations, the latter, correspondingly, was seen as an impediment to the urgent need to create a biologically strong and collectively functional nation.[53]

Both in suggesting methods for detecting social problems and in his self-reflexive anxiety over degeneration, Hajós mapped out a version of Hungarian eugenics as a social discourse on health. Within this framework, Hajós examined the standardization of medical practices, further extending his prophylactic agenda to include the wide Hungarian society, which he conceived as racially vigorous but endangered by a number of environmentally induced factors. Hajós' social critique was infused with medical and eugenic suggestions. Countering degeneration, Hajós argued, would be both a qualitative and quantitative process engaged with national protectionism. As such he offered a reading of the emergent eugenic culture that moved beyond the perimeters of individual health towards an inclusive vision of the racial collective.

The eugenic agenda here is clear. Consistent with the language of degeneration at this time, Hajós spoke for many of the anxieties over urban development and its attendant chronic poverty and social unrest. Public health reforms were, he insisted, to be implemented along scientific principles and, equally important, were to reflect a new understanding of society's eugenic requirements. Furthermore, Hajós' rhetoric of social defence reveals the concern with establishing modern ideas of hygiene; above all, Hajós conflated national health with its submission to scientific expertise. Both social and biological regeneration were deemed indispensable expressions of an advancing medical knowledge. Ultimately, these forms of health regulation, protectionism and treatment – which Hajós continuously invoked – sanctioned diverse forms of expert surveillance and state control. Eugenics thus galvanized what was already a long process in the subjection of the individual and society to the modern state, a development Michel Foucault has termed "the State control of the biological".[54]

As the national body was codified biologically, however, controlling the life of the population meant more than mere guardianship. The state was also responsible for ensuring that health policies were suitably pursued. Such interventionist agenda was endorsed during the International Congress on Anti-Alcoholism held in Budapest between 11 and 16 September 1905. Alcoholism was discussed in relationship to civil law, education, sexuality and hospital treatment. Participants included Taav Laitinen, Head of the Medical Service of Finland; Wilhelm Weygandt, a professor of psychiatry at the University of Würzburg; the Swiss psychiatrist and eugenicist August Forel; the Austrian physiologist and neurologist Rudolf Wlassak; Leó Liebermann, a professor of hygiene at the University of Budapest;[55] and Rusztem Vámbéry, a professor of law at the University of Budapest. These experts stressed the importance of the state for their social and biological agendas, claiming that it was the state's duty to guarantee not only civil and political liberties, but also to defend the population against its own social and biological degeneration. By endorsing state intervention in the life of the individual and the community, the participants – as health experts and scientists – made a more active eugenic commitment to ideas of social and biological improvement.[56]

Addressing the Medical Association in Budapest on 3 March 1906, the neuropathologist Gyula Donáth called upon Hungarian physicians to engage vigorously with combating the nation's physical degeneration.[57] Drawing on the wide-ranging eugenic literature, particularly authors such as Alfred Ploetz, Auguste Forel, Wilhelm Schallmayer and Alfred Grotjahn, Donáth discussed the harming effects of narcotics, claiming that moderation and self-control were equally moral and social goals. Eugenics was advertised as the remedy for degeneration, and Donáth saluted the *Archiv für Rassen- und Gesellschaftsbiologie* for promoting corrective social and medical measures. In unfolding the contents of his lecture, Donáth devoted particular attention to alcohol, seen as one of the most destructive elements of social and family life. Charting the fragility of the modern individual – both physically and psychologically – Donáth set the increased consumption of alcohol and other narcotics against a host of changes associated with the rapidly modernizing European societies. Modernity was seen as fostering nervous and mental illnesses, "neurasthenia, hysteria and progressive paralysis",[58] and ultimately national and racial decline.

This merging of alcoholism with sexually transmitted diseases, and the coalition between medical and social reform, informed how certain categories of individuals were regarded as eugenic problems. These individuals were discussed in terms of their alleged regressive behaviour, social and reproductive problems.[59] Alcoholics, in this instance, had to be disciplined, and the eugenicists entrusted the state with this civilizing mission.[60] In building a nexus between the individual, the society and relations of power, Donáth echoed the work of other European eugenicists. In the opening decades of the twentieth century, degeneration (in its multiple forms) came to be widely discussed and these debates exploited the central role played by the human body as a source of eugenic meaning. Such a reliance upon the protectionist role of the state called for a certain degree of limitation on the existing political economy and individual liberties in order to enact the necessary eugenic remedies. Sterilization, in this context, was increasingly viewed as one such remedy for counteracting the alleged increase of "degenerate" individuals.

The intellectual channels through which sterilization was articulated in Hungary at the beginning of the twentieth century were diverse but clearly interconnected. The issue that emerged first, and most urgently, was that of authority. Who should decide whether a person should be sterilized or not? This was a question raised as early as 1902 by Pál Angyal, a professor at Pázmány Péter University in Budapest, which he answered from the perspective of criminal law. Most of his supporting arguments emphasized the importance of the legal system in regulating the health of both the individual and the community, and as a solution to social problems.

A related consideration was that of a comprehensive surgical methodology. In this context, Angyal considered castration and ovariotomy as two efficient methods to prevent and punish crime. To substantiate his opinion, Angyal drew upon the legal and medical precedents for sterilization, both diachronically and historically, beginning with Roman criminal law and concluding with modern Hungarian legal history.[61] Angyal seized upon what he assumed to be his professional and civic responsibilities, and discussed sterilization as a legal procedure. He insisted that the social value of public health – as expressed in the protection of the family and society from criminals – be officially recognized.

Derived from the traditions of criminal law, Angyal's argument was reinforced by Rusztem Vámbéry and the legal experts József Illés and Dezső Buday, who advanced the social and biological necessity of regulating marriage through state intervention. According to Vámbéry, for example, such a task demanded a number of changes to both expert knowledge and legal practices.[62] Discussing marriage in terms of its social value, Buday further positioned it at the intersection of criminal law, sociopolitical value and private morality.[63] For Illés, on the other hand, marriage was more than just the simple coexistence of two individuals: it was a central ingredient of the social and national life that needed defending.[64] These authors combined the social and biological management of the population with a progressive national pedagogy. The protection of the family consisted of a complex assemblage, both social and biological, which operated within different areas of individual and collective life.

Yet medical instrumentalities were also competing with these legal theories of social improvement.[65] The attitude of criminologists and legal theorists like Vámbéry, Angyal and Buday soon converged with that of physicians, like Manó Szántó, who in his 1905 *A fakultatív sterilitás kérdéséről* (*On the Question of Voluntary Sterility*) centred the debate around contraception and on whether it was recommended for medical reasons (if the health of the mother would be endangered by the birth of another child) as opposed to non-medical, social purposes. Exploring the former, Szántó considered that fertility, pregnancy and abortion not only have physiological consequences for a woman, but they are also shaped by social consequences. Moreover, Szántó acknowledged that contraceptive interventions (of which he enumerated several, for both women and men) had multiple meanings, and that women's health was not always the result of negotiations between women and medical professionals; it often depended on economic and social circumstances. By paying closer attention to the ways in which women exerted personal control over their health and reproduction, Szántó pointed to the limitations in modern medical knowledge existing in Hungary, alongside the necessity for a closer supervision by the state of sexual reproduction. Concerned about maternal and infant mortality, Szántó wanted to regulate women's access to "traditional" techniques of birth control, arguing instead for modern reproductive technologies through public health and welfare programmes.[66]

Reflecting these concerns, in his 1906 *Születési és termékenységi statisztika* (*Statistics on Births and Fecundity*), the statistician Géza Illyefalvi-Vitéz analysed how social conditions determined reproductive behaviour, explaining fluctuating birth rates as a consequence of both changes in fertility and the "rationalization" of reproduction by wealthy families. He warned that the gap between private and public health care was widening. As a result, Illyefalvi-Vitéz envisioned an interventionist programme aiming to simultaneously protect, regulate and emancipate the population in accordance with new ideas of eugenics.[67] These ideas included insightful and progressive conceptions of social and medical insurance for workers, the need for a more equitable role for women in society and, not least, the reinforcement of the role of medicine in protecting society.

At the beginning of the twentieth century, new sterilization techniques were adopted by the medical profession including some, like vasectomy, which were also sanctioned eugenically. Although genitourinary surgeons were using vasectomy in connection with prostate operations by the early 1890s, it was Harry C. Sharp, a physician at the Indiana Reformatory in Jeffersonville, Indiana, who famously performed the first vasectomy for eugenic purposes in 1899. His declared intention was to prevent institutionalized patients from reproducing and thus burdening the state.[68] Whether Angyal and Szántó were aware of Sharp's sterilization experiments is not clear, as both authors failed to mention either vasectomy (for men) or salpingectomy (for women) in their studies. Nor did either seem to be aware of Robert R. Rentoul's 1903 *Proposed Sterilization of Certain Mental and Physical Degenerates*, a book which enjoyed considerable circulation among supporters of sterilization in both Britain and the USA. In contrast to Szántó, who focused on contraceptive methods by "voluntary sterilization", Rentoul understood vasectomy and salpingectomy as two surgical procedures whereby sexual desire is retained but there remained "no power to impregnate". He also called for the introduction of "compulsory sterilization" for "lunatics, epileptics, idiots, confirmed criminals and inebriates, and habitual vagrants".[69]

As the above suggests, a far more complex eugenic culture was coming into being in countries like England, Germany and the USA than Hajós, Angyal and Szántó had realized. In combination with various theories of degeneration and heredity, this new culture would eventually replace the nineteenth-century one within which Hungarian ideas of social and medical control were first developed. Eugenics emerged at precisely the point at which philosophies of social reform and medicine were focusing on "defective" individuals and attempting to construct detailed typologies for their diagnosis, treatment and control. Donáth's anxiety about the effects of alcoholism on society and race as well as Vámbéry and Buday's legal underpinnings of family protectionism were similar expressions of this transformation.

As narratives of social and biological degeneration began to travel across geographical and cultural boundaries at the beginning of the twentieth century, they also reached Hungary. The authors mentioned above did not neglect these narratives, but adopted them to Hungary's own social and medical problems associated with criminality, insanity and moral degeneration. To address these problems, they collectively underlined the need for a new eugenic strategy aimed at reaching the scientific and political leaders. The intellectual and cultural regimes in which scientific discourses about eugenic improvement would be formulated in the following decade in Hungary would reflect this attempt to place the discussion about the racial qualities of the population within a more engaging political matrix.

Reproductive Hygiene and National Biology

Whereas Teleki and Donáth invoked Ploetz and other German eugenicists, Balás and Hajós cited an assortment of British authors, including Francis Galton. These two eugenic traditions, German and British, were essential to the birth and the development of Hungarian eugenics. Take, for example, Galton's 1904 lectures on eugenics delivered to the Sociological Society of London,[70] which were immediately translated and published in the *Archiv für Rassen- und Gesellschaftsbiologie*. Interestingly, Ploetz chose to translate eugenics not as *Eugenik* or *Rassenhygiene*, but – using Alfred Grotjahn's term – as *Fortpflanzungshygiene* (reproductive hygiene).[71] Engaging with Galton's definition of eugenics as "the science which deals with all influences that improve the inborn qualities of a race; also with those that develop them to the utmost advantage", Ploetz explained that this definition of eugenics only referred to creating the optimal conditions for the successful production of offspring. As such, eugenics was only a division of racial hygiene, which in addition to reproduction also concentrated on the conservation of the race's highest qualities. Ploetz's definition of racial hygiene was thus broader than Galton's eugenics, and predicated upon all means necessary for the preservation of the racial community.[72] Eugenics, from this perspective, was concerned with promoting the well-being of the individual, whereas racial hygiene was mostly about encouraging the fulfilment of the individual within his or her racial community.

Hungarian eugenicists valued both Galton and Ploetz's theories. And it was not only conservative aristocrats like Teleki who were willing to engage with the fashionable British and German eugenics. In 1906, József Madzsar presented an abbreviated version of the *Archiv*'s translation of Galton's lectures to the Hungarian public. Madzsar (Fig. 1), like Ploetz, preferred to refer to Galton's eugenics as "reproductive hygiene".[73] The

Figure 1   József Madzsar (courtesy of the Semmelweis Museum, Library and Archives of the History of Medicine, Budapest)

ability to control reproduction, Madzsar believed, endowed eugenics with a dominant voice, in addition to authoritative knowledge about the social and biological role of the family. Eugenic reproductive technologies and procedures were, in fact, seen at the time as symbols of power and authority over the body. Accordingly, certain social domains, such as the protection of mothers and infants, as well as pregnancy, birth and abortion, became intensively eugenicized.

As one of the most important leftist intellectuals of the period, Madzsar was to play an important role in the eugenics movement over the next decade, and his articles and activities will be discussed frequently. But his 1906 review in *Huszadik Század* deserves a special mention, as it illustrates another important feature of Hungarian eugenics: its varied ideological composition. As in other European countries, eugenicists in Hungary belonged to diverse social and cultural milieus, and although they often endorsed opposing political philosophies, these differences made their eugenic beliefs no less influential. Instead of simply agreeing upon a singular definition of eugenics, various models were produced in accordance with their authors' different sociopolitical backgrounds.[74]

Indeed, the first elaborate discussion of eugenics in Hungary was penned not by Teleki, Balás, Hajós, Donáth or Madzsar, their initial interest in eugenics notwithstanding, but by a neo-Thomist philosopher, Gyula Kozáry. Far from this being an isolated occurrence, Kozáry had already published two studies on heredity, in which he engaged with evolutionary ideas and the corresponding possibilities that modern technologies of health and hygiene held out for improving national culture.[75] In 1906, he began serializing his history of eugenics in the Hungarian Philosophical Society's journal, *Athenaeum*, ultimately published in four parts, the last in early 1907.[76] A brief look at the arguments put forward in his study exposes the methodological foundations underpinning Kozáry's eugenic worldview.

Kozáry employed a diachronic perspective, identifying the first expressions of eugenics in classical Greece, particularly in the Laws of Lycurgus, the legendary lawgiver of Sparta, and in Plato's *Republic*. It was, however, Galton's writings – with which Kozáry was fully acquainted – that provided the central reference point for his eugenic ideas.[77] In casting a critical eye over the emerging literature on eugenics, Kozáry also participated in wider discussions on Darwin's theory of natural selection and its potential implications for racial improvement.

Engaging with John Berry Haycraft's 1894 *Darwinism and Race Progress* and Otto Ammon's 1893 *Die natürliche Auslese beim Menschen*, Kozáry thus drew attention to the importance of eugenics in combating social and racial decline. And, following Galton's theory of social classification, Haycraft and Ammon themselves drew attention to differential fertility rates, arguing that modern societies were consequently prone to racial degeneration. Protecting the weak and the socially unfit were therefore deemed inadequate measures for ensuring the race's future. Following Vacher de Lapouge's 1888 *L'hérédité dans la science politique*, Kozáry stressed the need for unremitting competition and reflexive social selection among superior racial elements within society.[78] This combination of biological statistics, social anthropology and organicist theories of degeneration helped to broaden the intellectual basis of eugenics.[79]

Kozáry's approach to Lapouge's ideas, moreover, reflects the pivotal place anthropology occupied in discussions on race and eugenics at this time. Lapouge himself made this connection explicit when claiming that anthropology was "destined to revolutionize the political and social sciences as radically as bacteriology has revolutionized the science of medicine".[80] By the beginning of the twentieth century, the concept of race also embodied a specific set of meanings in relation to health and disease. Eugenics, with its proclivity for racial metaphors, provided the idiom of choice among a large number of social scientists, biologists, philosophers, physicians and politicians, enabling them to canvass a new biological order. A racial language once identified with social and physical anthropology thus dovetailed with the hierarchical scheme of social and biological differences that eugenics increasingly aimed to address.

Kozáry did not favour one intellectual option over others, for he sought to provide as wide-ranging a theoretical overview as possible. Aptly, Kozáry placed his discussion of eugenics within the conceptual framework that had been established by Galton, Ploetz and Wilhelm Schallmayer.[81] Following the latter, especially his 1905 *Beiträge zu einer Nationalbiologie* (*Contributions to a National Biology*), Kozáry linked eugenics with direct state intervention and with a commitment to controlling social and racial reproduction. A new relationship between science and politics ensued – one purportedly based on medical reasoning – in which social hygiene and public health morphed into a superior form of eugenic protectionism. This would, of course, be ideally suited to a new Hungarian body politic:

> Social hygiene should deal not only with personal hygiene, but also with heredity and racial hygiene. Public health should rise to the level of becoming national biology (*Nationalbiologie*), which first serves society by regulating the health of the people and second by encompassing racial services and institutions (heredity or the hygiene of procreation). The duty of the latter would be to prevent the population's hereditary degeneration and as much as possible to guide hereditary selection.[82]

Cast in these terms, eugenics provided an innovative terminology for an increasingly obsessive preoccupation with the hereditarian foundation of the nation. One main task was thus "to secure the health conditions necessary for performing various social duties as well as to safeguard, and to a certain extent, guide and influence the quantitative and qualitative development of the national body".[83] For Kozáry, moreover, it seemed that eugenics' focus on social and biological reproduction, rather than racial identity, was indicative of the larger conceptual dilemma preoccupying the authors he discussed in his article. Eugenics, he believed, was in fact underpinned by the language and logic of race. In deliberately moving towards a racial understanding of identity, Kozáry felt it necessary to add that

Concepts of race and people are not the same. People is a larger concept characterized by the unity of language; it comprises usually individuals belonging to several races and cross-breeds thereof. Race means ancient, externally recognizable ties of blood, and does not correspond to linguistic groups. Race is the first solid and constant, common and internationally shared trait of peoples; it means consanguinity.[84]

This depiction of race was – as already emphasized above – specific neither to Kozáry nor to Hungary. Within early twentieth-century discussions of nationalism and imperialism, this orientation, however, went hand in hand with an intellectual celebration of racial superiority. Eugenics, socialism and Aryan racism, personified especially by authors like Ludwig Woltmann in Germany, helped solidify Kozáry's final arguments. The ideal of "pure Germanness", and the conviction that "the German race" had provided the greatest achievements in Western art and culture, ultimately sanctified Kozáry's belief in European racial hierarchies: if Prussia was "Germany's leader", Kozáry declared, "Germany [was] Europe's, and Europe, the leader of the whole world!"[85]

Kozáry's approach to eugenics reveals the ways in which it became entangled with the dominant political ideas of the age. Yet, his arguments about race and the dominance of the Germanic element in Central Europe formed but one of the various strands pursued by supporters of eugenics in Hungary. As we have seen, Hajós and Madzsar were, by contrast, attentive not to racial struggle but to social and medical determinants of health, insisting on the idea of planned reproduction and a rational society. They, like other authors gravitating around *Huszadik Század*, adhered to eugenics on account of their medical expertise no less than their social and biological convictions. Eugenics expressed not only the need for rationalized forms of social inventory and control but also defined biological representations of the national community and its role in safeguarding the future of the Hungarian state. Authors like Teleki and Balás therefore placed eugenics at the confluence of political demography and population policy, retaining an interest in racial development and the nationalist competition among the countries in the region.

Medical Foundations

Notwithstanding the significant role played by anthropology, sociology, philosophy and demography in shaping the emergence of Hungarian eugenics, it was medicine – in all its various permutations – that became fundamentally associated with ideas of social and biological improvement. As in other European countries at this time, medical knowledge and eugenics were interrelated, fusing the vision of a healthy Hungarian nation with physicians' daily experiences of providing healthcare and social assistance. These ideas informed the eugenically protectionist agenda proposed by medical journals in Hungary like *Orvosi Hetilap* (*Medical Weekly*), *Gyógyászat* (*Therapy*), *Magyar Orvosok Lapja* (*Hungarian Physicians' Journal*), *Klinikai Füzetek* (*Clinical Review*), *Budapesti Orvosi Újság* (*Budapest Medical Journal*) and *Egészség* (*Health*). Other periodicals, like *Unsere Gesundheit* (*Our Health*), launched in 1898 by Sándor Szana in Temesvár (Temeswar, Timişoara), and *Volksgesundheit* (*National Health*), founded in 1902 by Heinrich Siegmund in Mediasch, attempted to reconcile the regenerative ideology of modern hygiene and eugenics with traditional morality in the respective Hungarian and Saxon communities of the Banat and Transylvania.[86]

Analogies between individual and collective welfare foregrounded medicine as the central producer of a healthy national culture. In 1902, in recognition of medicine's emerging social responsibility, the Social Museum (Társadalmi Muzeum) was established, based on the models provided by the Social Museum (Musée Social) in Paris and the Museum of Industrial Hygiene (Gewerbehygienisches Museum) in Vienna; it later became the Museum and Institute of Social Hygiene (Társadalomegészségügyi Intézet és Múzeum).[87] Within the wider matrix of social and eugenic reform advocated by health officials and welfare experts in Hungary, the Social Museum's declared purpose was thus to study "social, health and economic conditions of the industrial and agricultural workers, of collecting scientific material on social diseases, and wakening the interest of the public for a systematic campaign against social diseases and other evils connected to them".[88]

The concept of a "social museum" was launched at the 1889 Universal Exhibition in Paris. According to Janet R. Horne,

> The idea of creating a permanent social museum emanated from the basic museological function of the universal exhibitions themselves. By extending the Encyclopedic tradition into the visual public realm, these exhibitions became temporary outdoor museums for collecting, classifying, and displaying the physical results of modern technical, commercial, and social knowledge.[89]

In Hungary, the Social Museum combined social knowledge with an examination of economic factors, the health and work conditions of industrial and agricultural workers, protective measures and the prevention of accidents.[90]

The Social Museum also established its own bulletin, *A Társadalmi Muzeum Értesítője*, in 1909, in order to publicize its wide-ranging activities in the field of social hygiene, preventive medicine and public health. Behind this broad agenda there were new conceptions of urban and rural hygiene. As such, the Social Museum took its place alongside new scientific and public welfare associations, like the Society of Social Sciences, the National League for the Protection of Children (Országos Gyermekvédő Liga) and the Hungarian Society for Child Study (Magyar Gyermektanulmányi Társaság), which embraced this modern vision of public health, social hygiene and eugenics.[91] While the establishment of the Social Museum did not immediately usher in a nation-wide eugenic programme, it certainly constituted a significant addition to the practice of thinking about the relationship between individual health and social problems caused by economic development and industrialization.[92] In 1906, in recognition of these mounting concerns about urban poverty, infectious diseases, social welfare and housing, the periodical *Fajegészségügy* (*Practical Eugenics*) was published as a supplement to the prestigious *Orvosok Lapja* (*Physicians' Journal*) (Fig. 2).

Just two years after the publication of the *Archiv für Rassen- und Gesellschaftsbiologie* commenced, and three years before the first issue of *The Eugenics Review* was launched, *Fajegészségügy* became the second eugenic publication in the world. Yet compared to its German and British counterpart, it did not conceptualize eugenics in relation to race and racial protectionism but in terms of modern strategies of health and hygiene. Its editor, the social hygienist Henrik Pach, was a strong supporter of improving health conditions in the workplace through the provision of adequate social insurance and medical protection.[93] A modern state, he believed, required a modern system of welfare.

Pach actively propagated this vision in the pages of *Fajegészségügy*. Medical awareness and morality, he further claimed, were intrinsic to eugenics. Yet eugenics, for Pach, defined the problem of social and biological improvement in quite different ways than it did for Galton and Ploetz and, indeed, Kozáry. In a word, it was not concerned with "race". Eugenics articulated the need for new forms of health care and provision, extending the role of morality, familial duty and social productivity in assuring the nation's health. Specifically, Pach included the following factors in determining eugenics' fields of activity: (1) tuberculosis and infant mortality; (2) sexually transmitted diseases and prostitution; and (3) alcoholism and criminality. These factors

1906 deczember 20. Az „ORVOSOK LAPJA" 51. sz. melléklete. 1. szám.

FAJEGÉSZSÉGÜGY

A BETEGÁPOLÁS, GYERMEK- ÉS MUNKÁSVÉDELEM, IPAREGÉSZSÉGÜGY, TÜDŐVÉSZ ÉS NEMI BETEGSÉGEK, LAKÁSÜGY ÉS MUNKÁSJÓLÉTI INTÉZMÉNYEK SOCIÁLIS TÖREKVÉSEINEK TUDOMÁNYOS KÖZLÖNYE.

MEGJELENIK HAVONTA EGYSZER. * SZERKESZTI: PACH HENRIK Dʀ. * MINDEN JOG FENNTARTVA.

Az olvasóhoz!

A sociális eszmék hódító útjokon immár ezen ország határaihoz is értek. Őket feltartóztatnunk többé nem lehet, sőt nem is szabad! Feladatunk inkább a hatalmas árt e *fajérzés és erkölcse* emelte gátakkal irányítani és odahatni, hogy földünkre csak termékenyítőleg hasson!

Átmeneti, izzó korszakot él a mai nemzedék, melynek örökké változó alakulatai előre sejtetik a szebb és jobb jövő nagy feladatait és áldásait.

Ez alapvető munkálatokból a magyar orvosi tudomány és annak áldozárjai, a magyar orvosok is, kell hogy kivegyék a részüket! És a mint biztosra vehető, hogy a komoly búvárlat és kutatás mint eddig mindenha, úgy ezentúl is magán fogja viselni az új idők bélyegét, mert hiszen az orvosi tudomány csodálatos életrevalóságának titka abban rejlik, hogy mindig a kor vezérlő eszméihez simulni akart és tudott is: sajna ép oly kétséges, vajjon az anyagi gondok, a lenézés és mellőzés nyűgét békésen viselő, magyar orvosi kar idejekorán, a küszöbön álló társadalmi munkálatokból is kiveszi-e részét; vajjon azon magasztos feladatok és nagy problémák, melyek most a *faji egészségügy* körül megolandók, velünk és általunk dűlőre hozatnak-e?

Ép ezért indokoltnak, sőt szükségesnek tartjuk, hogy e *lapunk*, melynek tartalmát, színvonalát és irányát eddig mindig csak az orvosi gyakorlatra, tehát az élet nehéz küzdelmében már álló magyar orvosok nagy többségére való tekintetek szabták meg, ezentúl egy havonta megjelenendő *mellékletet* adjon ki, mely a *faji egészségügynek* kialakulásához, fejlesztéséhez és biztosításához szükséges ismereteket, törekvéseket és berendezéseket tartalmazza és terjessze.

A faji egészségügy szolgálatába álló eme lapunk tehát tartalmazni fogja:

a) a *tüdővész és gyermekhalandóság*, e két faji létünket leginkább veszélyeztető, *sociális betegség* sikeres leküzdésére alkalmasnak bizonyult törekvések és berendezések tudományos szemléjét;

b) tekintettel leend a *nemi betegségek és prostitució* elhárítására ajánlt és bevált óvintézkedésekre;

c) figyelemmel kiséri az *alkoholismus* elleni védekezés minden mozzanatát, nemcsak a hozzájuk fűződő *egészségügyi* hanem a *gazdasági és bűnügyi* vonatkozásion szemmeltartásával, végül

d) különös gondot fordít a *lakás és iskolaegészségügy, munkásbiztosítás, munkásvédelem* és *betestelhárítás* valamint az *iparegészségügy* eddigelé kevéssé mivelt és ápolt, de annál fontosabb szakkérdések *tudományos* ismertelésére.

Egyszóval lefoglalja a maga részére mindama társadalmi *intézmények* tudományos és tárgyilagos méltatását, melyek ép úgy az egyed mint a nemzet legértékesebb javának, az *emberi egészségnek* biztosítására szükségesek és alkalmasak.

Óhajunk, hogy rövid, de mindig a dolog velejét kimerítő közleményeinkben gyors útbaigazítást találjon a korával haladni kivánó, magyar gyakorló orvos!

Az „ORVOSOK LAPJA"
szerkesztősége.

Iskola-egészségügy.

× **Iskolabetegségek.** Habár mi a XX. század orvosai csak hirből ismerhetjük az elődeinktől annyira dicsért *házi orvos* állás örömeit és búját: a testi és lelki változások, melyeket «hű családjaink» *gyermekei* az iskolába való lépéssel, majdnem kivétel nélkül, felmutatnak, oly jellemzők és feltünők, hogy azoknak a mai iskoláztatás rendszerével való okozati összefüggéséhez kétség alig férhet. Teljesen figyelmen kívül hagyva itt a vidéki, falusi iskolák hihetetlen berendezését és tanmódszerét, joggal azt állíthatjuk, hogy a városi iskola ügyünk is, egészségügyi tekintetben még sok kivánni valót hagy. A ki az iskolaépülteink czélszerütlen sokemeletes berendezését ismeri, a ki tapasztalja, hogy kényelmességi szempontból a muzeumok, gyüjtemények sőt az igazgatólakások is földszinten vannak, a tantermeket pedig az emeletekre helyezik — mert oda már kevesbé hatol az utcza zaja!? — ez így vézna is roszszul táplált gyermekeink napjában többször sok lépcsőn fel és lefutni kénytelenek, az ugyan nem csodálkozhatik a sok szívbajon, melyeknek csirája sok gyermek az iskolában szerzi. Pedig már ama tapasztalati tény, hogy a legtöbb fertőző betegség — az ugynevezett gyermekbetegségek, — morbilli, scarlatina, dyptheria és az ezekkel járó következmény lek — első sorban a szívet támadhatják meg, követeli, hogy az emeleti tantermeket lehetőleg kerüljük.

Amúgy is nyilvánul az iskolaegészségügy iránti érzéketlenség, hogy a *tanítás* kezdetét, csupán hagyományos tekintetből, télen is reggeli 8 órára teszik. Vajjon mit szóljunk továbbá a nálunk divó helytelen vakáczió beosztásához?

Mi más mint az iskolatermek rossz világítása okozója ama feltünő sok rövidlátó gyermekeinknek? És vajjon ha már ők vétkesek a szemeink romlásában, miért idegenkednek az iskolaorvosok alkalmazásától, kiknek módjában volna a kezdődő bajt idejekorán rendelt szemüveggel, ha nem is jóvátenni, de legalább enyhíteni? Avagy még nem tudják, hogy középiskoláinkban átlag minden hatodik deák rövidlátó, hogy idővel mi is a «pápaszemesek birodalma» kétes értékű hirnévre pályázhatunk?

Mentségül feltesszük, hogy az iskola vezetésére döntő befolyással biró tek. és érdemes iskolaszéki tagok ezt nem tudják, de akkor miért nem hivják meg a hatósági orvost gyűléseikre? Ha ezt tennék és ezt saját gyermekük érdeke is javalja, akkor megtudnák, hogy az egészségügy követelményei szerint, az ablakok nagysága legalább a tanterem alapfelületének ötödrészével egyenlő legyen; hogy a balról felülről jövő világosság a legjobb és hogy a napvilágítás a legjobb világitó forrásnak mondható.

Jó és kiadó világítás mellett nemcsak a rövidlátók, hanem a görbehátuak (skoliotikus) száma is apadni fog.

Figure 2 The first issue of *Fajegészségügy* (1906)

reflected Pach's primary focus on fields as diverse as school hygiene, housing, employment insurance and work protection, accident prevention and industrial hygiene.[94]

Throughout its brief existence – having changed its title to the more appropriate *Szociális Egészségügy* (*Social Hygiene*) in 1908 – *Fajegészségügy* championed the introduction of health reforms aimed in particular at tuberculosis and alcoholism, as well as modern ideas of social hygiene.[95] For its contributors, and especially for Pach, the new science of eugenics reconciled social hygiene with health protectionism by emphasizing the medical dimension of each. But the intellectual overlap between social hygiene and eugenics was, in fact, an imperfect one. For the contributors to *Fajegészségügy*, hygiene was an instrument of social sanitization and of improving the population's collective health.

This view was shared by other supporters of eugenics in Hungary. For example, the secretary of the branch of the Institute of National Association of Public Health in Versec (a city in the Temes county), Péter Buro, further emphasized the question of hygiene and social responsibility when discussing eugenic prophylaxis. In his 1907 article on "Társadalmi és faji egészségtan" ("Social and Racial Hygiene"), Buro interpreted eugenics as a preventive health strategy dealing with both the individual and the race.[96] Hygiene was the essential component of this strategy. "The task of hygiene", Buro argued, was "not only to prevent disease but also to maintain and strengthen health."[97]

Inspired by the ideas of sanitary improvement and health prevention popularized by the German hygienist Max von Pettenkofer, Buro placed hygiene at the confluence of social and racial protectionism. Eugenics was, for him, broadly congruent with sanitation and social hygiene. It was also, however, directed at protecting the nation's racial body. "Racial hygiene [*fajhigiéne*]", he asserted, "deals with the question of the hereditary disposition of racial characteristics".[98] To strengthen this racial feature of eugenics, Buro preferred to employ the more potent term *fajhigiéne* rather than *faji egészségtan*, which he used in the title of his article. In itself this emphasis is unsurprising (both *higiéne* and *egészségtan* translate as hygiene), but for the general discussion on eugenic terminology at the time in Hungary, Buro's use of *fajhigiéne* is significant. He is the first author to use the term explicitly.

In widening his intellectual sources to include not only hygienists like Pettenkofer but also authors like Darwin, Spencer and Malthus, Buro associated social progress and preventive medicine with the eugenic goal of human betterment.[99] He drew no simple dichotomy between social (nurture) and racial (nature) improvement, and his insistence to keep the two together did not reflect a methodological uncertainty, but rather an awareness of the complex relationship between hygiene and eugenics. Through this association, eugenics became linked to a wide range of medical and social issues that preoccupied social and political reformers in Hungary at the beginning of the twentieth century. In Pach's case – as discussed earlier – this included the prevalence of contagious and venereal diseases, high infant mortality rates, social assistance, child welfare and the protection of mothers. In fact, precisely to the extent that it was conventionally reiterated, the alliance between medicine and eugenics served continually to overcome the distinction between the social and the biological.

This was also the strategy employed by the editor of the capital's main medical newspaper, *Budapesti Orvosi Újság*, the paediatrician Ferenc Torday in his article on the "health of the future generations".[100] Torday insisted that the nurturing of healthy children was one of the best ways to ensure the future of the nation against the danger posed by contagious diseases like tuberculosis. He championed modern ideas of health and urged the state to intervene to reduce the infant mortality rate. The relationship between contagion and social pathology reconfigured ideas of child welfare and the protection of mothers, as much as it raised awareness about the effects of alcoholism, unhealthy living conditions, lack of hygiene and poor nutrition on the nation's racial strength. The future of Hungary rested, in Torday's view, on the establishment of a modern system of health and hygiene.

When considered in the overall context of the early twentieth century, eugenic ideas of social and biological improvement recommended by medical journals such as *Egészség* and *Fajegészségügy* may appear different than those advocated by *Huszadik Század*. Yet, the important point is that all of these authors shared a modern conception of national welfare based on health, hygiene and eugenics. In turn, these diverse intellectuals harmonized their professional expertise with an assortment of public commitments to eugenics. Ultimately, they all conceived eugenics to be an overarching set of modern ideas of health and hygiene, offering wide-ranging solutions to society's alleged degeneracy.

Of particular importance in this context was the organization of the 16th International Congress of Medicine in Budapest between 29 August and 4 September 1909. With over 5000 participants from countries as diverse as Turkey and Paraguay, the congress provided an ideal forum for legitimizing Hungary's claim to scientific pre-eminence in the field of medicine and public health.[101] It was also the largest medical congress to date, spread over as many as 21 sections covering, among other branches of medicine, anatomy, embryology, experimental pathology, gynaecology, psychiatry, hygiene and tropical diseases.[102]

The distinguished Austrian bacteriologist and the director of the Institute for Hygiene in Munich, Max von Gruber, delivered a lecture on "Inheritance, Selection and Hygiene".[103] The major issue discussed was whether modern hygiene worked against the Darwinist theory of natural selection allowing those of lesser "racial value" to increase numerically and thus become a threat to the healthy population.[104] The concept of genetic inferiority (*Minderwertigkeit*) as a master trope in Gruber's eugenic theory gravitated around alcoholism and sexually transmitted diseases, both of which were blamed for causing degeneration of both the reproductive abilities and of the germ-plasm.[105] Eugenics, Gruber insisted, was the only way to alleviate the "biological grievances of human society and to forge a better future for mankind".[106] If Francis Galton coined the term eugenics, it was Wilhelm Schallmayer and Alfred Ploetz who first suggested efficient methods to prevent racial decline, Gruber suggested.

Assuring his Hungarian audience that he did not intend to leave them with "dark phantasies" about the future of the race, Gruber called for unity among those "who understood the importance of eugenics for the future of mankind", while emphasizing the need to disseminate the "understanding and love of eugenics" to the general public.[107] The establishment in 1907 of the Research Laboratory for National Eugenics at University College London, and of the International Society for Racial Hygiene (Internationalen Gesellschaft für Rassenhygiene), reinforced Gruber's hopes that the eugenic "movement for regeneration" (*Bewegung für Regeneration*) may one day succeed. The closing argument left the audience in no doubt about Gruber's commitment to the relationship between medicine and eugenics. British Prime Minister William Gladstone's prophecy that the "physician would become the protector of the race" would only come true, Gruber insisted, when modern medicine and hygiene performed its primary task: removing the agents that contributed to the weakening of the race or those who were already degenerate.[108] Eugenics was both a prerequisite for, and an extension of, any long-term medical programme to improve national health.

As the chairman of the Munich Society for Racial Hygiene (Münchner Gesellschaft für Rassenhygiene), Gruber was the first eugenicist of international standing to have spoken publicly in Hungary. In the history of Hungarian eugenics, his lecture and the personal connections he later developed with Hungarian eugenicists like Teleki would help construct a second tradition, complementary to – but distinct from – that set out by Galtonian eugenics. Whereas Galton had proposed a statistical method of controlling human reproduction, Gruber responded with the racial commensurability of eugenics. Each nation faced the spectre of degeneration, and the eugenicists who met in international congresses to address these issues offered solutions compatible with their national sociopolitical realities. The ideal of a healthy nation demanded not only the rejuvenation of the race but also the creation of a better society. In formulating a narrative of both racial and social improvement, eugenicists glorified the mentally and physically healthy, while warning against those deemed inferior.

Intersecting Eugenic Traditions

The articulation of eugenics constituted a powerful source for many early twentieth-century European discourses on racial degeneration, and Hungary was no exception. This new eugenic vocabulary attracted a rather eclectic segment of the Hungarian elite; geographers, sociologists and philosophers joined physicians and biologists in searching for the new scientific rationale for a nation purportedly affected by demographic decline and racial degeneration. Some Hungarian authors, however, felt that Western social and biological sciences had not been appropriately applied to local realities,[109] and that, moreover, not enough emphasis was put on explaining Hungary's dominant racial element and its subsequent national development.[110]

Engaging with this new nationalist ethos, some members of the Society of Social Sciences – including the well-known zoologist István Apáthy, the social economist Jenő Gaál, and the legal historian Ákos Timon – left the society and established their own association in 1907, fittingly named the Hungarian Association for Social Sciences (Magyar Társadalomtudományi Egyesület).[111] The consequences of this parting would prove to be significant for the development of eugenics in Hungary, not least on account of the association's own journal, *Magyar Társadalomtudományi Szemle* (*Hungarian Review of Social Sciences*), which consistently championed "Hungarian" ideas of social and biological improvement. The creation of the Hungarian Association for Social Sciences reflected how science, ideas of social reform and nationalism were closely intertwined in Hungary; it also mirrored Apáthy's insistent faith in eugenics as a *national* science.

In distancing itself from the Society for Social Sciences' alleged cosmopolitanism, the Association for Social Sciences wanted to protect what it deemed to be Hungarian national values. From its first issue in 1908, the Association's own journal, *Magyar Társadalomtudományi Szemle*, drew attention to the nationalist potential of eugenics, laying out the grounds for a new style of biologized politics. Apáthy, for instance, advocated the need for "national education" predicated upon Darwinist notions of evolution and selection. Societies, he consistently argued, were living organisms conditioned by the natural cycles of growth, maturity and decay.[112] Károly Balás, on the other hand, continuing the lines of demographic arguments exposed in his previously mentioned 1905 *A népesedés*, reviewed the importance of modern population policies (like neo-Malthusianism) for the political control of the national body.[113] Complementing *Huszadik Század*'s efforts to enlighten the Hungarian general public in eugenic matters, *Magyar Társadalomtudományi Szemle* added a nationalist fervour to the latter's cultural and political ambitions.

As preoccupations with Hungarian national identity increasingly refined the fundamentals of eugenics, there were additional developments, and two in particular are worth mentioning: the Turanic Society (*Turáni Társaság*) and the journal *A Cél* (*The Target*), both established in 1910.[114] Pál Teleki was elected the president of the Turanic Society, with the economist Alajos Paikert as vice president, alongside the esteemed Orientalist Ármin Vámbéry and Count István Széchenyi's son, Béla – a passionate explorer – as honorary co-presidents.[115] Together with *A Cél*, which was fashioned as a "social, economic, literary and sport review" and adorned with the Latin dictum *Mens sana in corpore sano* (Fig. 3), the Turanic Society spelled out the emergence of a new racial discourse, one closely attuned to eugenic technologies of biological improvement. Gradually, as will be discussed in the following chapters, this insistence on Hungarian racial identity shaped eugenic narratives during the First World War. Moreover, the conceptual interaction of nationalism with biological theories of evolution would have broader implications for the understanding of eugenics, especially when cast in terms of the dichotomy between the Hungarian nation's inherited characteristics and those of other ethnic groups living in the country.

The question, ultimately, was not which theory of heredity and evolution had pre-eminence in Hungary but, as Karl Pearson maintained,

What, from the scientific standpoint, is the function of a nation? What part from the natural history aspect does the national organization play in the universal struggle for existence? And, second, what has science to tell us of the best methods of fitting the nation for its task?[116]

46

Figure 3   The first issue of *A Cél* (1910)

Whatever conceptual strategy they pursued, Hungarian authors agreed that a connection between sociology, medicine and anthropology, viewed as national sciences, and eugenics – described as the most efficient mechanism to ensure the nation's social and biological improvement – needed to be immediately established. While *Fajegészségügy, Egészség* and *A Társadalmi Muzeum Értesítője* advocated this development through public health and sanitary reforms, *Magyar Társadalomtudományi Szemle* championed the protection of the racial elements of the Hungarian nation.

Huszadik Század, on the other hand, was conceptually more versatile, showing interest simultaneously in Ploetz's insistence on the protection of the race as well as in Galton's ideas about a population's "reproductive health", which he quantified and measured in terms of fertility statistics and infant mortality figures. Galtonian eugenics was favoured, however, as demonstrated by Jászi's letter to Galton and the subsequent translation of his Herbert Spencer Lecture on "Probability, the Foundation of Eugenics", which was delivered at the University of Oxford in 1907.[117] This translation was an important moment in the history of Hungarian eugenics, one cementing the connections with British culture and science which journals like *Huszadik Század*, for example, had progressively established since 1900.[118] Treading the well-known path of social and educational reforms necessary to support individual abilities, talents and faculties, Galton dwelt on the practical details related to the widespread adoption of eugenics. He cautiously suggested, however, that the time for legislating eugenics had not yet come; that it still needed to secure a favourable public reception and, ultimately, general acceptance. To this end, Galton was certainly right to argue that, for eugenics to be accepted by society at large and not just by a few individuals, the eugenicist needed to offer a theory of human development able to reconcile science with a strong sense of commitment to public education and training. He did not fail to grasp how profound eugenics' impact would be once the teachings of what he called a "new religion" were absorbed and accepted by society and the state. "When the desired fullness of information shall have been acquired", Galton declared, "then, and not till then, will be the fit moment to proclaim 'Jehad,' or Holy War against customs and prejudices that impair the physical and moral qualities of our race".[119]

Galton's eugenic philosophy concentrated on public awareness and scientific investigation, and these also became the core principles guiding eugenicists in Hungary and elsewhere during the first decade of the twentieth century. However, there was no effortless transition from treating eugenics "as an academic question"[120] to the subsequent engaging in the "Holy War" of its practical application to society. Ultimately, it was the development of scientific knowledge surrounding biological improvement that would determine the widespread reception of eugenics. Another reason for this impending achievement was heightened intellectual investment in the new disciplines of sociology, anthropology and demography, which, alongside medicine, attempted to explain – and counteract – the risks of social dislodgement and ethnic co-mingling in early twentieth-century Europe. By stressing the importance of these diverse intellectual sources underpinning the envisioned social, racial and national improvement, it becomes possible to recognize the ideological importance of eugenics, especially its all-pervading assertion that biological readings of the individual and of society were essential for building a strong, dynamic and healthy nation. As will be demonstrated in the next chapter, the public debate on eugenics that was organized by the Society of Social Sciences in 1911 unequivocally celebrated this diversity of views, clearly demonstrating the degree of interaction and cross-pollination achieved by Hungarian scientists by the end of the first decade of the twentieth century.

2
Debating Eugenics

During the process of absorption and adaptation to local Hungarian realities, eugenics disclosed a notable conceptual flexibility. Building upon this discussion, this second chapter aims to provide an inventory of the elements that constituted these foundations. Tied to ideas of public health, social hygiene and national welfare, eugenics aimed to exert biological control over the population. This ability underpinned the cultural framework in which ideas of social and biological improvement developed, providing the foundations for the first public debate on eugenics convened in Hungary in 1911. Detailing the eugenic views expressed on this occasion, and introducing their authors, therefore allows us to see this Hungarian debate on eugenics as a link in a chain of programmatic eugenic activities occurring simultaneously in Europe and the USA during the first decade of the twentieth century.

This debate on eugenics is important for three reasons. First, it created an auspicious environment for eugenics to interact with various scientific disciplines, from biology and medicine, to demography and sociology. As the previous chapter intimated, after 1900 eugenics was increasingly adopted by segments of the Hungarian intellectual and scientific elite, but this acceptance was never uncritical or monolithic. Evolutionary ideas permeating the Hungarian cultural landscape at the beginning of the twentieth century were by no means univocal, even if they tended to produce a yearning and identification with the increasingly articulated visions of social and biological improvement already present in the works of other European eugenicists.

Second, the debate demonstrates the level of scientific sophistication achieved by Hungarian eugenicists at this time, in addition to their reception of European and North American trends in heredity and genetics. During the first decade of the twentieth century Hungarian physicians, biologists, anthropologists and sociologists struggled – like their counterparts in Britain and Germany – with the public relevance and importance of heredity.[1] Debates on the nature of inheritance, like those between biometricians and Mendelians that caused many a methodological schism among eugenicists in Britain, were widely echoed within the Hungarian scientific community.[2]

The third reason why this Hungarian debate is important in the history of international eugenics concerns its wider political and social message. Eugenics was seen as a tool to decode social and biological predicaments, such as overpopulation, prostitution or declining birth rates, predicaments thought to be impeding the progress of the Hungarian state and society. This emphasis on community and society rather than race distinguishes the Hungarian debate from similar events at the time. Organized by the Society of Social Sciences and publicized by *Huszadik Század*, this debate on eugenics was as much scientific as sociopolitical. As such, it illustrates more than just the ambition of a few Hungarian eugenicists coming to terms with new developments in the social and natural sciences. Ultimately, the debate represented a successful attempt to enlighten the general public and the scientific community about the aims, methods and content of eugenics.

Biometrical Foundations

Debates between Lamarckians and Darwinians as well as between biometry and Mendelism were at their peak in the 1910s, and their convoluted ramifications echoed in Hungary too.[3] The biometric foundations of eugenics were blessed by none other than Galton himself, whose article on biometry was – as previously mentioned— translated into Hungarian in 1907. As elsewhere, Hungarian eugenicists established a relationship between traditional preoccupations with health and hygiene on the one hand, and a new social politics on the other, which derived its potent vitality from the ostensibly regenerative ethos of modern science.[4] Hoping to consolidate their authority within Hungarian intellectual circles, eugenicists approached *Huszadik Század*, a journal which – they reasoned – had already demonstrated its commitment to popularizing ideas of social and biological improvement. In 1910, as a consequence, three seminal articles were published in *Huszadik Század* which acted as a prelude to a more sustained debate on eugenics.

The biologist Lajos Dienes (Fig. 4) authored the first. Entitled "Biometrika" ("Biometrics"), this commentary briefly introduced Francis Galton and Karl Pearson's statistical and experimental studies, as well as

Figure 4 Lajos Dienes (courtesy of Katalin Czellár)

the intimate relationship between social statistics and eugenics. Dienes was sceptical of the extent to which an environment could influence biological evolution. He repudiated Lamarckian assumptions that social and biological qualities, such as "the child's behaviour or learning ability" as well as physical impairments, like "the myopia of school-children",[5] were determined by the environment, and instead entrusted their aetiology to heredity. The role of mathematical equations and statistics in identifying "the biological processes paramount to the functioning of society" was also vigorously defended. In keeping with this view, Dienes understood eugenics as a branch of biometry; albeit an important instrument of social control, whose main goal was to improve the physical, moral and intellectual character of the population. Dienes's second aim was to describe biometry's relevance to the study of heredity. Praising Galton and Pearson, he wrote, "the results of their method are entirely innovative; these results reveal the validity of things we could have thought of, but could never have demonstrated in the past". Although grounded in a novel approach to heredity, Dienes admitted that biometry "will probably not lead to a complete theory of society".[6] Despite this hesitation, Dienes optimistically noted that once the general public and academic communities were properly informed, biometry would be accepted as a scientific theory. As this was a short article, Dienes concluded by pledging to return to the topics of biometry and eugenics with more systematic observations.

József Madzsar wrote the second article, entitled "Gyakorlati eugenika" ("Practical Eugenics"). Although it echoed Karl Pearson's *The Problem of Practical Eugenics* (1909), Madzsar's text was less speculative. As in Dienes's case, this account was meant to be introductory. Nevertheless, Madzsar was less ambiguous than Dienes when it came to biometry's importance. He was rather more vocal in his criticism of the role environment and society played in shaping eugenic improvement. Furthermore, Madzsar insisted one should not anticipate that the population's biological improvement would come from "technological progress, better public health conditions, proper education and the change in the economic situation".[7]

As a first step, Madzsar sought to offer an assessment of the individual's hereditarian value according to Pearson's biometric epistemology. "By employing the exact methods of biometrics", he asserted, "it was possible to demonstrate how certain maladies, for instance tuberculosis, deaf-muteness, and mental disorders", differed in their range of variability.[8] Next, Madzsar criticized state institutions for their attempts to improve the health of the population through social assistance. These he felt to be working against "the goals of natural selection". Madzsar's conception of "practical eugenics" appropriated Galton and Pearson's biometrical and statistical evaluation of hereditary determinants, and opposed other visions of public assistance and medical reforms based on humanitarian principles. Unhesitatingly, Madzsar declared that "the present form of social charity is even more dangerous because in most cases it obstructs the suppression of elements which are most burdensome and dangerous for society and it encourages their proliferation".

Moreover, Madzsar did not stop at simply criticizing policies of "social charity"; he went still further and suggested radical policies of "negative" eugenics, including sterilization. Invoking Plato's argument that the "disabled should be banished from the state", Madzsar maintained the necessity of employing this practice in contemporary Hungarian society. Eugenicists, he continued, should resist their "pseudo-humanism" and "at least pursue the goal of preventing the proliferation of the unfit and promoting the proliferation of the fit". In such a context, the state was to be invested with eugenic privileges, and Madzsar placed consistent emphasis on the tight connection between eugenic politics and state interventionism. The role of heredity in mythologizing the national body was a paramount one. Indeed, according to Madzsar, "if the state has the right to deprive citizens of their freedom, of their life even, it undoubtedly has the right to sterilize as well, especially when this can be performed without any other unpleasant consequences for the individual". There were thus a "multitude of queuing tasks" facing eugenic reformers but – as Madzsar hoped – "through the scientific organization of the research into heredity and through the establishment of eugenic laboratories we will have to create a scientific basis for all those tasks".[9] Madzsar therefore demanded that the modern state purify its national body by eugenic means. Ultimately, the state was the repository of the nation's biological qualities, and needed to act correspondingly as their vigilant custodian.

There was more. Madzsar linked "practical eugenics" to social control and political discipline, seen as a means of enabling "worthy" individuals to flourish in a eugenically planned society. In his conclusions, unsurprisingly, he positioned eugenics within the field of politics, rather than within that of medicine and science. Quoting Galton, Madzsar predicted that

> a eugenic religion will take shape in the realm of ideologies. This religion will forbid all forms of sentimental charity which are damaging to the race; it will enhance kinship and increase love for the family and for the race. In brief: eugenics is a manly, promising religion which calls upon the noblest feelings of our nature.[10]

According to Madzsar, this synthesis of biological and mathematical sciences and politics reflected more than just concerns with future generations' deterioration in health; it also endowed eugenics with the important task of improving the population's racial quality and quantity.

In the third article "Eugenika" ("Eugenics"), the natural scientist and one of the translators of Darwin into Hungarian, Zsigmond Fülöp, expanded upon some of the arguments outlined by Madzsar. Having established eugenics' historical genealogy by tracing it back to the ancient Greeks, Fülöp then turned to general considerations on the revolutionary theory of natural selection. In fact, it was within the scientific revolution brought about by Darwinism that he placed the emergence of eugenics. As such, Fülöp's definition of eugenics was simply "the application of Darwinism to society with the scope of improving the qualities of the race, especially social qualities".[11] Since Darwin had demonstrated that the mechanism of natural selection was based on inheritance, variation and selection, Fülöp considered that selection in particular assumed the pivotal function of social and biological engineering. Equipped with such knowledge, the eugenicist was able to record, classify and accordingly decide the future of degenerate and deviant individuals.

Fülöp was also impressed both with how Galton reconciled his theories of heredity with Darwinism and, in addition, with how methodical biometry was. "To achieve the tasks of eugenics", Fülöp noted, "it is not enough just to establish which biological factors work in society. To understand the role of these factors we need to establish *their mathematical definition*".[12] Galton had usefully employed such precision, Fülöp believed, in developing his theories of ancestral inheritance and filial regression. Popularizing eugenics was just as important as the scientific research itself, and Fülöp correspondingly praised the pioneering work undertaken by *Biometrika* and the English eugenicists affiliated with the Eugenics Laboratory at University College London, especially Karl Pearson and David Heron.

However, British eugenics was not the only source drawn upon by Fülöp in order to demonstrate his intellectual dexterity. Engaging with contemporary German literature, particularly Wilhelm Schallmayer's 1905 *Beiträge zu einer Nationalbiologie* (*Contributions to a National Biology*) and Robert Sommer's 1907 *Familienforschung und Vererbungslehre* (*Family Research and Heredity*), Fülöp also analysed other forms of eugenic policy necessary to improve the race. "Eugenic research must be extended", he argued, beyond revealing merely the hereditarian past and future of an individual, but to observe the "new races resulting from the mixing of different nations and races" as well. Exploring eugenics' impact on the contested landscape of race, Fülöp insinuated that modifying existing racial boundaries could have positive results, including the creation of a new type of "superior social value".[13]

In a similar manner to Madzsar, towards the end of his article Fülöp turned to the practical application of eugenics. Based on his Darwinist beliefs, and following Galton, Fülöp identified two ways for improving society: "negative" methods, centred on impeding the proliferation of degenerate and unfit individuals; and "positive" methods, based on encouraging healthy individuals to reproduce. The pursuit of "positive eugenics" was, however, hindered by factors which, according to Fülöp, had to be addressed first if the racial improvement of Hungarian society was to succeed. Fülöp produced a range of policy proposals, including the abolition of conscription, even of war as such, as it periodically maimed and killed the finest male individuals. Furthermore, mental asylums should be closed down, as they only allowed the mentally ill to survive in modern society. Urbanization, moreover, was blamed for causing degeneration, while religious celibacy was attacked for hindering both marriage and procreation by the educated elite. Conversely, degenerate infants should not be permitted to live, and marriage among family relatives was to be forbidden. Finally, Fülöp enumerated two additional critical factors contributing to the weakening of the Hungarian body politic: the one-child practice and increased emigration to the United States.[14]

Not surprisingly, too, Fülöp accused the existing social system of "pseudo-humanism" and "charity", which advanced the proliferation of the unfit and portended the "biological ruin of the race". In response, he proposed radical solutions, such as giving society "ideological and emotional training" to help it accept "negative eugenic measures". Fülöp's eugenic vision was highly interventionist, and the state was again invoked as the only credible, efficient means of introducing eugenic policies. Echoing Madzsar, Fülöp posed the following question: "if the state has the right, for the good of society, to execute the criminal, why should it not have the right, for the equally important biological development of society, to obstruct the reproduction of degenerates?"[15] The state, then, was to exercise complete control over the individual in the interest of improving the racial health of society. Ultimately, what Fülöp advocated was a national community subjected to regular eugenic close scrutiny.

Although he sympathized with Galton and Pearson – most notably in their approaches to popularizing eugenics, such as the establishment of research teams and societies as well as lecturing – Fülöp was not persuaded that this form of propaganda was, in and of itself, enough to make eugenics "the religion of the future". He even accused Galton of "religious idealism". The only way to achieve a eugenic transformation of society, Fülöp contended, was neither to encourage technological progress nor to support existing social institutions, but to radically change the social system itself. "Only then", he asserted, "would a healthy and unwavering sociopolitical eugenics be possible, namely the sacrifice of one's personal interests in favour of collective interests".[16]

These three articles collectively identified some of the problems pertaining to the scientific nature of eugenics, in addition to the difficulties of its practical application. In the wake of Galton's death on 17 January 1911, the stage was set for a wider public debate.[17] It also helped that Dienes, Madzsar and Fülöp were all well connected to progressive intellectual circles in Budapest. Eventually, personal connections as well as an increased public interest in biologically informed welfare policies would yield these men's desired outcome: a public debate on eugenics. More importantly, these articles – and the debate that followed – also made clear that eugenic ideas of social and biological improvement in Hungary did not simply celebrate scientific trends elsewhere in Europe, but also drew heavily on personal networking and the increased professionalization of Hungarian eugenicists.[18]

Lectures on Eugenics

On 27 and 31 January 1911, Oszkár Jászi sent an invitation to the Association for Social Sciences and the Royal Medical Association of Budapest (Budapesti Királyi Orvosegyesület), respectively, publicizing the forthcoming debate on the issues of eugenics (*eugenika*), racial improvement (*fajnemesítés*) and racial hygiene (*faji-higienia*).[19] Three speakers were announced, namely József Madzsar, Lajos Dienes and Zsigmond Fülöp, who were asked to expand on the arguments expressed in their recently published articles. The invitation also specified a set of questions for each participant to consider in the debate:

> Can racial degeneration be prevented and the physical and mental health of the race be improved? Can we accelerate or abate the effects of natural selection? Can it be possible that the development of society conforms to the laws of heredity? Which are the social and physical conditions for eugenics [to emerge]? What trends in social and population policies (e.g. neo-Malthusianism) are the consequences of eugenics?

In the first lecture on 1 February 1911, entitled "Fajromlás és fajnemesítés" ("Racial Degeneration and Racial Improvement"), Madzsar outlined a comprehensive eugenic programme for Hungary. From the outset, he observed the increasing level of physical degeneration among the male population as a result of conscription, illness and poor living and hygiene conditions. As revealed in his previous article, Madzsar was a supporter of the most rigorous form of eugenics, thoroughly disapproving of what he perceived to be public and official indulgence towards symptoms of social and biological deterioration. He pointedly remarked that degeneration was detrimental to both society at large and to the country's military strength. Producing "healthy individuals is not only the army's prerogative; to have as many healthy and as few individuals in poor health as possible is the essential prerequisite of our entire culture".[20]

Madzsar was particularly alarmed by the lack of social commitment and cultural participation by those he deemed "socially and racially weak". "We need only to consider how those in poor health do not engage with culture", he noted acrimoniously.[21] Invoking the physical elimination of undesired offspring as practised by the ancient Greeks, Madzsar eulogized their ideals of fitness and racial health. "We have grown too accustomed", he charged, "to the Christian imperative to pity the wretched, while, in the meantime, we have abandoned the pagan love of the beautiful and the healthy".[22] This denunciation of Christian charity was a bold statement, reflecting Madzsar's commitment to a society governed by biological rationality and eugenic responsibility.

In the process of scrutinizing the population's racial value, Madzsar also addressed the continued popularity of eugenic literature. He competently presented a conceptual synthesis of various works on eugenics, including Francis Galton's 1889 *Natural Inheritance*, Karl Pearson's 1907 *The Scope and Importance to the State of the Science of National Eugenics* and Leonard Doncaster's 1910 *Heredity in the Light of Recent Research*. Largely agreeing with these authors, Madzsar explained his position as follows:

> First of all, and following the example of the Galton Eugenics Laboratory, we need to work on the science of eugenics, examine the rules of heredity, collect good and bad pedigrees, thus creating a proper theoretical foundation for its later, practical implementation. In order to discover and adopt biological views everywhere in our practical social policy, we need to reformulate our entire thinking according to the accumulated data. It is neither enough to deal with economic aspects and external circumstances, nor to enhance education and individual hygiene: we need to consider the future generation as well.[23]

Madzsar then elaborated on some of the ideas presented in his previous article on practical eugenics, of which the most important concerned degeneration, the control of reproduction and marriage. Madzsar's view was that the eugenic integrity of the nation was threatened by social and biological deterioration, and that the proportion of constitutionally weak and mentally disabled Hungarians was increasing. Transposing these anxieties to the population's perceived decline in racial quality served to reinforce Madzsar's central argument, namely that for any form of social and biological engineering to succeed it required the control of reproduction. In due course, "the institution of marriage" would need to be reformed. Echoing familiar tropes from neighbouring European discourses on eugenics, Madzsar suggested that marriage between the mentally disabled caused profound social and biological instability. Eugenic principles should, in turn, be employed to regulate such marriages. The creation of "a new biological aristocracy of fit families" – one based not on social class, but upon hereditary qualities – was what Madzsar hoped to achieve in Hungary.[24]

To this end, Madzsar's eugenic goals included an attempt to improve social policies in Hungary, most notably in the spheres of medical care and social assistance; second, to educate a responsive audience in the virtues of eugenics, especially its heuristic and social potential as a source of individual liberation; and finally, to develop and clarify a policy of eugenic reform. Aware of the distance separating the profound transformations called for in his eugenic doctrine and the existing sociopolitical context in Hungary at the time, Madzsar explicitly acknowledged that "eugenics is still a utopia today, but it might not be such a distant issue; after all, all theories have been utopias at a certain moment".[25]

Galton and Pearson's contributions to eugenics' consolidation as a scientific discipline, as well as the relationship of eugenics to the statistical analysis of fertility rates and family histories, were further explored by Lajos Dienes in his lecture "A fajnemesítés biometrikai alapjai" ("The Biometrical Basis of Racial Improvement"), delivered on 8 February 1911.[26] Dienes accepted Galton's ideas regarding heredity's dominance over environment, and insisted on the biometrical approach's importance to understanding heredity through statistical techniques. According to Dienes, differences in physical traits, health and intelligence could be explained only through the statistical study of natural selection within the population. Dienes thus combined Pearson's population approach to Darwinist variation with Galton's hereditary conception of society as a way of clarifying his own concept of eugenics.

Biometry, Dienes elaborated, was not only mathematical and scientific scholarship par excellence, less exposed to gratuitous speculation, it also convincingly explained how hereditary variability and correlation influenced natural selection. "In any case", Dienes noted, "the most important fact is that the effect of biological factors upon society can be identified and compared quantitatively". Based on this assumption, a much larger programme of eugenic rejuvenation could be envisaged for Hungary: "We cannot fail to realize the importance of these statements for the assessment of social factors and for the elaboration of our future social policy, and it is impossible not to discover the urgent need for continuing these investigations."[27] Highlighting the importance of social factors in the implementation of eugenic programmes nurtured the hope that biological and social degeneration could be countered, if only the proper policies on public health and preventive medicine were applied widely enough. Incorporating biological and medical sciences into social reform was challenging, Dienes argued, but possible. The making of a eugenic culture presupposed, in fact, that the boundaries of the social and its corresponding biological framework were continually under medical surveillance.

The last speaker to address the Society of Social Sciences was Zsigmond Fülöp, on 14 February 1911. In "Az eugenetika követelései és korunk társadalmi viszonyai" ("The Demands of Eugenics and the Social Conditions of Our Age"), Fülöp portrayed eugenics more as a social than a biological discipline, and this was especially the case with the branch of eugenics intending to deal with collective improvement.[28] Fülöp did not share Dienes's enthusiastic support for quantitative eugenic policies, and instead offered his support to the qualitative aspect of biological improvement. "The best definition of eugenics I could think of, for the purpose of my lecture", he remarked, "is that eugenics is a demographic policy which takes qualitative rather than quantitative aspects into consideration."[29] This warning was needed to illustrate Fülöp's next argument; namely, that the main objective of eugenics was to determine which individual qualities were hereditary and which were not. Only then could eugenicists act accordingly.

Fülöp's attention then turned to the arguments raised in previous lectures, most notably the issue of degeneration. Fülöp too was adamant in his insistence that degeneration constituted one of the most important issues facing eugenics, both in Hungary and elsewhere. Thus, "as rapid degeneration is an undeniable fact", the question preoccupying eugenicists became "whether we need to initiate a eugenic social policy, or in other words: is there a need for the creation of a human race physically and psychologically stronger and more valuable than the present one? The answer to this question is repeatedly yes."[30] It is within the context of considerations such as these that Fülöp assessed the role of eugenics in social welfare and public health.

The social and biological improvement of society were inextricably linked, as illustrated by Fülöp's anticipated eugenic agenda, one framed by the following questions:

1. Does eugenic research offer enough positive results which could in turn lead to the practical realization of eugenics? 2. What are the most appropriate tools for this realization? 3. Can the identified

goals of eugenics be realized within the present social framework, or should we modify our social conditions first?[31]

In answering these questions, Fülöp engaged in a critical evaluation of biometry and eugenics. First, he refined his approach to statistical methods in the study of heredity. "Not even the most perfect *statistical* research", he believed, "can elucidate *the causes of variability and heredity*". Fülöp further explained his position on this issue thus:

> I think that, from a eugenic point of view, it is much more important to determine whether acquired features are hereditary or not than to perform statistical research, because we will have to give eugenic social policy an entirely new direction if it turns out that we need not deal with such acquired characteristics or with the impact of the environment, but only with hereditary features instead.[32]

Fülöp's lecture offered a similarly unorthodox perspective with regard to the relationship between heredity and practical eugenics. Both had important ramifications for revolutionary changes to the existing social order. On the one hand, Fülöp questioned the connection between research into heredity and "eugenic social policy" while, on the other, he emphasized the importance of Darwinism in articulating artificial selection. "In the beginning", Fülöp claimed, "it will suffice for us to know the most important rules of general biology, which play an important role in the evolution of species, rules which have been applied to the breeding of animals and plants for thousands of years."[33] In recent times, however, eugenic selection had been stymied, mostly due to improvements in the medical sciences. While commendable in terms of intellectual achievements, modern progress did not, Fülöp believed, constructively contribute to racial improvement. Eugenics was called upon to rectify an unhealthy physical environment, coupled with the gradual deterioration of the nation's physical body. Fülöp accordingly argued for the introduction of eugenic legislation to improve the social environment while remaining faithful to a political philosophy based on national protectionism.

Fülöp's preference for qualitative eugenics was an attempt to establish the primacy of nature over nurture in the subsequent public debate. Biological identity was central to individual self-representation, which was surely strong enough to overcome cultural division and social and national estrangement. These alleged "ills" happened as a result of the anomic experience of modernity. Fülöp developed his critique of modernity into an articulated eugenic commentary, one which chastised the existing social and political elites for their complete ignorance of "biology and sociology".[34] Rather than limiting the power of the state, he believed, eugenicists needed to rely on its coercive force in order to induce the collective transformation of the nation along biological lines. Fülöp's eugenic vision, therefore, yielded to state interventionism. In a modern eugenic state thus defined, the national value of biological improvement would be officially recognized.

Fülöp's eugenic interventionism diverged from the form of eugenics advocated by Galton and his supporters, which was decidedly noninterventionist and individualist.[35] Mindful of important structural differences between societies in Britain and Hungary, Fülöp was fearful that in his own country eugenics would be undermined without state resources, or eventually rendered ineffective in the wake of constant ethnic and social changes. In addition to its potential for channelling social transformation, eugenics should also prompt the creation of a new "national ethics"; namely, an evolved form of social solidarity that ensured the racial continuity of future generations. In reference to Schallmayer, Fülöp concluded that, with the emergence of a new morality and ethics based on biology, the old order – sanctified by religion and outdated political philosophies – would finally be replaced. Only when this secular religion predominated would eugenics triumph.

Engaging with Eugenics

These debates over the meaning of eugenics illustrate the emergence of a scientific language of social and biological improvement in early twentieth-century Hungary. But these debates should also be placed in broader social and cultural contexts. How did contemporaries react to the public lectures delivered by Dienes, Madzsar and Fülöp? Was there any interest in eugenics outside the community of physicians and evolutionary biologists? As the Society of Social Sciences anticipated, reactions to the lectures were immediate. They formed the first public discussion of eugenics within the Hungarian scientific community, aptly called "Eugenika vita" ("The Eugenic Debate").

Published by the *Huszadik Század* between February and September 1911 under the title "A fajnemesítés (eugenika) problémái" ("The Question of Racial Improvement—Eugenics"), this public debate had a clear aim: to clarify the scientific and social challenges associated with various interpretations of eugenics. A varied terminology was employed simultaneously – *eugenika, eugenetika, fajromlás* and *fajnemesítés* — a clear indication that there were doubts over whether the Galtonian term *eugenics* was sufficient to encompass the participants' competing perspectives on social and biological improvement.[36]

The list of contributors included individuals from diverse professional groups, such as lawyers, psychologists, veterinary surgeons, physicians, sociologists, social hygienists and natural scientists. István Apáthy and Leó Liebermann were also in attendance. Sessions were chaired by the president of the Society of Social Sciences, Gyula Pikler, and the vice president, Ervin Szabó. Considering this wide and heterogeneous mix of participants, it is perhaps unsurprising that the debate was characterized by scientific arguments concerning the application of eugenics to society, in addition to those considering how theories of evolution could substantiate demands for state intervention.

While some observations simply recapitulated arguments put forward by the three speakers, most participants aspired to be innovative. Apáthy, for example, prepared two sets of questions, one about social biology and the other about practical eugenics. Some of them were just rephrasings of the original questions which the three initial speakers – Madzsar, Dienes and Fülöp – engaged with in 1910, particularly with respect to the practical application of eugenics; the importance of biometry for the study of eugenics; the relationship between neo-Malthusianism and eugenics; and the imperative of social reform. Other questions, however, raised the issue of the role played by natural and sexual selection in modern societies; racial degeneration; marriage and racial protection; eugenic morality and, finally, the role of the state in fostering it.[37]

Two dominant themes emerging from Apáthy's questions were extensively discussed during this first public debate on eugenics: combating social and racial degeneration; and the creation of a new social order based on biological and eugenic principles. It was generally agreed that heredity was paramount in shaping any individual's life, but difference of opinion persisted on two key questions. First, were alleged social afflictions, such as alcoholism or sexually transmitted diseases, hereditary? Second, could the transmission of hereditary traits be affected by external influences, like environment, education and so on? On 4 April, in his contribution to the debate, the Doctor in Chief of the Nagyvárad Garrison, René Berkovits, explicitly connected the issue of degeneration to heredity:

> To employ a purposeful and practical form of eugenics we need to determine the relationship which exists between two major issues: the heredity of acquired characteristics and degeneration. The matter of

degeneration appears under very different light depending on whether we acknowledge or deny the heredity of acquired characteristics.[38]

Berkovits further believed that this imbalance had to be remedied by appropriate social and family policies. The solution, as seen by supporters of this view, was one of artificial selection. The issue was less about improving the performance of the race as a whole than encouraging the reproduction of individuals with "good heredity" while discouraging the existence of those designated as "unfit". According to Berkovits, "it is of utter necessity to prevent the proliferation of degenerates; this is the only way to extinguish haemophilia, inherited defective constitution, endogenous mental diseases, hereditary antisocial behaviour, and so on". Moreover, an interventionist health policy was simply insufficient. What was also needed, Berkovits concluded, was to establish "a eugenic commission to research the specific matters and come up with a clear recommendation for the legislation; the negative part of this legislation, namely the list of those who should be banished from marriage, has already grown enough to be formulated".[39]

Berkovits was not the only participant to connect degeneration with modernity's effects on Hungarian society, or to advocate negative measures for eugenic segregation. In his comments from 11 April, the physician Imre Káldor, for instance, suggested that social transformations induced by industrialization and urbanization contributed to a weakening of human resistance to a variety of degenerative factors, ranging from spiritual corruption to loss of physical prowess and decreasing birth rates. Racial degeneration was thus associated with modernity, Káldor continued, as more and more modern medical instruments were invented to correct, and thus normalize, hitherto conspicuous physical deficiencies. Underlying "defective" hereditary traits would, however, re-emerge again in the next generations, medical interventions to protect the body notwithstanding. Káldor was not, however, against modern social assistance in general and argued that a proper eugenic policy and careful social selection could, in fact, balance the effects of racial degeneration.

Káldor did, however, reject "positive" eugenics as "unscientific, because we cannot determine those features that are worth cultivating from society's point of view". "Negative" eugenics was "practically more manageable", especially in the case of "natural born criminals and the mentally afflicted" for whom, Káldor reasoned, sterilization was the only solution if these individuals were to be reintegrated into society. Accordingly, "the real eugenic movement is the one which strives to eliminate the natural causes inflicting human degeneration".[40] Káldor focused on the environmental causes of degeneration, but also believed that there was only one solution for "undesirable elements": eugenic sterilization.

Similarly, in his intervention from 28 March, István Apáthy endeavoured to offer a more detailed analysis of eugenics and its relationship to degeneration. Insisting that the scientific context required terminological clarification, Apáthy constantly employed the term *fajegészségtan* for eugenics rather than *eugenika*, despite other participants in the debate having predominantly used the latter.[41] Apáthy's eugenic ideas will be discussed in detail in the next chapters. It is worth mentioning here, however, that he understood racial degeneration to be "a malady of the species, a malady which can be cured by extremely special methods". Well versed in hereditarian theories and a supporter of schemes of preventive hygiene, Apáthy adopted both as sources of inspiration for the "special methods", which he advocated to combat the "malady of modern degeneration". Thus, he continued, "hygiene is concerned with the improvement of public life and public health; eugenics struggles with certain maladies which endanger not only the survival of isolated individuals but the survival of the entire species". The science of hygiene therefore needed to adapt to challenges posed by eugenics – especially the notion that benevolent hygienic schemes resulted in the multiplication of the "unfit". The relationship between eugenics and hygiene was thus a substantial one, as Apáthy explained:

> The endeavours of these two sciences are in many aspects similar; furthermore, the improvement of hygiene itself is one of the methods employed by eugenics. Yet the latter has also borrowed methods which first originated in sociology, on the one hand, and in ethics, on the other; and, finally, it adopted an entirely specific biological method as well. This method is the deliberate selection of those elements that protect the race and the impediment of the reproduction of certain individuals who might have a damaging effect on the future generation.[42]

As with previous speakers, Apáthy also realized that the main obstacle to applying eugenic principles was to be found in navigating between the interests of the individual and the powers of the state. Apáthy outlined a new set of priorities for eugenics upon which the Hungarian modern state was to be built. Connected to this, existing biological degeneration and deteriorating social conditions only strengthened

Apáthy's conviction that the private sphere ought not to succumb to excessive individualism and egoism. On the contrary, eugenic virtues were embedded in collective social morality, with the state overseeing the nation's biological capital. "It is a great mistake", Apáthy noted, "to believe that the interest of the individual is in conflict with state interest, or that private interest fights public interest regarding the problems of eugenics."[43]

Apáthy advanced specific examples of observational and experimental evidence to support his theoretical perspective, ranging from cytology and animal breeding to the impact of Malthus on Darwin's thinking. As long as "the task of producing the next generation" was left to "the best individuals in each generation" society would be protected against degeneration. While acquired character was adaptive, evidence of hereditary inheritance would support the claim that a "healthy environment" was decisive for the improvement of social conditions. Pervasive "social illnesses" such as prostitution, homosexuality and alcoholism could therefore be cured if the environment was eugenically cleansed of "negative influences". In this context, then,

> the goal of eugenics is to call attention to the fact that each social order, each habit, fashion or morality which acts against the selection of the best individuals in fact undermines the future of the entire society, and that it trades evolution, the salvation of future generations, for the pleasures of the moment, for the individual comfort of the present generation.[44]

Eugenics should thus appeal beyond professional scientists to all those concerned with racial preservation and the social reconstruction of Hungarian society. In conclusion, Apáthy reiterated this point:

> Eugenics will be satisfied even with the possibility to prevent the proliferation of less desired individuals and to increase the proliferation of desirable individuals. Neither the morality of racial improvement nor state intervention will make the accomplishment of these requirements easy. The two might initiate, in more favourable social conditions, a series of changes which might lead to the creation of a happier, better and more noble human race.[45]

Clearly, Apáthy demanded well-defined eugenic standards for both public and private life. For he, too, feared a corrupt and decadent modernity, one that could fatally affect Hungarian national morality and, ultimately, the body politic. He, like other participants, used an eclectic methodology to support his arguments, combining ideas about culture and biology with projects for social improvement, and moral precepts with schemes centred on hygiene and public health.

While it cannot be denied that ideas of social and biological degeneration played an important role in shaping the Hungarian debate, eugenics represented something far more complex than merely a counter-discourse. Indeed, at a more general level, the overall goal was to provide the basis for a creative political biology, one capable of developing and pursuing the ideal of a healthy Hungarian nation. Consequently, this approach to eugenics resulted in the development of a language of scientific expertise that sought to forge a consensus on national welfare. A new social and biological solidarity was heralded during this public debate, and rarely was eugenics to be more frequently invoked in Hungary as a vehicle for the radical transformation of existing social and biological conditions than at the beginning of the 1910s.

Practical Eugenics

Social and biological strategies to improve the nation scientifically in accordance with eugenic principles were enunciated passionately during this first Hungarian debate on eugenics.[46] All those discussing the issues took a keen interest in contemporary social problems, and their views on eugenics emphasized broader ideas of biological rejuvenation of society brimming with the positivist psychology and various philosophies of social progress in Hungary at the time.[47] The source of this transformation, they professed, was a resolute belief that eugenics, underwritten by hereditarian laws, had to be extended into every area of social policy and public health – if not even further.

Practical eugenics remained a cardinal theme. Eugenic methods were invoked in order to reinforce the importance of immediate policies on social and biological improvement. Yet, the recognition of these positive features did not convince all participants that "negative" eugenic practices, such as sterilization and segregation, were unreservedly required. In his remarks from 18 April, the lawyer Zoltán Rónai, for instance, countered Madzsar and Fülöp's negative arguments about social charity; a predisposition he also associated with German eugenics. "There is no contradiction", Rónai maintained, "between altruism and racial hygiene, as advocated by Ploetz, for instance."[48] Progressive social policies and the new science of eugenics were not antagonistic. Incorporating eugenic arguments about the health of future generations was supported by those who – like Rónai – were alarmed by the increased threat posed by alcoholism, prostitution and the spread of venereal diseases. He further noted that "society stubbornly clings to alcohol-consumption and prostitution, and [that] the unavoidable conflict is not between the protection of the weak and the future of the race, but between the assault on the weak and the future of the race instead".[49] This led Rónai to doubt whether even negative eugenics could be effective:

> The interdiction to marry and artificial sterilization, through analgetic surgical intervention, of those with venereal diseases, alcoholics, and other degenerates threatening the future of the race has little consequence. The syphilitic infection or the germ-intoxicating alcoholism can also intervene postnuptially, and it would hardly be possible to sterilize the huge numbers of syphilitics and alcoholics, or those degenerated under the influence of syphilis and alcohol.[50]

To succeed in its battle against degeneration, Rónai argued, Hungarian society needed urgent change, both socially and politically. "The ultimate goals of eugenics", he believed, "could only be realized in a socialist society." Accordingly, "eugenicists must strive to encourage not a new nobility, but a new culture, and in this respect we can say that the real ennoblement of the human race can be achieved only together with the real ennoblement of society".[51]

Rónai's socialist inclination was not unique among supporters of eugenics in pre-1914 Hungary.[52] In fact, the majority of those participating in the debate were socialist sympathizers, if not outright activists, like Madzsar, for example. As Diane Paul and Michael Schwartz, among others, have eloquently written, it was common for Marxists and Fabians in Britain, and social democrats in Germany, to entertain eugenic ideas and to suggest a fusion between socialism and eugenics.[53] Projects of biological engineering squared neatly with socialist attempts at transforming humankind. In many ways, socialist thinkers were quicker to grasp the importance of eugenics in shaping projects of national rejuvenation than other political ideologies. The emerging scientism of the early twentieth century, for its part, no doubt drew socialism and eugenics closer together. Interestingly, however, their fusion was not necessarily based on positive eugenic methods, but rather negative ones as well, including sterilization and social segregation.

But the motives behind support for negative eugenics in Hungary were largely biological. Rónai concurred with other eugenicists that schemes of racial improvement derived more generally from the natural sciences, biology and medicine than from the social sciences. This is not to say that the form of biological improvement advocated by eugenics on the one hand, and the social transformation envisioned by social sciences on the other, were completely separated. On the contrary, as we have seen in the first part, eugenics and sociology often became entwined. As Dezső Buday explained,

> Scientific research relates to the questions of eugenics in two ways. One is the biological method which analyses the rules of procreation and the heredity of the individual based strictly on a biological basis; the other is the sociological approach which determines the effects of natural selection, of sexual selection and of heredity on the basis of wider research and biometrical statistics. Both methods are equally necessary.[54]

Admittedly, there was a degree of conceptual overlap between the two, for the language of eugenics and politics equally charted the importance of biological laws in shaping the character of the individual and the race. Even those eugenicists less inclined to adopt a political language – like the biologists László Detre and István Apáthy, for example – encoded their speculations about heredity with metaphors applying eugenics to contemporary society. There were some, however, who were openly critical of eugenics' ambition to interfere in the scientific realm, or who opposed eugenicists' interference in the medical sphere, one hitherto reserved for social hygiene.

Underpinning this debate was a deeper argument over health improvement and social environment. Diseases and impoverished social conditions in Hungary seemed to determine the life of the individual more than heredity. For eugenicists supporting this view, nature rather than nurture provided the key to understanding how genetic material was transmitted from generation to generation. Disagreements persisted, however, and the willingness of prominent participants in the debate to accept a biological definition of eugenics should not obfuscate the fact that other contributors adhered, in fact, to a social interpretation. In this context, on 11 April, the histologist Tibor Péterfi criticized the careless use of the term "race" employed by eugenicists. His was one of the most sustained charges made against eugenic representations of the national body: they lent themselves to racial exaggerations. Péterfi wondered if this desire to create a healthy society should be undertaken without connecting it to ideas of race preservation? Nation-wide hygienic programmes were one solution, he suggested. Not surprisingly, then, Péterfi believed that eugenic strategies to protect and improve society owed less to "the laws of biology and phylogeny" and rather more to "the means and results of the hygienic and medical sciences".[55] Péterfi also had some reservations about the cult of the state promoted by some eugenicists. Claiming that inclusive state control would only lead to "the most boisterous absolutism", Péterfi endorsed the racial improvement of future generations through sustained social welfare and public health policies.[56]

In the session organized on 9 May, Leó Liebermann (Fig. 5) advocated a similar idea, while questioning the practical application of eugenics. Remarking that some of the greatest minds of European culture, including Immanuel Kant, had fragile constitutions and suffered from hereditary diseases, Liebermann retorted that "it would be far-fetched to try and produce the artificial or the drastic elimination of the weak or of those who have a poorer constitution for the furthering of the cause of eugenics".[57] To those demanding "the artificial termination of the reproductive ability" for those deemed "unworthy", Liebermann pointed out that many of the initial North American and Australian colonists were

Figure 5   Leó Liebermann (courtesy of the Semmelweis Museum, Library and Archives of the History of Medicine, Budapest)

not always of "worthy stock". "It is not necessary to demonstrate the worth of their descendants", Liebermann continued, "in order to prove that the civilization of the world would have indeed suffered great harm if these anti-social elements would have been destroyed only because they did not find their place in the old Europe."[58] In turn, Liebermann ridiculed the proposition that society could only be improved through negative eugenics. Dividing the world into racially worthy and unworthy, he argued, not only endangered the nation's biological survival, but its intellectual and cultural legacy as well.

This more flexible and multi-layered conception of eugenics also appeared in Sándor Doktor's contribution to a previous session convened on 21 March. In complex ethnic societies like Hungary at the beginning of the twentieth century, national communities were a symbiosis of both biological and cultural heritages. Doktor rejected excessive biologism, and instead offered a fluid conception of heredity that allowed for the retention of multiple inheritance factors. Illustrious Hungarian politicians like Petőfi, Széchenyi and Kossuth, for instance, have "contributed more to the improvement or the degeneration of the race with their acquired intellectual wealth than with what they have passed on to their offspring".[59] Eugenics was only the most recent symbol of emergent theories on heredity, not their ultimate manifestation. Next to eugenics, Doktor placed culture and education as central factors in determining the nation's historical destiny.

Sharing the session with Liebermann and Dezső Hahn, the feminist Vilma Glücklich added a new dimension to this interpretation of eugenics as a means to facilitate comprehensive public health reforms and social progress. She pointed out that women's emancipation was being greatly neglected by male eugenicists designing schemes for social and biological improvement. She then insisted that proper sexual education and widespread dissemination of basic hygienic knowledge were required before the eugenic perfection of the race could be achieved. Modern eugenic ideas of reproduction were indeed needed – Glücklich further argued – but only in a society in which women's responsibility for childbearing and child-rearing was not abused for economic and political purposes. To empower women as mothers was as important as the eugenic transformation of society itself, Glücklich insisted.[60]

Adding to the concerns raised by Doktor and Glücklich, Zsigmond Engel had already linked the practical application of eugenics to the protection of children in his session of 18 April. A qualitative population policy, able to accompany quantitative measures, was vital for a new eugenic mentality; moreover, this task was a national duty for both the state and the eugenicists. Modern child welfare, Engel insisted, should not be based on crude Darwinist principles, which viewed "the assistance of the poor and the protection of children" as leading to "racial decline". It should, instead, be based on a corresponding improvement in social relations, appropriate legislation and the principles of moral tolerance.[61]

Taking the discussion a step further, one of Hungary's foremost venereologists, Dezső Hahn, firmly rejected what he considered to be scientifically flawed approaches to eugenics. He assessed the impact of biometrical research critically, proposing a much more restrictive interpretation. Indeed, he overtly rejected "negative" eugenic methods: "in some American states certain recidivist criminals or alcoholics – only men – are violently and artificially, by way of operation, prevented from procreation; well, this is nothing but a distasteful caricature of eugenics".[62] A more critical reception of eugenic ideas was needed instead, Hahn contended:

> the foundations of eugenics do not appear in such a light, and the great and most important task of serious eugenic research is to clarify its basic assumptions with the tools of objective and scientific criticism, and not to waste its power on the utopian discussion of various means to practical accomplishment.[63]

Hahn did not subscribe uncritically to eugenics. Artificial selection, whether in the form of negative or positive eugenics, was "practically impossible to realize" – he noted. Environmental factors, too, were not to be neglected. Even Richard Dugdale's classic study of the hereditary nature of anti-social behaviour, *The Jukes: A Study in Crime, Pauperism, Disease, and Heredity* (1877) – which he quoted approvingly – failed in persuading Hahn to modify his neo-Lamarckian view that a proper social environment was as important in shaping the individual character as heredity. Ultimately, however, he nonetheless confessed his support for eugenics, and considered that it had essentially two practical tasks: combating social degeneration and creating a "racial ethics which would introduce the feeling of responsibility for the race besides healthy selfishness for the conscience and acts of man".[64]

Hahn did not abandon his belief in the necessity of eugenics as such. He shared many of the views advocated by other participants during the debate. The general tendency, however, was to consider eugenics to be of fundamental importance to Hungarian society, not only in assisting schemes of public health and social hygiene, but also in shaping new forms of political biology. The corollary to this view was the elevation of eugenicists to a new status, that of national shamans protecting the well-being not just of the individual but of the nation itself. This inter-mingling of eugenics, morality and national valour encapsulated the spirit of the debate, as both Lajos Dienes and Zsigmond Fülöp noted in their concluding remarks, presented to the Society of Social Sciences on 16 May. Perhaps unsurprisingly, the debate raised more questions than it answered. Dienes rounded on the criticism voiced against biometry, particularly by Fülöp and Rónai, highlighting two fundamental short-comings in their method. The first was their limited understanding of statistics and its importance for eugenics; the second, their disguised predisposition towards Lamarckism.[65] Biometry, Dienes concluded, was the scientific methodology that best supported eugenic research.

Fülöp, on the other hand, recommended that eugenic politics was never meant to be applied uncritically, and was only valid when sub-stantiated by "the empirical results of biology". However, a still broader purpose behind this debate on eugenics, he stated, was to discuss those principles and values able to inspire collective action against what all speakers perceived to be the illness gnawing at Hungarian society: social and racial degeneration.[66] Fülöp, like the other contributors, truly believed in the transformative power of eugenics. While differing over the social and political ramifications of the science of heredity, all participants were ultimately in agreement that the Hungarian nation's protection fell within eugenic responsibility.

Eugenics as Social Reform

What, then, was the impact of this debate on eugenics on the general public? Considering that the journal *Huszadik Század* had a circulation of *c*.3000 copies per issue and that the opening lectures and the debate were published over six issues, one can safely surmise that this was the first time that eugenics had enjoyed such wide national circulation and recognition in Hungary.[67] In organizing such debates on various cul-tural and social issues, the Society of Social Sciences professed, in fact, a wider ambition, namely "to enlighten not only the legislative organs but also individual citizens about the nature of social truths".[68] For one of the organizers, Oszkár Jászi, therefore, eugenic ideas of biological improvement were closely connected to progressive values and notions of building a new modern Hungarian society. It reflected his vision of a pragmatic Hungarian cultural politics, one guided by the twin princi-ples of science and progress.[69] Eugenics served this group of progressive intellectuals with new ideas of social and biological improvement, reflecting their own vision of a modern nation state.

Different models of practical eugenics emerged during the debate, and all were based on three core principles: the importance of heredity in shaping the destiny of individuals and national communities; the link between biology, medicine and national health; and finally, the connection between science and political power. Universal in its inspiration, the progressive intent of eugenics was nevertheless held together by a specifically national understanding of the science of heredity and its application to society. During this debate, Hungarian eugenicists formulated collective agendas, conceived as totalizing cultural and social responses to any number of scientific and practical topics, spanning heredity, biometry and Mendelism along with degeneration, sterilization and marriage counselling. Even when they did not share the same ideas about eugenics, they did agree that the improvement of the nation's health and Hungary's transformation into a modern state were closely interrelated.

Ultimately, the debate on eugenics convincingly demonstrates that eugenics was equally appealing and challenging for many Hungarian scientists. Some assumptions formulated during the debate on eugenics were based on local, specific Hungarian experiences; others were drawn from similar debates in Germany and Britain.[70] Karl Pearson's programme of "national eugenics", for instance, was one such external influence. "Every nation has in a certain sense its own study of eugenics", Pearson argued,

> and what is true of one nation is not necessarily true of the second. The ranges of thought and of habit are so diverse among nations that what might be at once or in a short time under the social control of one nation, would be practically impossible to control in a second.[71]

It was in this context that the participants in the debate elaborated the necessary concepts they needed to identify the social and medical problems characterizing their society and which they hoped they could solve through the development of a specifically Hungarian form of eugenics.

At a time when theories of biological destiny enthralled political elites and intellectuals alike, eugenics offered the possibility of comprehensive national renewal, combining scientific dogmas with scientific utopianism that derived from Galton's description of eugenics as a "new religion". Once eugenics had succeeded in attracting a critical mass of supporters, two dominant viewpoints emerged with respect to its proposed application to Hungarian society. One category of eugenicists argued that economic reform and the improvement of social conditions should be directed by the precepts of race protectionism. Advocates of this approach to social and biological transformation advocated the need for selective breeding policies designed to prevent those individuals with "negative" characteristics from social interaction and, ultimately, reproduction. Such a radical position did not go unchallenged, however. There were also supporters of eugenics who defined their goals as primarily directed towards public health reforms and medical improvements. For this group, eugenics was part of a broader hygienic *Weltanschauung* that included diverse welfare strategies such as improving the living conditions of the urban underprivileged classes in the suburbs of Budapest and other Hungarian cities, child protection, social assistance for young mothers, the prevention of sexually transmitted diseases and the social reintegration of prostitutes. As will be discussed in the next chapter, however, both groups attempted to legitimate eugenics as a compellingly modern technology of social and biological engineering.

3
At a Crossroads

The debate on eugenics occasioned by the lectures Madzsar, Dienes and Fülöp offered to the Society of Social Sciences between February and May 1911 highlighted diverse theories of social and biological improvement. A crucial realization in itself, it also substantiated the claim that Hungarian eugenics could not be viewed in isolation, but in terms of a set of ideas and practices which travelled from one individual to another, and from one country to another. This debate marked a major landmark in the history of eugenics in Hungary, but is it possible to shift the scholarly focus from a single cultural moment to a network of connections between this society and other societies at that time, and no less between eugenicists in Hungary and abroad?

Until now, a central role has been assigned to the Society of Social Sciences and its journal, the *Huszadik Század* in the diffusion of eugenic ideas in early twentieth-century Hungary. Nonetheless, this role must be constantly reviewed. Eugenics was not a fixed set of ideas, but a nexus in constant change, subjected to diverse and often conflicting intellectual influences. As seen, this conceptual elasticity allowed for biologically determinist and environmentalist visions of social reform to co-exist side by side. Eugenic agendas thus included improvements in the health of the population, as well as concerns with racial reproduction and family protection.

Like other eugenic movements emerging after 1900 in Europe and elsewhere, the Hungarian one came to be validated on the basis of its modern cultural and scientific methodologies. What started out in the 1880s in Britain – as Galton's erudite project to understand and historically locate the hereditary improvement of noteworthy families – had become by the 1910s a widely shared methodological vocabulary for a number of scientific disciplines that competed in offering remedies to various perceptions of alleged national and racial decline. This synchronized process of scientification with nationalization also marked eugenic developments in Hungary. As a result, a tension arose between those who viewed eugenics as a narrative of social reform and those who placed it within the political tradition of the nation as the ideological legitimation of the status quo. The eugenic vision of a modern Hungarian state created by these ideological differences became particularly evident in the years preceding the First World War. By 1914, Hungarian eugenics stood at a crossroads.

Spreading the Word

The scientific merits of eugenics continued to be discussed after Hungary's first public debate ended on 16 May 1911, and not only in Budapest. Eugenics may have been the latest scientific fashion but it was certainly not confined to the capital. During the summer of 1911, members of the Society of Social Sciences travelled to provincial towns, offering free lectures on Darwinism, evolution and eugenics. The jurist and future minister of Foreign Affairs in the short-lived communist republic, Péter Ágoston, for example, spoke of "Darwin's Impact on Modern Thinking" in the north-eastern towns of Debrecen and Nyíregyháza, and in Lugos (Lugoj; Lugosch), a town in the Banat region. Dezső Buday, on the other hand, discussed the demographic effects of the "one-child" practice in his lecture in the town of Nagyvárad (Oradea; Großwardein). More eugenically inclined members of the Society of Social Sciences also took their cultural mission seriously. While Zsigmond Fülöp explored the importance of "scientific reasoning" in Miskolc, Oszkár Jászi investigated the relationship between "Nation and Race" in his lecture in Nagyvárad. In Erzsébetfalva, József Madzsar discussed the "Causes for Racial Degeneration", while in Moson (north-western Hungary) – he was invited to the opening ceremony of a Workers Gymnasium – and in Lugos, he lectured on "Racial Hygiene" and "Racial Improvement", respectively.[1]

These lectures confirmed that eugenics mattered and that it should have a public voice.[2] It was not only local officials and scientists who garnered support for the speakers and their ideas – as in Miskolc, when the local section of the prestigious Association of Hungarian Doctors and Naturalists (Magyar Orvosok és Természetvizsgálók Társasága)[3] co-sponsored the meeting – but also members of the wider public interested in the educational work promoted by the Society of Social Sciences. For many in the small provincial towns this was their first encounter with eugenics, as well as with the type of social and biological reform promoted by this group of progressive intellectuals. While the main debates on eugenics would continue to take place in Budapest, the fact that the Society of Social Sciences had supporters in provincial towns across Hungary helped in disseminating eugenic ideas to a diverse public.

These cultural and scientific tours in the countryside were seen at the time as a necessary requirement for the popular acceptance of eugenics. Hungarian eugenicists further stressed the difficulty of finding an accessible language to express their ideas, arguing that imposing ideas of social and biological improvement upon an unprepared general public would be ultimately counter-productive. They tried to maintain a balance between conveying ideas plainly in order to gain popular endorsement, and advancing complex theories and scientific terms. They all agreed that eugenics was predicated upon a mastery of technologies of health and social reform that were judged to be scientific, modern and beneficial to the progress of Hungarian society. Yet how best to convey these visions of a new concept of a healthy, race-centred modern state remained open to debate. As their publications increased in frequency, Hungarian eugenicists became more aware of the need to capitalize on the intellectual impact of their work.[4]

The outcome was a deliberate and determined pursuit of a eugenic praxis tailored to reflect local, Hungarian realities. Quite appropriately, then, in the capital of Transylvania, Kolozsvár (Klausenburg, Cluj), the ethnographer Antal Herrmann gave a lecture on 1 February 1913 to the Tourist Association of Hungarian School Teachers (Magyar Tanítók Túrista Egyesülete). Herrmann's lecture drew directly on eugenics in order to portray domestic tourism as a means of building the nation's physical vigour and spiritual strength. It was mostly in the East, in Transylvania, Herrmann believed, that authentic Hungarians were to be found. Transylvania's well-defined rural communities and mountainous geography provided unparalleled opportunities to study the endurance of distinct ethnic characteristics.[5] "Educating the youth towards the eugenic ideal" was the key catalyst in the formation of a healthy national community in Hungary, Herrmann noted.[6] The task before Hungarian eugenicists now was how to appreciate and celebrate this national authenticity through a programme of social and biological improvement.

Not all eugenicists, however, were attracted to national themes. Lajos Dienes, for instance, was one author who consistently promoted a more universalist interpretation of eugenics, one first addressed in his article on biometrics (published in the *Huszadik Század* of the previous year) and then developed in his February lecture to the Society of Social Sciences. Like József Madzsar and Zsigmond Fülöp, he benefited directly from the publicity surrounding the debate organized by the Society of Social Sciences. As Dienes's involvement with eugenics continued, he expanded the biometrical framework in order to encompass a much wider range of eugenic issues. In a lengthy article entitled "Eugenika" (Eugenics) published in *A Társadalmi Muzeum Értesítője* (*Bulletin of the Social Museum*) in two instalments between May and August 1911, Dienes established himself as one of the most refined Hungarian eugenicists of his time.[7]

Having dutifully expressed his conviction that eugenics was the science of the future in his opening passage, Dienes then provided a detailed discussion of Galton's contributions to the study of heredity and human improvement. Consistent with the "need for eugenics" was Galton's insistence on the influence of heredity on social and biological environments.[8] According to Dienes, setting the health of the future generation in a hereditarian framework provided a rationale for social and biological improvement. It also highlighted the correlation between medical statistics, particularly as applied to birth and mortality rates, and the importance of demographic management alongside population planning and social welfare.[9]

Another significant feature of eugenics was that it brought biological pragmatism into the sphere of social reform in Hungary. Dienes spoke earnestly of the need for eugenicists to understand current social issues, and to investigate the relationship between society and its biological foundations.[10] This emphasis on nature and heredity over nurture and environment was a recurrent eugenic claim. But nature and nurture did not function as neatly exclusive categories, and Dienes admitted that the boundaries between them have always been porous. Not until the rediscovery of Mendel's laws was biological determinism reformulated. Mendelism endorsed the idea of genetic differences, Dienes believed, but it was Galton's biometrical "method of applying exact statistics to biology" that accentuated hereditarian hierarchy within society.[11] Mathematical and statistical models such as Pearson's correlation coefficient,[12] moreover, could provide – Dienes argued – a more convincing account of the process of hereditary transmission of physical characteristics from parents to their offspring.

Just how certain diseases like syphilis and forms of physical and moral degeneration like alcoholism and prostitution were inherited, if at all, remained to be determined. Above all, biometrics provided the main themes and ideas that were to guide statistical and descriptive studies of

eugenics. But Dienes envisioned eugenics to be more than just a repository of new statistical accounts of hereditary variation; rather, it was to
provide a unified ground for scientists to engage in the battle against
social and biological pathologies – thus preventing the "degeneration
of the family".[13]

A central theme that emerged from these observations was society's
failure to defend itself against perceived degeneration. Correspondingly,
scientists needed to provide detailed explanations for a range of social
and biological illnesses affecting the nation as well as recommending
appropriate eugenic measures in response. According to Dienes, modern
institutions devoted to eugenic research were central to this programme
of biological enlightenment, and he praised the work carried out by
one such institution, the University of London's Eugenics Laboratory.[14]
Anxieties about the nation's health had deepened with the onset of
modernity at the end of the nineteenth century, Dienes noted. Faced
with the challenges of both social and biological degeneration, modern
welfare projects looked less credible without a sustained examination of
mortality and birth rates, in both healthy and "pathological families".[15]
The fertility of the "unfit", as far as eugenicists were concerned, applied
to those with mental and physical disabilities as well as to those living
in poverty. In illustrating the latter point, Dienes reproduced lengthy
details about fertility rates in several "manufacturing" English towns
(including York, Manchester, Bradford and Leeds), first presented by
Pearson in his 1909 *The Problem of Practical Eugenics*. The data obtained
was then used to argue that "unfit" individuals had higher birth rates
than "fit" individuals. This statistical reasoning about fertility, social
efficiency and the working classes was made to illustrate the importance
of economic productivity to the eugenic ideal of a modern, healthy
nation. Had not Pearson himself declared eugenics to be "a doctrine of
national welfare" and "a branch of national economy"?[16]

Granted, Pearson was writing in the heavily industrialized English
context, but Dienes understood his concerns. The justification for combating the nation's alleged degeneration, and the anxieties it inspired,
ultimately rested on intricate statistical and eugenic arguments, which
actually set the boundaries between those who were deemed beneficial
to society and those who were not. What was demanded, in short, was
a broadly conceived eugenic programme. "Was", Dienes asked in the
concluding part of his article, "a eugenic social policy possible?"[17] It was
not only possible but also necessary, Dienes concluded. He made a final
strong plea to his contemporaries regarding both the effects of degeneration and that only an appropriate social policy would ensure that

the "nation's physical and mental qualities improved".[18] Undeniably, Dienes continued, the government and other institutions in Hungary had made considerable progress in fighting tuberculosis, venereal diseases and alcoholism, together with improvements in social insurance and the protection of mothers and children.[19] But more state intervention was needed.

Clearly, harnessing notions of collective progress to improvements in medical care and public health had evoked the greatest enthusiasm among progressive Hungarian social reformers. As a leading figure in this group, Dienes was aware that a "eugenic social policy" was slow in the making, as knowledge about the "application of biology to social processes" was still in its infancy. Though slow, certain eugenic strategies for social improvement were nevertheless possible, Dienes argued. Extending the period of confinement in asylums was one such strategy; restriction of marriage was another. On the positive side, Dienes recommended "awarding economic benefits to healthy families".[20] The latter, however, was not simply a matter of redistributing state welfare. As a "social policy", eugenics was predicated upon an understanding of "the biological processes in the population".[21] A synthesis of demography, statistics and eugenics was necessary, Dienes concluded.

It is noteworthy that Dienes appreciatively returned to the activities of the Eugenic Record Office and the Eugenics Laboratory at the University of London, again suggesting that the popularization of eugenics and its institutionalization were interconnected. The acceptance of eugenics was ultimately the result of sustained public engagement, and the question of agency was an important one, Dienes insisted. Scholarly debates and the wide circulation of eugenic literature were identified as the channels through which eugenic ideas were discussed by educated elites, alongside the organization of public events such as health and hygiene exhibitions. Without their own organization, Dienes noted, eugenicists in Hungary had to rely on other institutions, like the Social Museum, in order to promote ideas of social and biological improvement.

Dienes championed the merits of British eugenics, but – as demonstrated earlier – this intellectual affinity characterized other Hungarian eugenicists as well, especially those gravitating around the Society of Social Sciences. The veterinary surgeon and sociologist Jenő Vámos provides further evidence of the continuing appreciation of Galton and his disciples in Hungary. In his article "Az alkalmazott eugenika" ("Applied Eugenics"), published at the end of 1911, Vámos enlisted Galton's mathematical and quantifying approach in support of his eugenic pedagogy.[22] "Galton's biometrical studies", Vámos maintained, "demonstrate that eugenics should not be confused with hygiene or social-politics, as many here in Hungary already do. Instead, it is to be considered a new science which incorporates all positive knowledge aimed at the definite ennoblement of human races".[23]

Eugenics was, for Vámos, intimately connected to both social and racial improvement. In many ways, he followed in the footsteps of other participants in the public debate organized by the Society of Social Sciences, especially in terms of entrusting eugenics with a social and medical mission. In other significant ways, moreover, Vámos underlined a hitherto criticized argument: the practical application of eugenics should also take into account the society's racial texture. Thus, "for eugenics, the improvement of the individual is only a means which, in effect, should lead to the improvement of the race".[24] Vámos consequently proposed a two-pronged eugenic agenda based on "*a priori* eugenic goals". The first measure was to facilitate "the natural selection of healthy individuals", while the second was to encourage the reproduction of individuals who were "fitter from a racial and sexual point of view". Healthy individuals were to be defined on the basis of a rigorous racial and sexual profile:

> If we want to improve the Caucasian race, those individuals are *a priori* suitable who bear the typical characteristics of the race, provided they serve a social purpose. On the other hand, for instance, such individuals who show characteristics contrary to their sexual nature – like the homosexuals – are *a priori* unsuitable, even if, in other aspects, they do befit a social purpose.[25]

Vámos endorsed such a challenging description of eugenics with an equally sophisticated theoretical argument, borrowed from his own area of expertise. In fact, Vámos's well-rehearsed methodology was based largely on a unified interpretation of biology and natural selection, one which asserted the universality of reproductive practices valid in the natural world. Such a view was widely disseminated after the publication of Charles Darwin's *Variations of Animals and Plants under Domestication* (1868), and it certainly influenced Galton's description of "stirpiculture", a term that fascinated him before the coinage of "eugenics" in 1883 and one that was popular among eugenicists in the United States at the end of the nineteenth century.[26]

Vámos identified six principles adapted from animal breeding able to provide a useful underpinning for human eugenics. The first was interbreeding: "The task of eugenics would be to facilitate, with the help of the proper means (i.e. the restriction of reproducing possibilities of those individuals characterized by detrimental characteristics), the consolidation of the good qualities, and the elimination of the bad ones." The second principle endorsed the belief that racial qualities were mainly inherited maternally. On this assumption, Vámos established a direct link between feminism and eugenics. "Feminism", he remarked, "found its perfect confirmation in eugenics".[27] As with other Hungarian progressive and leftist intellectuals, Vámos related women's cultural and social emancipation to economic prosperity, while, at the same time, insisting on their reproduction obligations for nation and race.[28] He thus predicted that "only when the woman is totally emancipated in spiritual and economic terms, will the goals of eugenics be realized".[29] Vámos similarly considered a more egalitarian society to be a precondition for implementing a eugenic programme: "The goal of feminism is the improvement of women, and the ennoblement of the race has to be grounded in the improvement of mothers."[30]

Women aware of their eugenic role, according to Vámos, were more likely to choose their partners carefully. Marriage, therefore, should only be allowed between compatible partners whose racial qualities were complementary. In this context, "the main goal of biometric measurements is to determine which qualities – and in what quantity – are transmitted by the fathers". The precision of biometry was also invoked in Vámos's fourth eugenic principle, which assumed that children inherited more from the parent of the opposite gender (i.e. boys from mothers, girls from fathers). Vámos then considered the relationship between environment and heredity, and argued that both played an equally important role in the history of a given race. Nevertheless, "creating proper living conditions is a major task for a eugenic social policy, although alone, is not considered eugenics". Finally, the sixth principle identified by Vámos was that of "saturation", meaning that, in "the case of multiple conceptions by the same father", it was the last child that inherited most of "the father's qualities".[31]

These Lamarckian analogies notwithstanding, and despite the fact that much of his eugenic language was more selectionist than hereditary in tone, Vámos hoped to improve the quality of the race both in a biological and a social sense. In his view, the modern conception of a healthy society was plainly speculative if it was not accompanied by pragmatic efforts to create favourable social and political conditions for racial improvement. The application of eugenic principles could not merely be determined arbitrarily or haphazardly but should be the work of suitably trained health officials and physicians. Echoing Madzsar and Fülöp, Vámos explained,

> I do not wish to elaborate here a detailed judgement about how this application would be imaginable. However, for those who, under the pretext of respecting the rights of the individual, want to impede the conscious application of eugenics, I have an argument. If the Church had, and in many places still has, the right – without respect to the interests of the individual – to prevent the procreation of desirable offspring for the race through the interdiction of mixed marriages, why would the future physicians of eugenics not have at least as many rights in favour of the improvement of the human race, and in favour of the individual too?[32]

With the consolidation of eugenics as a scientific discipline, progress could finally be made towards improving human society, so that, according to Vámos, "In today's society applied eugenics is not a utopia anymore, and it will be even less so in the society of the future."[33] Eugenics in Hungary, he suggested, should be both a *theoretical* programme, contemplating a new social and biological order, and a *practical* strategy, calling for the immediate and the interventionist management of society. This two-pronged strategy personified pragmatism and, in so doing, adapted international trends to the specific realities of Hungarian society.

Internationalization

Apáthy's anxieties concerning the development of eugenics as a "national science" in Hungary were not unfounded. He was aware of international efforts to expand the Society for Racial Hygiene,[34] but did not share Ploetz's *Grossdeutsch* and Nordic ideals. Perhaps unsurprisingly then, Max von Gruber's visit to Budapest in 1909 did not result in Apáthy or other Hungarian eugenicists joining an international organization under German leadership. However, this did not prevent interaction between Hungarians and eugenicists from other European countries. On the contrary, the internationalization of eugenics increased in the years preceding the First World War, as illustrated by the widespread impact of the Dresden Hygiene Exhibition and the London Eugenics Congress. These international events inspired eugenicists in Hungary, reinforcing their conviction – so clearly expressed in the public lectures on eugenics organized by the Society of Social Sciences – of the growing role played by eugenics and heredity within both the natural and social sciences, in addition to their application to modern societies.

The aforementioned International Hygiene Exhibition opened in Dresden on 6 May 1911. With more than five million visitors, the Exhibition attested to the importance of biological sciences and medicine in shaping a society and state.[35] Such a commanding representation of the authority of modern science was efficiently captured by the symbol of the exhibition: an omnipotent "eye". The exhibition was organized in five sections (scientific, historical, popular, sports and industry),[36] each divided into "special groups" devoted to a specific topic: from tuberculosis, alcoholism and cancer to ethnography and eugenics. The main organizer, the industrialist, Karl A. Linger, "took hygiene in a popular direction that contrasted to the rigid scientificity of the state-sponsored *Volkshygiene* movement or to the elitism and the professionalism of racial hygiene".[37]

An illustrated catalogue was prepared for the exhibition on eugenics; it contained a lengthy text from Gruber, in which he discussed the relationship between reproduction, heredity and racial hygiene.[38] Ploetz provided the analytical framework, attractively presented to the public through two exhibition panels, one explaining the relationship between racial hygiene and the other sciences, and a second delineating the areas covered by racial hygiene. Casting his net as widely as possible, Ploetz divided racial hygiene into two large groups. The first, which he termed "quantitative" racial hygiene, focused on: (1) the birth rate, both that of the general population and individual mothers; (2) the death rate; and (3) the birth excess in relation to the racial struggle for existence. The second group, described as "qualitative" racial hygiene, included: (1) selection, itself divided into (a) non-selective elimination through death or infertility; (b) selective elimination and the elimination of the "unfit" from the race; (c) counter-selective elimination through, for example, war; and (d) counter-selective selection such as the marriage between blind people. The other components of "qualitative racial hygiene" were: (2) reproductive hygiene (Galton's eugenics was considered as belonging in this category); (3) racial care in relation to reproductive strengths; and (4) racial care of physical and psychological abilities.[39] To present these ideas to the general public visiting the exhibition was a very effective way of popularizing eugenics within – and especially outside – Germany.[40]

The eugenic exhibition was also accompanied by public lectures, culminating in the meeting of the International Society for Racial Hygiene, which took place between 5 and 6 August 1911. Speakers at this significant gathering included Pontus Fahlbeck, Professor of Political Science at the University of Lund (Sweden), who lectured on the eugenic significance of declining birth rates and low fertility, and Hans Breymann, Chairman of the German Biographical and Genealogical Bureau, who urged a closer collaboration between genealogists and eugenicists. In addition to German and Swedish eugenicists, Ploetz and Rüdin were also successful in attracting British and American colleagues to participate. More importantly, one Hungarian eugenicist, Pál Teleki, was also in attendance.[41] Following Gruber's visit to Budapest two years earlier, Teleki's contact with Ploetz and Rüdin in Dresden in 1911 was another important episode in the internationalization of Hungarian eugenics.

Teleki claimed that he was the only Hungarian to attend the conference organized by the International Society for Racial Hygiene in Dresden.[42] This is quite possibly the case, but nonetheless there were other meetings devoted to eugenics in Dresden during the Hygiene Exhibitions, and other Hungarians attending them. Two in particular deserve to be mentioned: the Fourth International Neo-Malthusian Congress (IV. Internationaler Kongress für Neumalthusianismus) and the First International Congress for the Protection of Mothers and Sexual Reform (I. Internationalen Kongress für Mutterschutz und Sexualreform). Both events illustrate the widespread importance eugenics had acquired at the time, and just how wide-ranging the theme of social and biological improvement had become, since it included such issues as the control of reproduction, gender equality and the protection of mothers and children. These were topics that infiltrated into the central debate about the quality and the quantity of the population, a debate which often posited eugenicists in open conflict with neo-Malthusians about the latter's insistence on family limitations and contraception.[43]

The conference on neo-Malthusianism was held between 24 and 27 September 1911. Alice Drysdale-Vickery, at the time also President of the Universal Federation of Human Regeneration (Fédération universelle de la régéneration humaine) was appointed Chair.[44] Other participants included the German feminist and eugenicist Helene Stöcker; the neo-Malthusian enthusiast Charles V. Drysdale (Alice's son); the Dutch promoter of birth control Johannes Rutgers; and the Hungarian feminist Rosika (Rózsa) Schwimmer, who presented a paper on "Child Labour and neo-Malthusianism".[45] The main topics addressed at the conference centred on public health, economics, medicine, eugenics, ethics and feminism.[46] Unable to attend, August Forel and Wilhelm Schallmayer sent their papers on eugenics to be read in their stead.

Feminists and eugenicists like Helene Stöcker, for instance, were keen to use this occasion to popularize broader issues relating to motherhood, reproduction, sexuality and social protection.[47] As a result, the First International Congress for the Protection of Mothers and Sexual Reform began immediately after the conference on neo-Malthusianism concluded, on 28 September 1911.[48] Stöcker, in her capacity as the Chair of the German League for the Protection of Mothers (Deutscher Bund für Mutterschutz) was the driving force behind this event.[49] Again, eugenics featured prominently on the agenda, as the physician Eduard David, a Social Democratic member of the German Parliament, lectured on the "Protection of Mothers and Racial Hygiene".[50]

The congress was divided into two main sections, one on the "Protection of Mothers", the other discussing "Sexual Reform". Delegates from various countries, including Great Britain, France, Italy, Germany and Austria, all presented short reports. Amélie Neumann and Rosika Schwimmer represented Hungary – Neumann contributed to the panel covering "the protection of mothers through maternity insurance and child pension" while Schwimmer participated in that focusing on the "social condition of the unmarried mother and her child".[51] Both demanded greater educational, professional, economic and political opportunities for both married and unmarried women, buttressing their arguments with reference to specific Hungarian conditions.[52]

A resolution was passed on the last day of the congress, decreeing the creation of an International Commission for the Protection of Mothers and Sexual Reform (Internationale Vereinigung für Mutterschutz und Sexualreform, or IVMS). In a statement addressed to "all men and women from civilized countries", the "organic-spiritual perfection of the race" was presented. For this to be achieved, "maintaining the health of the race" was deemed imperative, together with modern ideas of marriage, reproduction and social emancipation.[53] Two Hungarian organizations joined the IVMS: the Feminist Association (Feministák Egyesülete) and the National Association for the Protection of Mothers and Infants (Országos Anya- és Csecsemővédő Egyesület). If the former campaigned for women's emancipation and political rights, the latter expressed their greater devotion to the biological welfare of the national community.

Rosika Schwimmer participated in both congresses. At the time she was known in the international feminist movement as a committed campaigner for women's rights in Hungary.[54] Schwimmer was equally interested in child protection and youth education[55] as well as eugenics. Like her friend and colleague, Vilma Glücklich, Schwimmer argued for the empowerment of women and the alleviation of their social and economic conditions. She also favoured negative eugenics as a means by which racial improvement could be achieved.[56] As other scholars have noted, Schwimmer's vision of reform for women was rooted in a modern conception of motherhood that also informed her advocacy of eugenics.[57] Feminism, in this context, provided Hungarian women with possibilities for education, social mobility and economic freedom. Although Schwimmer supported women's control over reproduction, she situated it within a larger eugenic goal of creating a healthy Hungarian nation. Schwimmer – like other feminists,[58] socialists and Social Democrats in Hungary – avoided any association with unsavoury racism, but nonetheless endorsed the confinement and even the sterilization of the "congenitally defective".

A preoccupation with race and racial improvement remained central to the discussions and papers on eugenics, neo-Malthusianism and sexual reform organized in Dresden on the occasion of the International Hygiene Exhibition. If eugenics was seen as intrinsic to a healthy life, race (in all its permutations) was still believed to be central to eugenics. In other words, feminists and social reformers explicitly linked campaigning on behalf of the family, women and children to the interests of the race or the dangers of national degeneration. This should not, however, suggest that there was a general consensus about the meaning and utility of racial typology. As discussed in Chapter 1, there was strong disagreement between sociologists like Max Weber and eugenicists like Alfred Ploetz over a racialized understanding of social improvement. A more sustained critique of ideas of racial superiority and inferiority, and their relationship with human development and culture, was offered at the First Race Congress, organized at the University of London between 26 and 29 July 1911.[59]

This congress was an impressive meeting, with over 1200 participants. Among those attending were anthropologists, including Felix von Luschan, Giuseppe Sergi and Franz Boas, the sociologists Ferdinand Tönnies and W. E. B. Du Bois, and the philosopher Alfred Fouillée. The third session dealt with "Conditions of Progress (Special Problems)". One Hungarian participant, Ákos Timon, Chair in Constitutional and Legal History at the University of Budapest, presented a paper on the "Theory of the Holy Crown, or the Development and Significance of the Conception of Public Rights of the Holy Crown in the Constitution".[60] Another Hungarian, Ákos Navratil, Professor of Political Economy at the University of Kolozsvár, contributed to the session on "Special Problems in Inter-Racial Economics".[61]

The congress was discussed and reviewed in Hungary, which served to add to domestic discussions about the role of race in society. Though generally sympathetic, Hungarian reviewers noted the lack of conceptual clarity characterizing some of the papers presented at the congress. But this form of academic socialization was commended, as it offered the Hungarian participants a chance to interact with renowned scientists; to be exposed to new ideas; and, equally importantly, to be offered the possibility to present theirs.[62] The progressive Catholic theologian Sándor Giesswein also attended the congress and participated in the subsequent discussions. At a meeting of the Association for Social Sciences on 29 November 1911 he offered a personal response to the views on race expressed at the congress.[63] Following Gobineau and Chamberlain, Giesswein accepted the existence of races, but his understanding of historical differences between various nations was based less on racial determinism and rather more on cultural and economic factors. "Why", he asked "are the Hungarians closer to the Germans than to the Romanians?" And "why", he continued, "were the Croats and the Serbs – considered to be the same nation, linguistically and ethnically – against each other?"[64] Ideas of racial inequality were as much cultural and historical as they were biological. It was misleading, Giesswein concluded, to favour only one factor over others. These international meetings further highlight the conspicuous circulation of eugenic ideas of social and biological improvement in Europe and elsewhere at the time. If the Hygiene Exhibition in Dresden and the First Race Congress in London were the most important international events dealing with eugenics and race in 1911, the First International Eugenics Congress and the Second International Congress of Moral Education were the main gatherings of 1912.

Organized between 24 and 30 July 1912 at the University of London, the First International Eugenics Congress brought together supporters of eugenics from a number of European countries as well as North America.[65] At the inaugural dinner on 24 July, the President of the Congress and the Chairman of the Eugenics Education Society, Leonard Darwin, proposed a toast to science. "We are scientific", he claimed, "or we are nothing".[66] The very purpose of the Congress was "to convince the public [...] that the study of eugenics is one of the greatest and most pressing necessities of our age".[67] The Congress's programme certainly impressed the Transylvanian lawyer and future legal scholar Zsombor Szász, who was in England at the time. Writing to Apáthy from a small village in Sussex on 8 July 1912, Szász wanted to know whether any member of the Association for Social Sciences would be attending the Congress; if not, he volunteered to write a summary of the papers presented.[68] The review was sent to Apáthy on 16 August 1912,[69] who published it immediately in the *Magyar Társadalomtudományi Szemle*.[70]

Szász's review offered thorough analysis of some of the main papers presented at the congress, including those of Leonard Darwin, Charles Davenport, Sören Hansen, Lucien March, C. W. Saleeby and Alfred Ploetz. Szász also noted some of the criticism expressed towards negative eugenic measures, most notably the report read by Bleeker van Wagenen, Chairman of the Committee on Sterilization of the Eugenic Section of the American Breeders' Association.[71] Certainly, the readers of the *Magyar Társadalomtudományi Szemle* would have found it difficult to accept some of the radical measures proposed by American eugenicists to reduce the number of "defectives". These included:

1. Life segregation (or segregation during the reproductive period); 2. Sterilization; 3. Restrictive marriage laws and customs; 4. Eugenic education of the public and of prospective marriage mates; 5. Systems of matings purporting to remove defective traits; 6. General environmental betterment; 7. Polygamy; 8. Euthanasia; 9. Neo-Malthusian doctrine, artificial interference to prevent conception; 10. Laissez-faire.[72]

In addition to guarding society from its alleged degeneration, eugenics was thus responsible for the moral education of individuals. All eugenicists were, in fact, dedicated moralists, believing that a healthy life was based on both physical and spiritual achievement.

This new eugenic morality advocated shared beliefs and methodologies with other international reform movements. One example, previously noted, was the connection with neo-Malthusianism (explored further below) and the movement for sexual reform; moreover, eugenics was also linked with the ethics reform movement. Originating in the Society for Ethical Culture, established by Felix Adler in New York City in 1876,[73] the ethics reform movement soon expanded to Europe. A German Society for Ethical Culture was created in 1892, followed by the Viennese Ethical Society in 1894. In England, the Moral Education League – considered by Pauline Mazumdar to be "the Eugenics Society's direct forerunner"[74] – was established in 1898. Taken collectively, Tracie Matysik has argued that the ethics reform movement claimed "a special role as the most fundamental of all categories of reform because it offered them a way to rethink what social life was possible in a modern context".[75] This narrative of belonging – concentrating on the *moral subject* – dovetailed with the eugenic vision of the individual and the national community as a *biological subject*.

With such insistence on both morality and belonging the ethics movement was in complete harmony with core eugenic principles of the day. James W. Slaughter, Secretary of the Sociological Society of London, remarked upon this close relationship in his paper, presented at the First International Moral Education Congress at the University of London between 25 and 29 September 1908. The concept of ethics was germane to the understanding of eugenic rhetoric on biological improvement. According to Slaughter, "Eugenics supplies a new moral principle. We have without doubt entered upon a new chapter in ethics based on knowledge and man's nature and conditions of his descent. This biological knowledge is demanding a corresponding sense of biological responsibility."[76] Rooted in biology, eugenics claimed a distinct moral authority to address issues that bore directly on both individual and community, both men and women.

Elaborating on this topic, the Bishop of Székesfehérvár, Ottokár Prohászka, linked eugenic morality to the notion of the Christian family and its role within the racial community. By the 1910s, Prohászka was already a central figure in the Christian Socialist movement in Hungary.[77] In his paper entitled "Ethical Co-operation of Home and School", he revived the importance of religious experience for youth education and national morality. Prohászka located the regenerative potential of eugenics within Christian morality and used it to reinforce the importance of instruction and schooling in the formation of a strong national community. "A pure and strong race", he concluded, "can only result from pure and well-disciplined family life, while the school must awaken and keep alive the desire for a pure life, which brings more happiness than impure and debilitating excesses".[78] The idea of human improvement was not some paean to the human spirit, Prohászka suggested, but was equally a re-evaluation of the importance given to the human body. The traditional Christian model of human corporeality – in which the body was devalued as fallen – remained central to Prohászka's interpretation of eugenics, which nevertheless conceded the renewed importance modern sciences had given to the establishment of a new physical vision of humanity. This was not eugenics as a secular religion, as Galton envisioned it, but eugenics as an appendix to religion.

Sarolta Geőcze also discussed Christian morality and the depiction of children as custodians of the nation's future in her paper on "Environment and Moral Development". Founder of the Hungarian

National Association of Christian Women Workers (Magyar Keresztény Munkásnők Országos Egyesülete), Geőcze appealed directly to biology in asserting a bond between morality and religion. "Properly speaking", she noted, "heredity itself does not influence [the child] merely as a biological agent but also as a family or race tradition; it is not only a bodily, but a mental disposition, strengthened by ever repeated examples".[79] Thereafter, Geőcze championed nationalism as the antidote against degeneration:

> The surging up of the national spirit is an elevating influence upon the young; so is a national war for independence. The Japanese or Polish child is ennobled by his patriotic feelings. But the children of oppressed degenerate races are themselves degenerate and prone to evil.[80]

Geőcze attributed this decline to a combination of cultural, environmental and hereditary factors. Like Prohászka, Geőcze largely viewed modern feminism as disorderly and subversive. To counter its effects, both Prohászka and Geőcze portrayed motherhood and youth education as a means of advancing Christian and family values.[81] Education was to be the medium through which a radical transformation in collective thought and national values was to be undertaken.

Other members of the Hungarian delegation, such as Sándor Giesswein[82] and Mór Kármán, the Secretary of the National Council of Education (Országos Közoktatási Tanács), insisted on the importance of ethical and character training.[83] This was a subject further developed by Ferenc Kemény, an esteemed pedagogue and member of the first International Olympic Committee, in his contribution to the Second International Congress of Moral Education held in The Hague between 22 and 27 August 1912.[84] In his paper, presented to the panel dealing with eugenics, Kemény affirmed the importance of physical education and hygiene in upholding and maintaining a healthy body. "The future", he declared, "belongs to the rational body-culture ennobled through hygiene."[85]

Such views echoed arguments over the relationship between morality, physical education and eugenics offered by the other Hungarian participants. Two of these, Kármán and Geőcze, reiterated their commitment to a national ethics, to be achieved through modern educational and welfare policies.[86] With renewed awareness, Geőcze launched yet another devastating critique of modern civilization, which she held responsible for the disintegration of traditional family, public brutality and widespread criminality. The legal expert, Pál Angyal was likewise preoccupied with the proliferation of criminality among Hungarian youth, and proposed various methods for combating it.[87] Angyal, like Geőcze, was particularly concerned with the impact of modern life on the young generation, though he insisted less on Christian morality and more on the improvement of the social environment, particularly when combined with modern penitentiary childcare and protective social hygiene.

This was a panel that also attracted contributions from John Russell, representing the Eugenics Education Society, who spoke on "The Eugenic Appeal in Moral Education"[88] and Caleb W. Saleeby, who discussed "Eugenic Education or Education for Parenthood".[89] Both highlighted the centrality of eugenic responsibility towards future generations, with Saleeby especially commending the advantages of eugenic marriages. According to these authors, moral education was but one aspect of a broader movement for social and biological improvement – one embracing not only eugenics, but equally child welfare, the protection of mothers, the reformulation of women's reproductive role, and a renewed interest in religion.

At these international congresses it is possible to see how eugenic ideas were discussed and circulated, especially beyond the immediate confines of a particular national tradition or profession. These organized meetings, moreover, facilitated the interaction between Hungarian educators, legal experts, social workers and religious leaders with eugenicists from other countries. For instance, in Dresden, Teleki connected with Ploetz and Rüdin, while Schwimmer networked with Stöcker and Drysdale; Prohászka and Giesswein conversed with Lombroso in London, while Geőcze and Angyal shared a panel on eugenics with Russell and Saleeby in The Hague. A more specifically articulated eugenic vision of social and biological improvement emerged from these encounters, a vision worth emphasizing here as it gained even greater significance when viewed at the intersection of biology, culture and religion.

Dissemination and Debate

Eugenics consistently advocated a renewed affirmation of the nation's social and biological improvement, even when couched in the language of universalism. Against this wider background of scientific internationalization, efforts by individuals and institutions in Hungary to publicize the significance of health, hygiene and eugenics were diverse though insistent. International events, such as the International Hygiene Exhibition in Dresden, the First International Eugenics Congress or the First and Second International Moral Education Congress, gave Hungarian experts an opportunity to engage with foreign medical cultures and to refine some of their ideas on health and hygiene. This, in turn, increased these experts' experience and fuelled their ambition to put eugenic ideas into practice.

One such opportunity was provided by a conference on public health convened in Budapest between 6 and 23 February 1912 by the Association for Social Sciences. In his introductory remarks, Jenő Gaál emphasized "the importance of hygiene in our time",[90] encouraging invited experts to reflect on the recent exhibition in Dresden as an example of how progress in medical knowledge – comprising modern conceptions of public health, social hygiene, preventive medicine and eugenics – strengthened the relationship between society and state. Gaál's was not just a call for the expansion of medical institutions, the adoption of modern legislation of social assistance and the protection of children and mothers, but was also one that considered these developments within the larger agenda of the nation's biological protection. "Hygiene", he pointed out, "can save the race from degeneration".[91] Ultimately, for him, hygiene functioned best as the modern state's comprehensive strategy of forging a healthy national community.

Thus defined, hygiene and public health offered participants a broader framework in which to debate their eugenic ideas. Some of them, like Leó Liebermann and László Detre, had also participated in the debate on eugenics organized by the Society of Social Sciences a year earlier. For others, like György Lukács,[92] Lajos Nékám, Augusta Rosenberg, Ödön Tuszkai, Ernő Deutsch and Ottó Pertik, this conference on public health was a new opportunity to discuss their first eugenic initiatives. Finally, for participants like Zsigmond Gerlóczy, Farkas Heller and Henrik Pach, this conference reaffirmed the role of the health expert in society.[93]

According to Lukács, government officials had long welcomed conferences on public health. As Minister of Religion and Education between 1905 and 1906, Lukács himself had adopted the cause of national health, introducing a new set of policies meant to improve hygienic conditions in Hungarian schools. He coupled these measures with establishing new educational opportunities for women, while offering increased financial support for childcare and wider social assistance.[94] In his contribution to the conference, he emphasized the need for better education through "regular popular health lectures" to be held not only in major cities but in the villages as well, so as to attract as wide an audience as possible.[95] Lukács further identified "the health of the mother" and "infant hygiene and child care" as related areas further testifying to the growing importance of health education.[96]

Other participants voiced similar concerns, aided by experiences from their respective areas of expertise. Lajos Nékám, for example, a specialist in dermatology, was concerned with sexually transmitted diseases, and insisted on coordinated government control and the introduction of stricter legislation.[97] According to Nékám, another area in need of eugenic intervention was state care for illegitimate children. He eschewed the social opprobrium attached to illegitimate children, particularly in poor areas. Child protection had to be regulated in the interests of the nation's racial future, not dictated by social and cultural norms.[98]

Much of this, of course, related directly to broader debates on social hygiene, public health and eugenics. Some participants directly advocated eugenic measures as means to protect the nation's health. Both Augusta Rosenberg, the vice president of the Federation of Hungarian Women's Associations (Magyarországi Nőegyesületek Szövetsége), and the pathologist Ottó Pertik, for instance, believed that the introduction of "health certificates" before marriage was an efficient way of combating the spread of venereal diseases and thus "protecting the family and the race".[99] These preventive eugenic measures, together with anti-alcohol and anti-tuberculosis campaigns, could play a key role in reversing "the degeneration of the race".[100] The overall quality of the race itself was deemed to be of paramount importance.

This eugenic vocabulary of racial degeneracy also characterized Ernő Deutsch's contribution. He linked social hygiene (*sociálhygiéne*) and racial hygiene (*fajhygiéne*) to both public health and "modern biological research".[101] According to Deutsch, effective protection of the national community resulted from an intricate web of medical, social, cultural and political strategies. As with other speakers, Deutsch endorsed general education and intensive health propaganda. In this respect, he praised the Social Museum in Budapest for its attempts to popularize the most recent technologies of public health and social hygiene.[102]

In his concluding remarks, Gaál returned to the wide-ranging impact that events such as the International Hygiene Exhibition in Dresden had had on the general public in Germany. In order to best translate ideas into reality, Gaál advocated the staging of similar events in Hungary. Just as this conference on public health had occasioned an exchange of eugenic views on health certification before marriage, repeated reference to the welfare of the population needed to prompt reflection on the central role of public health in Hungary. The eugenic transformation that these experts wished to effect in Hungary was not possible – they argued – without the patronage of a modern state committed to interventionist health policies.

The enthusiasm expressed by the Association for Social Sciences for the popularization of public health received further impetus from the Social Museum in Budapest. In the summer of 1912, this museum organized health exhibitions in Hungary's main provincial towns: Veszprém (25 August–6 September 1912), Kaposvár (3–14 November 1912), Székesfehérvár (21 December 1912–3 January 1913), Kolozsvár (1 May–12 June 1913) and Győr (14 August–10 September 1913). The number of visitors varied, but in larger places like Kolozsvár the exhibition attracted 42,160 people, while 24,579 and 30,168 visited the health exhibitions in Székesfehérvár and Győr, respectively.[103] The popularity of the exhibition with the general public confirmed the organizers' claim that the transformation of Hungarian society was possible through health and education.

The focus of the exhibition also changed over these months, from having five sections in Veszprém in 1912 (including Tuberculosis, Alcoholism and Child Welfare) to staging a complete 13-section programme in Győr in 1913 – extending to Organs and Functions of the Human Body, Occupational Diseases, Emigration, Housing, Social Work, Infectious Diseases, Tuberculosis, Alcoholism and Children's Hygiene.[104] The avowed aim of this exhibition was to place hygiene and health protection at the very centre of the national community. This sentiment was best asserted by Menyhért Szántó, Director of the Social Museum: "To preserve the life of the body, which is the temple of the soul, to pay due regard to our own health and that of our fellows, is a paramount duty, and to spread this feeling of duty has been the purpose of our exhibitions."[105] Yet even more importantly, Szántó believed that modern social reform presupposed the creation of a healthy national body.

Complementing this concerted effort to inform and educate the public was a subtle shift in the cultural and scientific language describing the relationship between eugenics and the modern Hungarian state. As noted earlier, the early twentieth century was a period when Darwinist metaphors of struggle and visions of society as a living organism prone to degeneration permeated the scientific marketplace in Hungary. In this context, eugenicists hoped their various programmes for a healthy nation could be more widely disseminated through the establishment of new publications and new professional associations. No journal devoted exclusively to eugenics existed as yet in Hungary. *Fajegészségügy* had been published briefly in 1906, but it covered mostly social hygiene and public health. Medical, cultural and feminist publications, such as *Egészség*, *Gyógyászat*, *Huszadik Század*, *Magyar Társadalomtudományi Szemle* and *A nő és a társadalom* (*Woman and Society*), often published articles on eugenics, yet the attention was mostly on developments elsewhere. Hungarian scientific journals like the renowned *Természettudományi Közlöny* (*Natural History Bulletin*), moreover, although extensively promoting Darwinism, were less interested in eugenics. Zsigmond Fülöp – a keynote speaker at the 1911 debate on eugenics – however, sought to redress this balance. In 1912, Fülöp became the editor of a new journal, fittingly named *"Darwin"* (Fig. 6), which aimed to raise public awareness about the interconnections between heredity, evolution and eugenics.

Fülöp also created a "Darwin" collection of books for the general public, which included Hungarian translations of Darwin himself, J. Lamarck, A. Weismann, W. Bölsche, as well as D. Diderot and Emil du Bois-Raymond.[106] Like *Huszadik Század* a decade earlier, *"Darwin"* claimed to speak for a new generation of Hungarian readers who had reached maturity in the 1910s and who were positivist, rationalist and secular in their outlook.[107] It was to this new generation and its vision of social reform that journals appeared to respond when publishing articles on eugenics, by Hungarian authors or, alternatively, translations from foreign authors. What was remarkable about Fülöp's idea for a "Darwin" collection in Hungarian, in short, was his intention to make available a number of authors identified as essential to better understand recent theories of evolution and heredity. Translations of eugenic texts were not particularly a fashionable practice at the time, since most educated Hungarians could read these texts in German, English and French. Indeed, Galton's "Probability, the Foundation of Eugenics" was his only text translated into Hungarian.

Another strategy in Hungary was to commission articles from authors already known to the reading public. As Darwinism and socialism were widely discussed among progressive Hungarian circles at the time it is no surprise that Fabian ideas were frequently discussed, particularly in *Huszadik Század*. Another journal, *Szociálpolitikai Szemle* (*Social Political Review*), edited by the sociologist Jenő Lánczi, secured an article from leading activists Sidney and Beatrice Webb in 1912.[108] Published under the rather misleading title "Poverty and Racial Beauty", the article located an intimate link between social reform and eugenics. The Webbs began by reviewed the impact of social and economic factors on race protectionism and racial fertility.[109] The British state was then criticized for its non-interventionist attitude and for not encouraging social and biological selection. Moreover, that the "fittest" members of the race

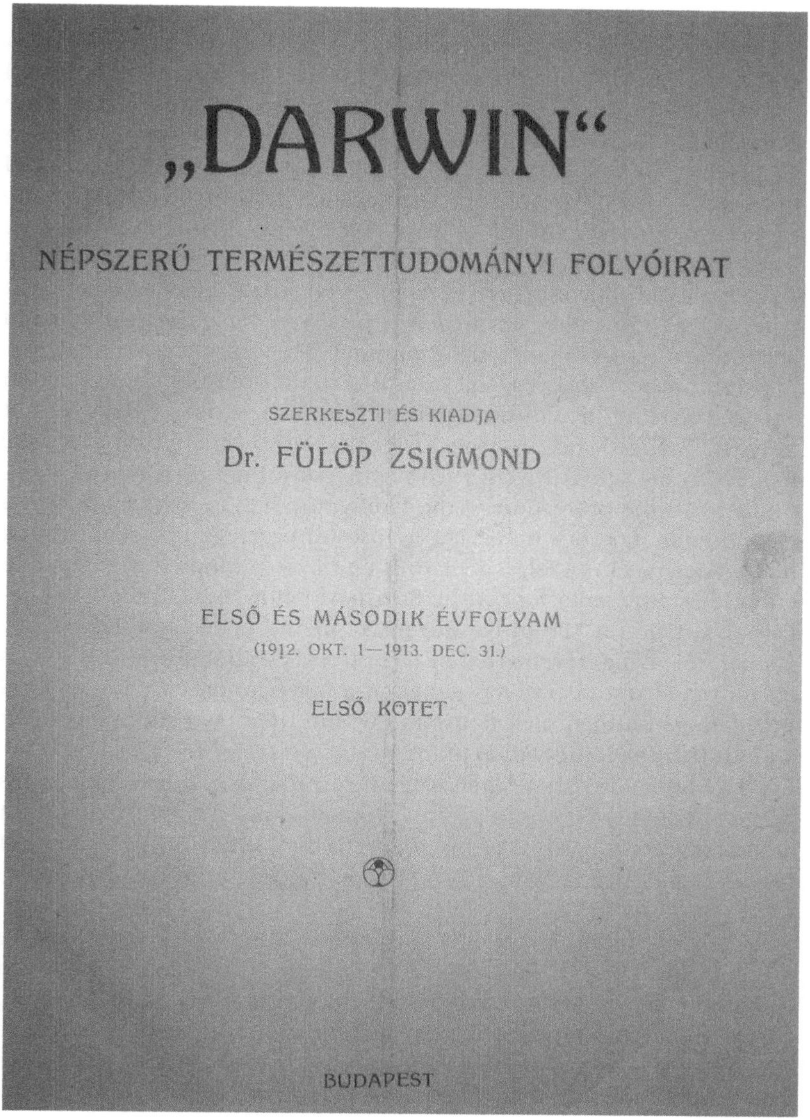

„DARWIN"

NÉPSZERŰ TERMÉSZETTUDOMÁNYI FOLYÓIRAT

SZERKESZTI ÉS KIADJA

Dr. FÜLÖP ZSIGMOND

ELSŐ ÉS MÁSODIK ÉVFOLYAM

(1912. OKT. 1—1913. DEC. 31.)

ELSŐ KOTET

BUDAPEST

Figure 6 The first volume of *"Darwin"* (1912/13)

would invariably always survive was chimerical, the Webbs believed. In light of modern social protection, the racial quality of the nation had not improved, but deteriorated instead. Institutional reform and concerted action on eugenics could, if applied properly, deter those who should not have children – for one reason or another – from doing so. "Next to biological heredity", the Webbs asserted, "there is social heredity".[110] For these campaigners, eugenics cultivated both. If it were to succeed, however, those who were perceived as "unfit" for work were considered "asocial" and "unhealthy" and needed to be prevented from reproducing. Following Karl Pearson's lead, the Webbs assumed that "the central problem of practical eugenics is to make the healthy child a valuable asset to the national economy".[111]

Thus, for the Webbs, eugenics was tied to the control of reproduction and perceived problems associated with human fertility. Yet it was not only the Fabians who connected eugenics with restrictive reproductive policies. Some activists went even further, positing that a eugenic commitment to the protection of the family, the nation or the race was a basic human trait, one necessarily transcending geographical or cultural boundaries. Such widely circulating opinions prompted a Hungarian ethnologist and collector for the British Museum, Emil Torday, to publish an article on "Primitive Eugenics" in the ephemeral *The Mendel Journal*.[112] Having travelled extensively across Central Africa, Torday was an informed researcher who aimed at a better understanding of local populations on their own terms and within their own cultural context and on their own individual terms.

While remarking that "the savage negro [sic!] is not less fond of his progeny than the European parent", Torday adduced another important argument: "fond as he is of his children, the welfare of his *race* is still *instinctively* dearer to his heart".[113] Correspondingly, Torday continued,

> When the infant is born, it is examined carefully; if it is weak or deformed, then in one way or another, it is no longer allowed to burden its own life nor handicap its race in the struggle for survival. This is the reason why one sees no cripples or other kind of defective persons in Central or West Africa; this is the reason why man there is a *man*, virile in habit, strong and lithe in body.[114]

As this suggests, Torday understood health and physical strength to be exclusively natural attributes. The implicit critique of European narratives of biological superiority is obvious here, even if Torday's description of "primitive eugenics" retained traces of colonial mimicry. The Africans' allegedly conspicuous physical attributes and untainted health directly invoked the Enlightenment view that indigenous peoples were "children of nature", whose very existence was, by the outset of the twentieth century, threatened by European conceptions of hygiene, disease and race.

This brief excursus beyond the confines of Europe and into the complex colonial setting of Africa provides supplementary evidence of how eugenics circulated across and in-between disciplines, as well as of the intellectual dexterity of Hungarian intellectuals and scientists. Furthermore, it also underscores the ways in which institutionalized forms of communication, embracing public lectures, conferences, academic congresses and journals, fostered professional interaction, conceptual and ideological affinities and, ultimately, the emergence of an international culture of eugenics. Bearing these considerations in mind, the interest shown by British neo-Malthusians in Hungarian practices of family control – in addition to viewing the relationship between these practices and eugenics more broadly in terms of demography and neo-Malthusianism – gains a deeper significance.

Demography and Neo-Malthusianism

It is hardly surprising that, at the turn of the twentieth century, emerging interpretations of eugenics often prioritized individuals over families. Pragmatically, this meant that diverse groups – including feminists to sexual reformers – engaged with eugenic thinking on social and biological improvement. Yet, in contrast, there was no consensus on the ideal family size; in fact, supporters of quantitative (large families) and qualitative (small families) eugenics often disagreed over which strategy was best suited to ensure the healthy biological future of the nation.[115]

In his 1912 *Neo-Malthusianism and Eugenics*, Charles Vickery Drysdale endeavoured to explain similarities and differences between the two.[116] While endorsing "negative eugenics", indeed equating it with neo-Malthusianism, Drysdale was nevertheless adamant in his condemnation of "positive eugenics". He described it as "brutal, unscientific and immoral", since it erroneously presupposed that "fit" individuals would "reproduce as much as possible in order to eliminate the 'unfit' by the struggle for existence", but also "because of the suffering it involves to women and children and to the whole community".[117] Neo-Malthusianism, on the other hand, was

> Race Control, concerned both with the quantity and quality of the race, and with the welfare both of the individual man, woman and

child, and to posterity; and it absolutely refuses to accept the proposition that the improvement of the race demands the sacrifice of women to passive and unlimited maternity.[118]

It was the last argument regarding the control of reproduction – in particular – that infuriated the eugenicists. Alfred Ploetz's view, expressed in his paper presented to the First International Eugenics Congress, exemplifies eugenicists' grave concern over birth rates, fertility, reproduction and the health of future generations.[119]

As elsewhere, of course, Hungarian eugenicists were divided over fertility welfare strategies. As discussed earlier, feminists like Rosika Schwimmer or Vilma Glücklich connected education and limitation of fertility with women's social and economic emancipation; others, like Sarolta Geőcze, disagreed, arguing that a woman's main role was to be a mother of the nation. For Geőcze, measures to curtail reproduction were both anti-Christian and anti-patriotic. The neo-Malthusian movement for birth control, while not lacking supporters, more generally remained disorganized in Hungary. Considering the widespread predisposition towards small families in certain rural Hungarian communities in Transdanubia, it comes as no surprise that the general attitude among eugenicists was essentially pro-natalist. Increasingly, the "one-child" practice was one that eugenicists, social reformers, public health officials and politicians deemed incompatible with the dream of a racially strong Hungarian nation.[120]

For passionate advocates of birth control like Drysdale, however, the "one-child" practice in Hungary remained a praiseworthy example of systematic population control – one exercised voluntarily by a community in harmony with its social and cultural ideals. Following his trip to Hungary in 1913, Drysdale enthusiastically asserted that

> All who are interested in social questions should certainly take the first opportunity of visiting Hungary. It would be difficult to find another such combination of backwardness and enlightenment, of rapid transition from eastern conservatism to ultra-progressive modernism, and of diverse races and creeds.[121]

Drysdale first published these reflections on the demographic situation in Hungary in *The Malthusian*. He considered the country "progressive from the Neo-Malthusian point of view" due to the voluntary "one-child" practice and, moreover, because "Hungary was most certainly the first country to give an official medical judgement in favour of prevention of conception".[122] Drysdale was referring here to a well-known

Memorandum submitted by the Alliance of Smallholders (Kisgazdák Szövetsége) to the Minister of the Interior in 1911, discussed below.

Throughout his Hungarian sojourn, Drysdale visited county asylums for infants and children in Debrecen, Nagyvárad and Szeged; as well as homes of foster parents in the village of Nagyszalonta (Salonta; Großsalontha); and even a reformatory school for boys and girls in Szeged. According to his impressions, he was especially well treated in Nagyvárad by his host, Menyhért Edelmann, a physician in charge of the local asylum whom he had met in 1911 at the Hygiene Exhibition in Dresden.[123] Drysdale also commended Hungary's child protection system, introduced in 1901, for its effectiveness, even if he raised objections on neo-Malthusian grounds that it allowed over-population and indulged social humanitarianism.[124] Yet few Hungarians – concerned as they were with demographic growth – would have agreed with Drysdale's suggestion that a "falling birth rate" was in any way beneficial to the nation.[125] There were, however, reasons to hope that in a country like Hungary, where the "one-child" practice was widespread, neo-Malthusian ideas could find a positive reception. Similarly, the emergent Hungarian feminist movement was another positive development, according to Drysdale, as it advocated birth control and the biological empowerment of women. Both Schwimmer and Glücklich were respected internationally for their efforts to inform public opinion in Hungary regarding the importance of neo-Malthusianism, but also for campaigning against women's exclusion from socially and economically productive labour.

There was further support for neo-Malthusianism and demographic management among Hungarian medical elites. When the aforementioned Alliance of Smallholders submitted a Memorandum to the Minister of the Interior in 1911, demanding that the state intervened against the "one-child" practice, the document was forwarded for appraisal to the National Council of Public Health (Országos Közegészségügyi Tanács). At the time, the Council was under the chairmanship of Vilmos Tauffer, Hungary's foremost gynaecologist and Director of the Institute of Obstetrics and Gynaecology at the University of Budapest's Medical Faculty.[126] A main source of concern expressed in the Memorandum was the contested custom of early marriages:

> Especially conducive to the Single Child System are *inter alia* the absurdly permitted marriage of 18 year old youths with 15–16 year old girls. What a danger for the race is implied in these early marriages! They simply ought not to be allowed by the laws. Such early

sexual life results in "female diseases", premature old age, sterility –
or, at least, in defective offspring.[127]

In response, the National Council dismissed these claims as "entirely
unwarranted. Early marriage can bring with it many social evils", it was
admitted "– perhaps also ethical and economical disadvantages, which
we will not consider – but never 'female diseases', premature old age,
sterility or defective offspring". By contrast, youthful qualities were to
be nurtured, not chastised: "It cannot be supposed that the marriage of
youthful persons (assuming they are physically fit, as should be medically ascertained) gives rise to hygienic evils."[128] There were no sound
medical reasons to object to early marriages, the Council concluded.

Yet on the issue of contraception, the Council sombrely agreed with
the Alliance of Smallholders' claim that "the practice of abortion pre-
vails to a horrible extent not only in the capital and the great towns,
but also in the country. This social disease devours the life force of the
people, for it is a source of much injury and life-long invalidism."[129]
Entirely eliminating contraception was realistically impossible although
strict control of prophylactics, alongside an improvement of living and
sanitary conditions, was endorsed. Another point of agreement was
the prevalence of high rates of infant mortality, "which our population has in its germ, so to speak". Although it was possible to entertain
hereditary arguments for these high rates, the phenomenon was more
likely on account of a combination of factors, including "unorganized
administration in hygienic matters" and "the poverty of the people".[130]

The Memorandum's conflation of marriage, contraception and infant
mortality was not accidental. Since rural families tended to be smaller
than urban ones, preserving the peasant Hungarian family became an
act of racial protection – not least since other ethnic groups within
Hungary were known for their higher fertility rates. For the Alliance
of Smallholders, celebrating natalism and the large family ideal was a
useful method to promote demographic growth, even if rather incon-
gruously, as the Council delicately remarked, this was not a strategy
similarly adopted by the landlords and aristocratic elites: "It is well-
known that the members of the propertied classes bring only a few
children into the world, in order that the standard of life of the children
should not fall below that of the parents."[131]

Viewed in this light, attempts to limit family sizes were less about
racial protectionism than about the preservation of class and social
status: "The whole neo-Malthusian practice owes its origin to the
propertied classes", the Council reminded the Minister of the Interior.

From this perspective, quantitative eugenic rhetoric yielded to a discriminatory limitation of fertility. "The very moment, however, that the working classes commence the adoption of this practice, the ruling classes proclaim all such conduct as immoral, which they, by their own conduct, have recognised as moral."[132] This unnecessary stigmatization of the peasant and working classes, the Council believed, only served to undermine the Alliance of Smallholders' wider welfare measures and strategies for racial improvement.

Ultimately, the National Council of Public Health rejected the Alliance of Smallholders' call for interventionist, state-organized natalist policies:

> The State possesses neither the power nor the means to prevent or diminish family limitations; for when, the working classes have realised that excessive reproduction puts a burden on their progeny, and have learnt the means of restriction, there is no law or power which can bring them back to renewed over-reproduction.[133]

In practical terms, the mere ideal of a numerically strong Hungarian nation was not a sufficient basis for a comprehensive demographic policy; substantive economic and social transformations were required beforehand. Vilmos Tauffer and other members of the National Council of Public Health were sympathetic to neo-Malthusian arguments despite believing that any inclusive conception of family limitation and controlled reproduction presently contained little applicability to Hungarian conditions.

Seen more broadly, the exchange between the Alliance of Smallholders and the National Council of Public Health illustrates the multifarious nature of eugenic experience in early twentieth-century Hungary, particularly in relation to political activism and nationalism. This fluid and diverse conception of eugenics and the internationalism it engendered is important to remember, however, as it never weakened the perceived national imperatives of social and biological improvement; quite the contrary. Taken as a whole, Dienes and Vámos's articles, the public lectures on Darwinism, evolution and eugenics offered by members of the Society of Social Sciences to local interested groups in various provincial towns, together with the itinerant health exhibition organized by the Social Museum and the conference on public health organized by the Association for Social Sciences, only strengthened the conviction that eugenics was completely compatible with a wider transformation of Hungary into a modern national state. If eugenicists associated with the Society of Social Sciences tended to import models of thought developed elsewhere, particularly Britain, they nonetheless adopted them to their own, cultur- ally specific problems. It is therefore not only the transmission of eugenic ideas, but the way in which these ideas were refashioned in the specific Hungarian context that needs to be further analysed.

New areas of eugenic inquiry had emerged through this intellectual syncretism, including child welfare, social protection for mothers, neo-Malthusian family planning, sexual reform, the introduction of moral education and increased population policies. The exhibitions, conferences and congresses discussed in this chapter helped to com- municate this nascent eugenic culture, both easing the transfer of ideas and facilitating a better public understanding of social and biological improvement. Increasingly, however, there was little room for the flex- ible principles of liberal-progressive eclecticism, which had dominated eugenic thinking in Hungary during the first decade of the twentieth century. As Prohászka, Geőcze, Lukács, Nékám and other eugenicists of similar orientation envisioned it, eugenics needed to become national and thus better suited to address the complex problems characterizing modern Hungarian society. This latter group understood eugenics less as a social movement for reform and emancipation and rather more as a larger biopolitical project that equated national progress with the healthy development of the Hungarian race.

4
Towards National Eugenics

One Hungarian eugenicist, even more devotedly than others, actively rejoiced in the progress of eugenic thinking in Hungary discussed in the previous chapter. His name was István Apáthy. As he understood it, eugenics was essential not only to reinforce scientific norms of health and hygiene, but to the very survival – let alone strength – of races and nations. It was this commitment to preserving racial individuality, which Apáthy shared with other Hungarian eugenicists, that nourished the emergence of a distinct and exclusive body of eugenic knowledge, and ultimately the transformation of eugenics into a national science with its own organization and dedicated members.

For any initiative to be successful, however, it required the work of more than one individual. As will be discussed in this chapter, in Hungary the establishment of a eugenics society commenced with an intense exchange of ideas between István Apáthy and the other Hungarian eugenicist of national and international reputation: Géza Hoffmann. Each proposed a coherent eugenic programme of social and biological improvement that could be applied to the Hungarian context. Ultimately, it was Apáthy's vision of eugenics that was adopted by the participants at the opening meeting of the Eugenic Committee of Hungarian Societies in January 1914.

István Apáthy's Conception of Eugenics

It is easy to detect in most Hungarian texts discussed so far the widespread influence of Galton's model of eugenics as well as the growing importance of German racial hygiene. During the first decade of the twentieth century, the eugenic vocabulary provided by these traditions marked and reshaped emerging Hungarian narratives of social and biological improvement. Apáthy, however, expressed concerns about what he perceived as uncritical imitation and superficial understanding of eugenic theories developed elsewhere. As discussed in the previous chapter, Apáthy reminded the Hungarian public that eugenics was essentially a *biological* science, following strict scientific principles; it entailed identifying, defining and naming the biological processes contributing to human improvement.

István Apáthy (Fig. 7) – a professor at the University of Kolozsvár and the dean of the Faculty of Natural Sciences – was a well-known physician, zoologist and histologist with an impressive scientific career. He was an authority on the structure of the nervous system[1] and was a key supporter of the new theories of evolution in Hungary at the beginning of the twentieth century. His academic achievements were both numerous and widely celebrated.[2]

Apáthy was born in 1863 in Pest. He studied medicine, zoology and biology under one of the first Darwinists in Hungary, Tivadar Margó, at Péter Pázmány University in Budapest, where he graduated in 1885.[3] Shortly thereafter, Apáthy went to Italy for four years, working as a

Figure 7 István Apáthy (courtesy of the Semmelweis Museum, Library and Archives of the History of Medicine, Budapest)

Research Fellow at the Zoological Station of Naples.[4] Upon his return in 1890 he was appointed professor and Head of the Zoological Department at the Franz Joseph University (Ferenc József Tudományegyetem), the Transylvanian university of Kolozsvár. In 1903 he was elected Rector of the university, a position he held until 1905.[5] Apáthy's national and international scientific authority was further established in 1909 by the creation of the Zoological Institute (Állattani Intézet) in Kolozsvár alongside his appointment as its Director.[6]

After 1900 Apáthy increasingly turned his attention to the impact of Darwinism on society and culture, particularly in Hungary.[7] He began promoting a vision of national eugenics that combined evolutionary biology, social Darwinism and Hungarian nationalism.[8] What made Apáthy's eugenic texts distinctive from that of other eugenicists in Hungary was not only his knowledge of biology, but also his ability to discuss a wide range of topics – from patriotism to organicist sociology and popular science.[9] For Apáthy, the connection between eugenics and nationalism was obvious, and was driven by his commitment to understand Hungarian society no less than to reform it. The nation's regeneration was to be achieved through the cultivation of Hungarian racial qualities and no one, according to Apáthy, had devoted more practical work to this goal than the great reformer of the mid-nineteenth century, István Széchenyi.[10] Alongside his veneration of Széchenyi, Apáthy also aligned his views regarding the beneficial nature of patriotic education and state intervention on national issues. In this context, a key principle, which Apáthy reaffirmed together with many other Hungarian nationalists, was a belief in the historical and cultural destiny of the Hungarian race in the Carpathian basin.[11]

In addition to his crucial role in establishing the Association for Social Sciences in Budapest in 1907, Apáthy was also active in the Medical-Natural Science Section (Orvos-Természettudományi Szakosztály) of the Transylvanian Museum Society (Erdélyi Múzeum-Egyesület) in Kolozsvár.[12] To reaffirm a fitting description offered by Carl Schorske in a different context, Apáthy "was not just a university scholar and teacher. He was a *praeceptor urbis* in the fullest sense of the term".[13] At the beginning of the twentieth century, Apáthy was thus recognized as one of Hungary's foremost scientists and was one of Kolozsvár's most pre-eminent residents. During the 1910s, he also revealingly came to personify central concepts of the eugenic movement.

This was particularly evident in Apáthy's contribution to the debate on eugenics as well as in his first major eugenic text published in 1911, entitled "A faj egészségtana" ("Hygiene of the Race").[14] From the outset,

Apáthy's article expressed his preference for the term *fajegészségtan*, suggesting that this was "practically what Galton means by *eugenics*". Pointing to a widespread conceptual confusion, he observed that the scope of eugenics was not "human breeding" but "racial improvement". In fact, he added, it would be appropriate to describe eugenics as "the hygiene of the human race".[15] There was a close connection between hygiene and eugenics, even if the former focused on understanding and improving the human environment while the latter emphasized the protection of the individual and the race. To some extent, Apáthy continued, eugenics was an environmentally oriented conception of hygiene and public health, predicated on the "deliberate selection" of those individuals whose heredity was deemed valuable, with the simultaneous rejection of those who "may be harmful to the next generation".[16]

Apáthy insisted in portraying eugenics as a *biological* project of national improvement. If nothing else was achieved, he remarked sharply, the debate on eugenics organized by the Society of Social Sciences would still be remembered for providing "some knowledge of biology to our public and writers dealing with sociological issues".[17] This was a warning for those who placed too much emphasis on the social basis of eugenic theories of national improvement. Such an adjustment to a more biologized interpretation of eugenics was needed, Apáthy believed, in order to make this new science more compatible with observable human nature, in addition to making it more attractive to the Hungarian scientific community. Moreover, and contrary to the extensive use of the word *faj* (race), Apáthy's approach to eugenics was not informed by the views of German eugenicists like Ploetz and Schallmayer, whose work he had read but did not endorse. He criticized Schallmayer for not having fully understood the relationship between selection (natural and sexual) and evolution, for example, but admitted that his 1903 *Vererbung und Auslese im Lebenslauf der Völker* (*Heredity and Selection in the Life-Process of Nations*) was "the best survey" of questions also preoccupying eugenicists in Hungary.[18]

Apáthy was not, on the whole, satisfied with the general understanding of how evolution fully worked in human societies, but he did agree that degeneration was not simply a convenient eugenic fiction manipulated by social reformers and progressive intellectuals.[19] Indeed, it was an indisputable fact of life as a whole. To succeed in countering degeneration, Apáthy argued, the modern state had to embrace the eugenic project of social and biological improvement. To do this, Apáthy recommended providing the Hungarian state with "eugenic morals" (*fajnemesítő erkölcsök*).[20] He was, however, cautious about attributing the exclusiveness of the eugenic project to the state. A "happier, better and more noble humanity" could only materialize, he concluded, if the state and society worked together, united by the common eugenic ideal of a healthy family and nation.[21]

The same ideas are repeated in Apáthy's 1912 *A fejlődés törvényei és a társadalom* (*The Laws of Evolution and Society*), a book containing some of his lectures, delivered in Budapest and Kolozsvár in 1909 and 1910, together with a key article on eugenics, published in *Magyar Társadalomtudományi Szemle* in 1911.[22] Apáthy's conception of eugenics was by now fully formed. It was a synthesis of Darwinism and Mendelism, one in which the hereditary legacy of the past was constantly re-enacted in the present. "The purpose of eugenics", he declared, "was to enlighten humanity"[23] regarding the importance of heredity. To think in eugenic terms was to accept responsibility for the nation's biological future. Eugenics, Apáthy maintained, also equally affected the individual in terms of his or her biological future. What may appear as a potential tension between the individual and the national community was reconciled in a narrative of eugenic improvement that required both to aspire to a new morality. This narrative was, in turn, underpinned by a nationalist emphasis on a strong Hungarian state and a quasi-Hobbesian interpretation of Hungary's multi-ethnic future. Accordingly, strengthening the national body was commensurate with ensuring its survival, within and outside the state. Evolution was the driving force behind not only eugenic improvement but also larger demographic changes in the racial and national composition of society, Apáthy insisted. "The only way we can live", he concluded, "was according to the laws of evolution."[24]

As suggested by Apáthy's book, it is not difficult to locate Darwinist instances of theories of racial conflict in the Hungarian political culture prior to the First World War.[25] An uneasy relationship existed between Hungarian claims to national hegemony and demands for cultural and political autonomy made by a number of ethnic minorities, particularly the Slovaks, Serbs and Romanians.[26] Underlying this relationship was the anxiety over what an unpublished 1912 paper – preserved in Apáthy's archive of the Association for Social Sciences – described as "Hungary's racial war".[27] The assertion of "the unity of the Hungarian state and the Hungarian political nation" was contrasted with the "persistent conflict between races".[28] In an increasingly nationalistic environment, the ostensible purpose of the Hungarian state was to ensure that the minorities' linguistic, cultural and religious rights were respected, but also that ethnic pluralism and the political *Ausgleich* with Austria would not affect Hungarian claims to political hegemony.[29]

Apáthy was thus fully aware of the ideological nature of eugenics. Importantly, he played a pivotal role in legitimizing eugenics in the eyes of the Hungarian nationalist elites and he did so in two crucial ways: through his writings and through the scientific activities organized by the Association for Social Sciences and the Transylvanian Museum Society.[30] Equally important, when compared with other eugenicists in Hungary discussed so far, Apáthy was far less inclined to follow eugenic models developed elsewhere. In a period when British and German eugenics dominated Hungarian debates on eugenics, Apáthy had the intellectual and scientific confidence to suggest that what was needed, first and foremost, was a specifically national eugenics, one commensurate with the collective interests of the Hungarian nation and race. In reality, however, the relationship between the national and international in shaping eugenic thought in Hungary was rather more dynamic than Apáthy admitted. As illustrated by Géza Hoffmann's international reputation, there can be little doubt that other Hungarian eugenicists found the eugenic experiences of other countries enlightening.

Géza Hoffmann and Practical Eugenics

Géza Hoffmann – the only Hungarian author to have received recognition in the literature on international eugenics – was born in 1885 in Nagyvárad. After completing his secondary education (*gymnasium*) in 1903, he enrolled at the Imperial and Royal Consular Academy in Vienna (former Oriental), in preparation for a career in the consular service.[31] In keeping with similar institutions in the late Austrian Empire, the Academy exuded an aura of ceremonial formalism, aptly captured in a group photo of Hoffmann and his colleagues from 1907 (Fig. 8).[32]

Hoffmann graduated from the Academy in 1908,[33] the same year Emperor Franz Joseph had ennobled his father, Hugó Hoffmann, for services to the Austro-Hungarian army.[34] According to William D. Godsey, who has closely examined the workings of the Austro-Hungarian Foreign Service before the First World War, "once a candidate had satisfied the various entrance requirements and performed acceptably on the diplomatic examination, he received, within a few days his appointment as a new attaché".[35] Hoffmann's appointment was at the Imperial and Royal Austro-Hungarian General Consulate in New York.[36]

At the time, Hungarians dominated the Habsburg Monarchy's diplomatic missions in the USA: in addition to the ambassador, László Hengelmüller, there were two additional senior Hungarian diplomats, Lajos Ambrózy and Moritz Szent-Ivány, at the Embassy in Washington.

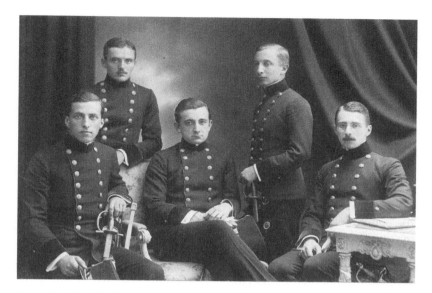

Figure 8 Géza Hoffmann (standing up on the right) in 1907 (courtesy of Pál Macskásy)

Considering the great number of immigrants from the Hungarian part of the Monarchy,[37] it was deemed practical to have Hungarian diplomats stationed in leading American cities, who would be able to "cultivate close relations with the local Austro-Hungarian colony".[38] Hoffmann's arrival in the USA also coincided with a period of intense interest in the life and working conditions of Hungarian Americans, exemplified by Ambrózy's tour of duty between 1907 and 1910.[39]

Not only abroad, but at home as well, Hungarian authorities were becoming increasingly concerned with emigration. In response, conferences were organized and studies commissioned to evaluate the social and economic effects of this phenomenon.[40] Some argued that economic benefits for the remaining population notwithstanding, emigration deprived the nation of some of its most hard-working and racially valuable members, especially as the majority of those leaving the country were young and healthy.[41] Nationalist arguments also began to appear in the media and official reports, warning that in some regions emigration favoured ethnic minorities – for an ever-increasing number were filling the void left by Hungarians who were moving abroad.[42]

Contributing to this emerging field of social studies of immigration and repatriation, Hoffmann published his first study on Hungarian-Americans in 1910.[43] This study was followed by a book entitled

Csonka munkásosztály: az amerikai magyarság (*A Mutilated Working Class: Hungarian-Americans*).[44] In both texts, Hoffmann used the political and cultural resources available to him as the cultural attaché to engage as directly as possible with the life of Hungarians in the USA. Using a range of primary sources – including interviews, newspaper articles and official documents – he provided a detailed survey relating to Hungarian-Americans' social, economic, cultural, religious and political life.[45]

Throughout his 1911 book, Hoffmann emphasized the social dexterity and economic acumen of Hungarian immigrants to the USA, as many distinguished themselves as civil servants, local politicians and religious leaders. The preservation of Hungarian culture, language and identity within the immigrant communities was also to be celebrated. Hoffmann similarly praised the role played by Hungarian women in maintaining the community's rituals and traditions, while at the same time pursuing modern education and professionalization. Commenting on the activities of other nationalities originating from the Austro-Hungarian Monarchy, Hoffmann noted that the Hungarians seemed to be the most passionate about their native land.[46] Yet, as they progressively integrated into American society, the more difficult it became to sustain their Hungarian identity, especially among the second generation.[47]

The issue of assimilation tested the new identity adopted by Hungarian-Americans, Hoffmann believed. Referring to Zsigmond Fülöp's 1910 article on eugenics published in *Huszadik Század*, Hoffmann reiterated the view that emigration was more than just an economic loss; it damaged the very future of the Hungarian race.[48] Most notably, the children of Hungarian immigrants were most greatly affected. "The generation born here" in the USA, he remarked gravely, "is lost for Hungary. It does not think of itself as Hungarian, not even Hungarian-American, but American; it felt Hungarian-American only for a short time, before it melted into American society."[49]

The personal emancipation of American-born Hungarians, together with education and social mobility, ultimately led to the abandonment of the Hungarian language and traditional Hungarian values. Assimilation, Hoffmann believed, simply assumed the adoption of an entirely new social and cultural identity. This was so much so that the body of the immigrant itself changed, prompting a new racial typology. In making this argument, Hoffmann relied on recent evidence provided by the distinguished anthropologist, Franz Boas in his report to the United States Immigration Commission. In this 1910 study published as *Changes in Bodily Form of Descendants of Immigrants*, Boas strikingly concluded, "the head form, which has always been considered one of the most stable and permanent characteristics of human races, undergoes far-reaching changes coincident with the transfer of the people from European to American soil".[50] Although Boas used this research to expose notions of racial permanence in favour of his theory of human biological plasticity,[51] for Hoffmann such "re-racialization" spelled the end of "specifically" Hungarian racial characteristics. "Hungarian Americans" would soon disappear into the racial melting pot of the USA, he believed. Hungarians who did not return home would gradually become American, in sensibility and even physiognomy.[52]

If Hoffmann's conservative and nationalist overtones were clearly discernible, so was his objectivity and analytical diligence, which one reviewer attributed to Hoffmann's having "learned the techniques of book writing in America" instead of Hungary.[53] More importantly for present purposes, Hoffmann's book also reveals his overarching preoccupation with the preservation of the Hungarian race, a theme that would reappear frequently in his future writings. Racial protectionism and eugenic incentives for Hungarian families were thus seen to be an important means for preventing emigration and a host of associated problems. For Hoffmann, crucially, it was in the USA that eugenics first matured into a practical programme of social and biological improvement, as confirmed by the wide application of sterilization laws.[54]

At the time, few would have guessed that Hoffmann would soon become an international authority on eugenics. In 1912 and 1913, he continued to publish on emigration and the "acculturation" of Hungarians in the USA,[55] now from the position of vice-Consul at the Imperial and Royal Austro-Hungarian General Consulate in Chicago. Also in 1912, he contributed his first article on American eugenics and marriage restrictions in German.[56] This article represented the beginning of his collaboration with the prestigious *Archiv für Rassen- und Gesellschaftsbiologie*, for which he repeatedly reviewed Hungarian books and reported on developments in eugenics, social hygiene and public health – such as the proposal on the introduction of health certificates advanced during the conference on public health organized by the Association for Social Sciences in February 1912, highlighted earlier in this chapter.[57] Likewise, this first German article marked the beginning of Hoffmann's interest in eugenic sterilization and other negative methods of racial improvement. By this time, eugenics had become more than a scholarly vocation for Hoffmann, it represented no less than a "new religion" – one which his wife, Paula, also shared.[58]

Few European eugenicists could have justifiably claimed a better knowledge of American eugenics prior to the First World War than

Hoffmann. He was not only its best overseas interpreter but also an enthusiastic supporter as well. His wholehearted devotion to eugenics became still clearer with the publication of Hoffmann's second book, *Die Rassenhygiene in den Vereinigten Staaten von Nordamerika* (*Racial Hygiene in the USA*).[59] This 1913 book established Hoffmann as a truly international specialist in American eugenics. Hoffmann assiduously assembled an outstanding collection of sources in order to highlight the presupposed concurrence between theories of eugenics and pragmatic social policies such as sterilization and immigration. He likewise endorsed negative eugenics policies on a practical level, while simultaneously emphasizing their potential theoretical importance for more general debates about the future of the nation and race.

"America is in no way radical", he declared at the outset of his book. "It is only rational to the point of sobriety." Therefore for Americans, eugenics was the expression of "pure, practical reason".[60] As a consequence, it was not surprising, Hoffmann noted, that the USA was the first country to introduce negative eugenic measures. Before guiding the reader through the eugenics labyrinth related to sterilization and immigration, Hoffmann provided an introductory chapter in which he discussed the basics of international eugenic thinking. Although "race-improvement" (*Rassenveredlung*), had existed from time immemorial, it was only through the "work of a brilliant thinker", Francis Galton, that the "science of racial hygiene, eugenics", had been initially and scientifically formulated.[61]

Both Darwin's theory of natural selection and Mendel's laws of heredity, which Hoffmann discussed briefly, provided solid foundations for the new doctrine of race improvement.[62] Furthermore, Hoffmann directly engaged with debates over the inheritance of acquired characteristics, especially as he viewed the latter in connection with "the fight against alcoholism, depravity and immorality as well as the preservation of the purity of family life", which were collectively deemed to be basic eugenic activities.[63] Why was the "improvement of the human race" (*Veredlung der Menschenrasse*) even necessary?, Hoffmann subsequently asked. His answer clearly illustrates Hoffmann's vision of social and biological degeneration:

> The incapable is sooner or later pressed against the wall; civilization protects the weak, the sufferer and the bad from downfall so that the inferior elements actually increase in numbers. Confirming a common observation, statistical investigations have shown that the population's lowermost stratum, including criminals and inferiors of all types, increases in size much more quickly than the population's capable elements.[64]

This allegedly higher fertility by those deemed "inferior" was an internationally representative eugenic argument. For Hoffmann, however, eugenics would "reach its goal not by killing but by preventing the birth of inferiors".[65] And who were these so-called "inferiors"? Here Hoffmann quoted a typology offered by Bleeker van Wagenen, in his previously cited report to the First International Eugenics Congress:

> Members of the following classes must be considered as socially unfit, and their supply should, if possible, be eliminated from the human stock: (1) the feebleminded; (2) the pauper class; (3) the criminal class; (4) the epileptics; (5) the insane; (6) the constitutionally weak, or the asthenic class; (7) those predisposed to specific diseases, or the diathetic class; (8) the deformed; (9) those having defective sense organs, as the blind and the deaf, or the kakaisthetic class.[66]

Which, then, were to be the "methods of racial hygiene"? At the forefront of scientific debate at this time were two principal methods: one "positive", referring to measures designed to improve and safeguard the qualities of the race; the other "negative", including methods designed to prevent the reproduction of those perceived to be "inferior" (physically, mentally, socially or racially).[67]

Having outlined this theoretical framework, Hoffmann then turned, in the second part of his book, to the (for him) familiar history of American eugenics. "Galton's dream, that eugenics will become the religion of the future, is fulfilled in America",[68] he noted enthusiastically. By contrast with other countries, the achievements of American eugenics were indeed impressive. Having incorporated eugenics into domestic politics, the USA was, according to Hoffmann, breeding "an ideal, world-dominating race".[69] To this end, he approvingly cited Woodrow Wilson's 1913 presidential address in light of its emphasis on national health and racial conservation.[70] Yet, in doing so, Hoffmann also highlighted the fact that eugenic policies aimed at strengthening the family were actually coupled with policies aimed at protecting "the purity of the race" against racial mixing.[71]

In addition to high politics and official rhetoric, Hoffmann further praised the broad dissemination of eugenic ideas in American schools and universities, the popularity of "Eugenics Clubs", as well as the interest in eugenics of major cultural institutions, such as the Eugenics Section of the Pittsburgh Academy of Sciences and Art.[72] Such a concerted national effort directly contributed to an increasing popularity of eugenics among the general public – as reflected by a range of marriage restrictions and control of reproduction introduced in a number of states, as well as by the growing number of eugenic organizations, alongside eugenically oriented medical and social institutions.[73] Hoffmann observed, perhaps unsurprisingly, that eugenics had progressively become an essential component of American society, one providing a vision of modernity to be successfully imitated elsewhere.[74]

The third part of Hoffmann's book was devoted to the control of marriage and its eugenic consequences. By connecting issues such as immigration, immorality and criminality to eugenic arguments concerning social and biological degeneration, Hoffmann powerfully articulated a specific relationship between the family and the state.[75] Eugenics both appealed to and embodied the ideal of a healthy racial community, one promoted by American politicians and social reformers alike. Extensive marriage restrictions had already been introduced in a number of states, including Connecticut, Indiana, Kansas, Michigan, New Jersey, Ohio, Utah and Washington.[76] Of course, while not all of these legislative efforts were successful in preventing the reproduction of "defectives" and "inferiors", these measures were nevertheless deemed to be eugenically progressive. In doing so, the strict hereditarianism of some American eugenicists – like Charles Davenport, for instance – was often combined with policies oriented towards improving both the social environment and the health practices aimed at combating diseases. The subsequent introduction of health certificates furthered this eugenic agenda, even if Hoffmann believed that "the time has not yet come" for their more general application.[77]

It was only in part four of Hoffmann's book that the "sterilization of inferiors" was directly addressed. Various surgical procedures were outlined, including castration and vasectomy, before Hoffmann proceeded to discuss the success of the initial eugenic sterilizations, such as those performed by Harry C. Sharp at the Indiana Reformatory in Jeffersonville, Indiana.[78] Hoffmann's survey was followed by a detailed description of leading categories of "inferiors" targeted for sterilization resulting from a number of medical, juridical, social or racial reasons.[79] Sterilization laws in various states were then described, beginning with Indiana (1907), continuing with California (1909) and New York (1912), before finishing with Oregon (1913).[80]

An equally detailed analysis was provided in Hoffmann's final chapter, which addressed immigration and a variety of regulations pertaining to its restriction.[81] These applied not only to individuals identified as "idiots, imbeciles, epileptics, blind and deaf", but also to those described as "anarchists and revolutionary socialists", as well as those who were "over 60 years old" and anyone classified as a "polygamist" or "illiterate".[82] Doubtless a wide-ranging programme for social and biological selection, Hoffmann had attempted to forge a new ideological alliance, one fully predicated on racial protectionism and the perceived regenerative power of eugenics.

A number of appendixes completed Hoffmann's book, providing a German translation of excerpts on marriage (Michigan 1905) and sterilization laws (first introduced in Indiana in 1907); followed by California, Connecticut and Washington in 1909; Iowa, Nevada and New Jersey in 1911; New York in 1912; and finally, Kansas, Michigan, North Dakota and Oregon in 1913.[83] In addition, Hoffmann translated fragments from the Federal Immigration Law of 1907 into German, while providing a comprehensive bibliography of almost 1000 titles on eugenics in the USA.[84]

Perhaps the single most important contribution by a Hungarian eugenicist to the history of international eugenics, *Die Rassenhygiene in den Vereinigten Staaten von Nordamerika* catapulted Hoffmann to the forefront of the European movement on eugenics and racial hygiene,[85] while simultaneously confirming his status as a leading authority on American eugenics.[86] In *The Eugenics Review*, Edgar Schuster strongly recommended the book to "eugenicists all over the world", for it "should be of great service to all who wish to undertake a serious study of eugenics, particularly in the case of what has been called restrictive or negative eugenics".[87] Hoffmann's support for negative eugenics was also remarked upon by Amey Eaton Watson, who, in her review published in *The Journal of Heredity*, considered the book to be "Of great service to all who wish to make a serious study of eugenics, especially of Negative or Restrictive Eugenics".[88] *The Bulletin of the American Academy of Medicine* was equally praiseworthy of Hoffmann's achievements: "We know of no other work which gives a more complete or concise statement of the progress of eugenics in the United States and every student of the subject should have access to the volume."[89] In a lengthy review published in the *Archiv für Rassen- und Gesellschaftsbiologie*, Fritz Lenz similarly praised the qualities of the book: "This is the first study that provides a reliable summary of the American eugenic laws", he concluded.[90] Just as importantly, according to Lenz, Hoffmann had established American eugenics as a model to be emulated. As Bernard Glueck, a Senior Assistant Physician at the Government Hospital for the Insane in Washington, DC, noted in his review of Hoffmann's book, "There was a time when the mere suggestion of the possibility of America teaching Europe anything in the biological sciences would have provoked a good deal of doubt; and it is very gratifying to see that the unexpected had happened."[91]

Moreover, the publication of Hoffmann's book also had a profound impact on the domestic development of the Hungarian eugenic movement.[92] René Berkovits, for instance, in a review article of the new directions in social biology published in *Huszadik Század*, praised the remarkable achievements of American eugenics as recounted by Hoffmann.[93] Similarly well versed in the literature on sterilization, Berkovits offered more than an enumeration of the book's main arguments. Invoking authors such as the Swiss psychiatrist Emil Oberholzer,[94] Berkovits highlighted the alleged "social and racial benefits" resulting from sterilizing "mental and physical defectives". He had long subscribed to Oberholzer's somewhat softer view that sterilization should be performed only after the consent of the family and favourable opinions from the judicial and medical authorities had been obtained.[95]

Hoffmann advanced an alternative model of eugenics to those advocated by Madzsar, Fülöp, Dienes on the one hand, and Apáthy on the other. Hoffmann's eugenic programme drew its sources, by contrast, from the practical application of hereditary principles as illustrated by American eugenics; it also articulated concerns over the conservative, aristocratic milieu of his social background and education. Unusually, Hoffmann came to eugenics from a social and diplomatic angle, and it is his contribution to German eugenics that is regularly recognized in contemporary scholarship.[96] Yet, Hoffmann aimed to create an official national eugenic discourse, one embedded in government politics that could, thereby, establish its intellectual and political reputation. In doing so, he adapted foreign models of eugenics to the domestic Hungarian context, while also providing eugenics in Hungary a degree of international visibility. If, before Hoffmann, eugenics in Hungary was scattered across academic disciplines – often hiding in the subtext of various intellectual debates on state and society – following his rise to prominence, eugenics in Hungary was no longer seen as marginal to these debates; instead, it became an identifiable cultural entity with its own society.[97]

A Nascent Eugenic Society

No eugenic movement was complete without its own organization. As mentioned previously, the first eugenic society in Europe was the Society of Racial Hygiene, established in Germany in 1905, followed by the Eugenics Education Society in 1907 and two years later, the Swedish Society for Racial Hygiene (Svenska sällskapets för rashygien).[98] The

French Eugenics Society (Société française d'eugénique) was established in December 1912,[99] while the Italian Committee of Eugenic Studies (Comitato Italiano per gli studi di Eugenica) first met in November 1913.[100] A similar tendency towards the institutionalization of eugenics can be observed in Central Europe as well. Polish supporters of eugenics in Posen (Poznań) – then in the German Empire – had already established a Society for Combating Sexually Transmitted Diseases (Towarzystwo ku Zwalczaniu Zakaźnych Chorób Płciowych) in 1903. A Eugenics Office was created at the Ernestinum, a hospital for the treatment of the mentally ill, in Prague in June 1913, by a group of Czech eugenicists, including Karel Herfort and Arthur Brožek.[101] Finally, in Vienna the Sociological Society (Wiener Soziologischen Gesellschaft) established a Section for Social Biology and Eugenics (Sektion für Sozialbiologie und Eugenik) in December 1913, with Julius Tandler delivering the inaugural lecture.[102]

These transnational and regional developments – together with the national publicity facilitated by the public debates on eugenics and public health organized by the Society of Social Sciences in 1911 and the Association for Social Sciences in 1912 – also encouraged eugenicists in Hungary to consider the creation of their own eugenic society. One early attempt belonged to René Berkovits. As mentioned in Chapter 2, Berkovits was one of the most active contributors to the public debate on eugenics. He was also familiar with developments in American eugenics, an interest that explains his letter to Harry H. Laughlin, at the time Superintendent at the Eugenics Record Office, Cold Spring Harbor, New York. On Christmas Day 1912, Berkovits wrote to Laughlin discussing his plans to create "eine ungarische eugenische Gesellschaft" in Nagyvárad.[103] Berkovits' initiative to create a eugenic society never materialized, but his letter once more illustrates the widespread reception of eugenics outside Budapest, especially in the provinces like Transylvania.

In 1913 Hoffmann was transferred from his post in Chicago to the Austro-Hungarian Embassy in Berlin. It was a move with significant consequences for the development of Hungarian eugenics. As discussed above, Hoffmann experienced the intense level of eugenic activity in the USA at first hand and was also familiar with eugenic societies elsewhere. To be sure, not all Hungarian eugenicists found Hoffmann's version of practical eugenics – with its strong emphasis on sterilization and segregation – to be an attractive proposition. Some advocated positive eugenics instead, focusing on environmentalism, public health and social hygiene. Yet all agreed that the need to establish a national eugenic society in Hungary was a pressing one.

Taking the initiative, Hoffmann sent a particularly revealing letter to Apáthy on 22 October 1913. There he outlined his suggestions for the creation of a eugenic society.[104] Hoffmann also informed Apáthy that, while visiting his hometown of Nagyvárad during the summer of 1913, he met various local notables interested in eugenics. These individuals were, in all probability, the same as those mentioned by Berkovits regarding the creation of a society for eugenics in Nagyvárad in 1912. Hoffmann had hoped to meet Apáthy in Kolozsvár in order to discuss these plans, but Apáthy was not available. Upon his return to Budapest, Hoffmann was, however, informed by Jenő Gaál, one of the presidents of the Association for Social Sciences, that they had already discussed the possibility of establishing a "eugenic section" under Apáthy's leadership. Encouraged by Gaál and Pál Angyal – whom he also met in Budapest – Hoffmann then wrote to Apáthy, offering his "foreign connections and [his] work, which was devoted to the cause of eugenics".[105]

Hoffmann had, in fact, already drafted his own plan for a "Hungarian eugenic society", which he also sent to Apáthy. Hoffmann's draft began with a short description of the main eugenic societies in Europe (Germany, Britain, Sweden, France and Austria) and the USA, followed by an outline of the main activities proposed for the Hungarian society. Hoffmann divided these activities into three groups: "scientific work", "advertising and information" and "practical work" – each, in turn, subdivided into sections. Using the data collected by the Society for Child Study and the Museum of Child Study (Gyermektanulmányi Múzeum) for eugenic purposes was for example considered "scientific work".[106] So too were the activities of various sections on diverse issues such as emigration, the "one-child" practice, birth rate, infant mortality, ethnic minorities, genealogy and so on. "Advertising" information referred to the organization of public lectures and the attempted popularizing of eugenics in newspapers and in academic journals. In this context, Hoffmann proposed the founding of a Hungarian periodical devoted exclusively to eugenics: *Eugenika: Fajegészségügyi Szemle*. It was to be published, preferably, as a supplement to the journal *Magyar Társadalomtudományi Szemle*.

Finally and most importantly, according to Hoffmann, the "practical work" of the proposed eugenic society presupposed persuading the government to adopt eugenic legislation. State involvement was essential for Hoffmann's eugenic project. Contrary to what one would have expected, however – considering the arguments put forward in his 1913 book on American sterilization laws – Hoffmann did not recommend the introduction of negative eugenic policies in Hungary. "I do not suggest", he underlined in the document, "the adoption of sterilization laws or marriage bans, but of laws which promote the quality of the population". Hoffmann's careful phrasing was deliberate. He certainly wanted to appeal to those supporters of eugenics in Hungary apprehensive about interventionist practices like sterilization. "Practical eugenic work" referred, therefore, to the establishment of "eugenic counselling centres" in Hungary, similar to those existing in Germany and the USA, which could both improve the health of the population and raise awareness about the racial quality of future generations.[107]

Furthermore, Hoffmann took it upon himself to liaise with Farkas Heller, Secretary of the Association for Social Sciences, concerning the financial considerations involved in the creation of such a society.[108] Yet Hoffmann's activities were not merely limited to the organizational, for he also proposed that a public lecture on eugenics be organized in November or December intended to serve as the society's launch. Hoffmann's intention, as he assured Apáthy, was not to "lead [the proposed society], but to put forward [his] modest knowledge of eugenics".[109] On the subject of leadership, Hoffmann made another suggestion: the president of the Hungarian eugenic society should be a "competent person", who also had "the time and the desire for such positive work". The personality who best matched this description was Pál Teleki (Fig. 9), the President

Figure 9   Pál Teleki (courtesy of the Magyar Nemzeti Múzeum Történeti Fényképtár)

of the Turanic Society and Secretary General of the Geographical Society (Magyar Földrajzi Társaság).[110]

On 2 December 1913, the Association for Social Sciences sent a letter to the Medical Association of Budapest, informing them of a meeting on eugenics scheduled for January 1914 in Budapest.[111] The need for a Hungarian eugenic society was explicitly formulated in the letter. Hoffmann, Teleki, Apáthy, Gaál, Angyal and other members of the Association for Social Sciences brought together hitherto disparate supporters of eugenics across political, cultural and religious boundaries. In the process, new alliances were forged, alongside a new eugenic worldview that was, from the outset, overtly nationalistic.

A Hungarian Eugenic Terminology

His 1913 book assured Hoffmann a unique position among Hungarian eugenicists. It was suggested immediately that his ideas needed to be made more accessible to the Hungarian public prior to the meeting on eugenics.[112] Hoffmann had already sent Apáthy a copy of his book on American eugenics alongside his letter of 22 October 1913. Correspondingly, Apáthy agreed to review the book for the *Magyar Társadalomtudományi Szemle* in January 1914.[113] As is often the case with those aiming at the same position within a movement, Apáthy sensed in Hoffmann a possible opponent. This review offered him the opportunity to demonstrate his superior theoretical understanding of eugenics once and for all.

Apáthy acknowledged Hoffmann's empirical thoroughness, but found his interpretation of eugenics to be problematic. As noted earlier, Apáthy's preferred Hungarian term for eugenics was *fajegészségtan*, not *eugenika* (eugenics) or *fajnemesítés* (race improvement) – two other terms used during the 1911 debate on eugenics. Occasionally he also used *fajegészségügy* (practical eugenics). "I was perhaps the first", Apáthy remarked, "to propose *fajegészségtan* as the Hungarian word for *eugenics*. *Fajegészségtan* may not perfectly reflect the linguistic meaning of the word 'eugenics', but it does tell us what the science of eugenics can in fact achieve."[114] *Fajegészségtan* was also preferred to *fajnemesítés* as it stressed the importance of hygiene rather than that of noble descent. In trying to disentangle the many meanings of the word *faj* (race) in Hungarian, Apáthy offered the following explanation:

> Within the general notion of species taken in the taxonomical sense, there is the subdivision of subspecies (*alfaj*); an even narrower definition is that of variety or *varietas* (*fajtaváltozat*), with which

a certain meaning of *fajta* (race; Rasse) overlaps. A narrow term is variety (*modificatio*) and an even narrower one is individual variation (*variatio*), which expresses personality and is nothing more than individual character. The common Hungarian language often does not distinguish between *faj* (species; race) and *fajta* (race), and every so often the word *faj* is used only to denote quality, selected origin, pure lineage, as in such expressions as *fajalma* (pure-breed apple), *fajbaromfi* (pure-blood poultry), *fajló* (pure-blood horse), and so on. *Fajegészségtan*, therefore, is fitting, because it deals with *faj* from the point of view of its innate racial quality.[115]

What, then, was Apáthy's definition of eugenics? In short, eugenics (*fajegészségtan*) was "*the science that studies the physical constitution of living organisms (anthropological and national development) from the point of view of racial improvement*".[116] Practical eugenics (*fajegészségügy*), on the other hand, was "the sum of those principles, institutions and regulations that ensure that eugenics' scientific findings can be put to practical use".[117] Nonetheless, *fajegészségtan* (eugenics) and *fajegészségügy* (practical eugenics) were paired concepts, according to Apáthy, existing in a mutually supporting relationship, similar to that between pedagogy and education. In other words, eugenics represented the *science* of racial improvement, while practical eugenics was its *modus operandi*.[118] Apáthy thus formulated an explicitly Hungarian eugenic vocabulary, one, he hoped, able to render the specific meanings of social and biological improvement for the Hungarian public. What he provided was, in fact, a "redescription" of eugenic ideas, "explained in a language which would sanction"[119] their application in practice. This was all part of the transformation of eugenics into a "genuine science" with its "special vocabulary, a journal, an association, congresses and lectures, and the support of leading academics".[120]

Eugenics (*fajegészségtan*) and practical eugenics (*fajegészségügy*) were simultaneously descriptive and evaluative terms, expressing the need for a normative eugenic vocabulary – one central to Apáthy's interpretation of a national eugenics in Hungary. Thus equipped, he then turned his attention to the object of his inquiry: the social and biological improvement of the Hungarian race. As noted earlier, Apáthy came to eugenics from biology and it was this discipline that endowed his thinking with its most important arguments. Contrary to the German eugenicists like Ploetz, Apáthy remained distrustful of anthropology, but he advocated a partnership with sociology. For Apáthy, eugenics was articulated through a variety of biological and sociological discourses regarding

the individual and the collective.[121] In this context, socialism – defined
not in its Marxist interpretation, as a historical phase of economic
development, but as a collectivist social philosophy – offered optimal
conditions for the practical application of eugenics.[122] "Individualism",
by contrast, "will always put the rights of the individual before the
requirements of eugenics."[123] According to Apáthy, commitment to
the nation's general welfare invariably transcended the personal interests of the individual: "eugenic measures often require a strong spirit
and strong faith in human evolution, and this is nothing but self-
abnegation".[124]

These theoretical guidelines were consciously devised to assist the
Association for Social Sciences in its attempts to establish a eugenic society. After a decade of uncertainties, Apáthy provided the most compelling eugenic project for Hungary's autochthonous traditions. Alongside
Hoffmann, Apáthy thus noted that the time had come for the "eugenic
question in Hungary" to be taken seriously, and he proposed the establishment of a "eugenic committee" (*fajegészségügyi bizottság*). It was not
only the Association for Social Sciences that advocated such a committee, but other medical institutions in Budapest as well, including the
Medical Association of Budapest and the National Association of Public
Health (Országos Közegészségügyi Egyesület).[125]

Following this exchange of ideas, events unfolded rapidly. As recommended by Hoffmann, on 14 January 1914 Apáthy offered Teleki the
chairmanship of Hungary's first eugenic society.[126] Teleki accepted on
16 January. As he enthusiastically informed Apáthy, had the Association
for Social Sciences not proposed this course of action, "the *Turanic
Society* would have had to, because only a race which was healthy could
become great and strong". Teleki further reminded Apáthy that he
had previously reviewed the first issue of Ploetz's *Archiv für Rassen- und
Gesellschaftsbiologie* in *Huszadik Század* and, perhaps more importantly,
that he was the only Hungarian to participate in the meeting of the
International Society for Racial Hygiene in Dresden in 1911. Yet he
claimed modestly to be nothing but a "dilettante" not a specialist like
Hoffmann – whose 1913 book on American eugenics Teleki had read
and apparently "really enjoyed".[127]

These concerted efforts aimed at popularizing eugenics had Apáthy
and Hoffmann's intended outcome.[128] On 24 January 1914, in the main
hall of the Royal Hungarian Society for Natural Sciences (Királyi Magyar
Természettudományi Társulat), the Association for Social Sciences,
together with the Royal Medical Association of Budapest and the National
Association of Public Health, met to discuss the creation of a "eugenic

section" (*fajegészségügyi [eugenikai] szakosztály*). Hoffmann was invited to deliver the opening lecture on the meaning and relevance of eugenics, but was unable to attend the meeting.[129] Another member of the Association for Social Sciences, Győző Alapy, read Hoffmann's text instead. The lecture was, furthermore, published in the February issue of *Magyar Társadalomtudományi Szemle* facilitating a long-overdue review of its main arguments.[130]

Hoffmann divided his lecture into three parts. The first offered an overview of Darwin's theory of natural selection and Mendel's hereditary laws, and in particular their impact upon the development of eugenics by Francis Galton, Alfred Ploetz and Wilhelm Schallmayer.[131] The significance ascribed by Hoffmann to these authors reflected conventional eugenic thinking in Hungary at the time. He aimed, however, at rather more than stating the obvious. This was Hoffmann's way of anchoring his ideas within a respected intellectual genealogy, in order to lend them academic credibility. Unlike Apáthy and other Hungarian eugenicists, moreover, Hoffmann had no systematic training in biology, sociology, anthropology or medicine. This different professional background did not deter him from articulating a convincing and engaging reading of the importance of heredity for eugenics.

Thereafter, Hoffmann spelled out the broad principles of his eugenic philosophy. First, he defined eugenics as "the science that examines the factors causing the improvement or the degeneration of the race". Second, he wanted to simplify existing eugenic terminology in order to dispel conceptual confusion where possible: "we have to decide how to name this new science in Hungarian". "Eugenics" (*eugenika*), he noted, "apart from being a foreign term", has its limitations for "it refers only to reproduction."[132] Following Galton, this was the term most Hungarian eugenicists used, as during the 1911 debate on eugenics. In fact, even *fajnemesítés* – viewed by Hoffmann as "the most appropriate Hungarian translation of the term eugenics"— was considered to be somewhat "naïve and ridiculous". In Hungarian, he added, *fajnemesítés* evoked animal breeding. Accordingly, Hoffmann recommended the terminology used by Apáthy in his earlier review: *fajegészségtan* and *fajegészségügy*, for these two idioms did not alter the meaning of eugenics in Hungarian, or render it more obscure. "*Eugenika*" remained in use in his lecture, however, as Hoffmann deemed it an effective word "to convey" the meaning of social and biological improvement "to a broader public".[133]

Finally, Hoffmann asserted eugenics' alleged ability to fashion the world anew. He appealed, however, to the general public not to be apprehensive about this claim. Eugenicists did not want, as was too often assumed, "a return to the cruel state of nature", nor to "ruthlessly eliminate the weak". Neither were they contemptuous of "Christian brotherly love" and "the current conceptions of social protection".[134] Hoffmann was irritated by what he regarded as this "misunderstanding", and his defence of eugenics became progressively more incisive. "Morality" and "philanthropy", he claimed, directly underpinned eugenic ideas of social and biological improvement. "What good would it do to us", Hoffmann asked, "if, on the one hand, we eliminated the weak but, on the other, did not improve the living conditions which ruin the healthy?" And again: "What good would it do to us, if we brought up a healthy new generation, but we did not care about its spiritual and moral life?"[135] Without "morality" and "philanthropy", humanity "would fall back into barbarity", Hoffmann believed.

As this suggests, Hoffmann had an unmitigated passion for eugenics. His aim in this article was not merely to set out the importance of heredity for social policies, or just to advocate measures of racial protectionism, but also to communicate the eugenic ideal for the modern Hungarian nation. Hoffmann repeatedly emphasized this vision of national regeneration as a core component of eugenics. While stressing the universality of the pursuit of happiness – "every single living soul had the right to the happiest life possible" – he also suggested, most tellingly, the appropriate method for avoiding "immense misery and suffering". It was not the elimination of "degenerates after they had been born", but keeping them from being born at all.[136] Thus expressed, Hoffmann's commitment was to improving the racial quality of the population through state interventionism and eugenic qualitative methods. These methods, or "instruments" – as he called them – were "the eugenic regulation of marriage, institutionalized segregation and statutory sterilization".[137]

Two additional elements of Hoffmann's eugenic programme need to be highlighted here: miscegenation (*fajkeveredés*) and eugenic marriage counselling. It is in connection to miscegenation that Hoffmann's racial proclivities were displayed most clearly. Like other eugenicists at the time, he believed that mixing "superior races" with "inferior ones" was to be avoided, and instead he advocated "the principle of racial purity". Furthermore, he argued that Hungary – where this form of "miscegenation was called assimilation or Magyarization" – was especially exposed to the detrimental effects of racial mixing. For Hoffmann, defending the purity of the race and eugenics strongly reinforced one another. In turn, he did not hesitate to criticize the policy of Magyarization from a eugenic point of view. With respect to marriage counselling – a proposal already discussed in the draft programme he sent to Apáthy on 22 October 1913 – Hoffmann recommended it in conjunction with racial protectionism as a means of cultivating a "eugenic responsibility" towards the interests of the community.[138]

The effectiveness of Hoffmann's eugenic programme stemmed from the boldness with which he asserted these eugenic values and principles. At the conclusion of his article, he earnestly urged his compatriots to initiate their own eugenic "investigations of the Hungarian race".[139] Undoubtedly, such research needed institutional support, most notably in the form of a eugenic society. He suggested that establishing such a society would not only signify an enlargement of the eugenic community, but that it would also go hand in hand with the education of the public regarding the importance of social and biological improvement. In this sense, Hoffmann envisioned the institutionalization of eugenics in Hungary as a collective endeavour: "Physicians and naturalists, jurists and politicians, teachers and priests, child psychologists and every philanthropist: we should all join hands and carry forward the work needed for the advancement of science and of the Hungarian race."[140] These assertions were firmly anchored in a coherent set of beliefs and assumptions about Hungary's continuing national improvement. Such nationalism closely connected to Hoffmann's eugenic thinking and, in many ways, flowed directly from it. To be sure, the symbiosis between eugenics and nationalism was in itself nothing original, but in 1914, and the years that followed, it nevertheless assumed special importance.

A Practical Eugenic Programme

The reading of Hoffmann's lecture was followed by Apáthy's succinct presentation of the aims and goals for "the future eugenic section".[141] Having already summarized his theoretical arguments in the review of Hoffmann's book, as well as in his previous articles and books, Apáthy now offered a vision of how eugenics could effectively be put into practice in Hungary. In his lecture, Apáthy advanced a practical eugenic programme, one whose declared purpose was

> the breeding and preservation of a new Hungarian-Turanic type. We do not want to breed Yankees or Germans. And we do not want to breed the abstract Man or the Human Ideal either. There is no abstract man; nor did the ideal human ever exist. For a Yankee the ideal human is a Yankee, as German is for the German and Chinese for the Chinese. Should Hungarians not have the Hungarian as their ideal Man?[142]

Contrary to expectations, Apáthy insisted, the "new Hungarian type" was not to be found among the "racially pure Hungarians", or – conversely – among the "parasitical, degenerate gentry", the "womanizers of Budapest" or "poor lads on the farms".[143] Yet, Apáthy refused to point out the exact ontological basis of this "new Hungarian type", arguing that this was precisely what Hungarian eugenics, with the assistance of modern biology, would successfully accomplish.[144] Creating the "new Hungarian-Turanic type" was as much about the *present*, namely the preservation of those racial characteristics perceived to be "Hungarian", as it was about the *future*, portending the introduction of a nation-wide eugenic programme of social and biological improvement.

Race, while not marginalized, was, however, to be incorporated into a more extensive interpretation of the nation, one that was less about reviving a mythical past and more about building a modern future. Eugenics reflected the yearning for national unity, whose centrality Apáthy did not hesitate to affirm. Similar to other Hungarian eugenicists, he hoped for a national future in which a strong Hungarian state commanded the obedience and fidelity of reverent minorities. This emphasis upon the unitary nature of the state was, of course, a common theme in early twentieth-century debates on the nation, in Hungary and elsewhere. In this environment, where nationalism played such a pivotal role, it was then only natural that eugenics – if it were to succeed as a movement in Hungary – could only be *national* in form.

Apáthy prefaced his interpretation of Hungarian national eugenics with a number of initial questions:

What physical form does the new Hungarian-Turanic type have? Which racial mixing, or more exactly, what group of hereditary characteristics contributes most to its creation? Which hereditary characteristic is it born with and which must it acquire from the environment and through education?[145]

Next Apáthy presented his practical eugenic programme for Hungary. This was divided into five clusters of ideas and practices, each with its own path and objective (see Fig. 10).

The first cluster was entitled "*praeparativus [sic!] eugenika (előkészítő fajegészségtan)*". As the name suggested, its purpose was to familiarize the general public with eugenic ideas. Accordingly, "social morality, economic policy and public education" in Hungary, Apáthy maintained, were to be brought in harmony with the core eugenic principles. A second cluster was devoted to "*praeventivus [sic!] eugenika (megelőző*

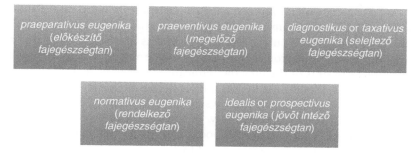

Figure 10 Apáthy's practical eugenic programme

fajegészségtan)", which aimed at preventing those deemed of lesser eugenic value from reproducing; in short, this was a "preventive eugenics" that promised the biological improvement of future generations. Achieving this objective, however, required eugenic management of not only the individual (anti-alcoholism and sexual abstinence, for example) and the family (child welfare and the role of women), but also "social life, social policy, public administration, legislation, domestic and foreign policy". Health values were thus framed in terms of eugenic norms, which together enforced an overtly biological vision of the Hungarian national community.

To separate "undesirable" from "desirable" individuals was the goal of the third cluster, fittingly termed *"diagnostikus or taxativus [sic!] eugenika (selejtező fajegészségtan)"*. As Apáthy, Hoffmann and others grew increasingly worried about the racial quality of Hungary's future generations, they aimed to employ eugenic strategies – positive and negative – attempted in other countries, from counselling centres and health exhibitions to welfare assistance for larger families, as well as the control of reproduction, marriage certificates and perhaps, eventually, sterilization. In this context, national protection dovetailed with eugenic improvement, which Apáthy hoped to have introduced in Hungary through political legislation and public support. Eugenic screening of the population assumed by the third cluster was to be naturally accompanied by state interventionism. A fourth cluster, therefore, was responsible for preventing "the reproduction of the least desirable individuals" while, at the same time, encouraging "the reproduction of those most desirable". A number of eugenic methods were proposed, including "medical examination before marriage, sterilization of degenerates, control of immigration" and so on. This, in short, was *"normativus eugenika [sic!] (rendelkező fajegészségtan)"*.[146]

Most eugenic programmes established elsewhere in Europe at the time emphasized similar directions for future research and active involvement with the wider public. What nevertheless distinguished Apáthy's eugenic programme from others was its wide range of activities. For example, the aims of the Eugenics Education Society were, by contrast, more modest. Established in 1907, it aspired, principally, to "set forth the national importance of Eugenics, in order to modify public opinion and create a sense of responsibility, in the respect of bringing all matters pertaining to human parenthood under the domination of eugenic ideals". The Society also endeavoured to "spread knowledge of the laws of heredity so far as they are surely known, and so far as that knowledge might affect the improvement of the race". Finally, attention would be devoted to "further eugenic teaching at home, in the schools and elsewhere".[147] The practical goals of the International Society for Racial Hygiene, as outlined by Ploetz in 1910, were slightly more ambitious. In addition to an "opposition to the two-child system" and the "establishment of a counterbalance to the protection of the weak by means of isolation, marriage restriction, etc., designed to prevent the reproduction of the inferior", Ploetz pledged "opposition to all germ-plasm poisons, especially syphilis, tuberculosis, and alcohol", the "protection against inferior immigrants" and, finally, the "extension of the reigning ideal of brotherly love by an ideal of modern chivalry, which combines the protection of the weak with the elevation of the moral and physical strength and fitness of the individual".[148]

Apáthy's eugenic programme was unusual in another respect: it promoted eugenics for the Hungarian nation and then applied it to the country's historical and geographical conditions. According to Apáthy's fifth, summative cluster, entitled *"idealis* or *prospectivus* [*sic!*] *eugenika (jövőt intéző fajegészségtan)"*, the main eugenic objective was to protect Hungary's multiethnic environment and the Hungarian race within it. To this end, the proposed eugenic society attempted to describe Hungary's "ethnic relations and the nationality question from the point of view of eugenics, and to draw attention to their eugenic significance".[149] In the highly charged ethnic context of 1914, in Hungary and elsewhere, eugenics was enlisted as one promising strategy meant to ensure the Hungarian nation's moral and biological renewal.

According to Apáthy, a new eugenic morality formed *the* essential prerequisite for any successful transformation of eugenics into a national movement. This complex process of achieving eugenic consciousness involved not only individual responsibility for one's race (the "one child" practice was, for example, described as a "moral scourge" and a "crime against the race") but also a "healthy life" in light of established sexual and gender norms. Any departure from these norms should be discouraged. As an example, Apáthy offered the new science of psychoanalysis, chastising Sigmund Freud and psychoanalysis for provoking "wildly sensual and sexual, erotic tendencies" and for spreading a "putrid worldview".[150] Maintaining a "healthy morality" occupied, of course, a central place in all eugenic movements, but seems to have held a special meaning for Apáthy, who used it as the necessary criteria for "the breeding and preservation of a new Hungarian-Turanic type".

Apáthy's eugenic ideas were not only innovative in design but also in terms of their declared goals. He aspired to create a modern Hungarian nation upon decidedly eugenic foundations. Like the British eugenicists, Apáthy believed that education was essential in disseminating the eugenic ideal, even if he favoured state intervention and championed the decisive role of the national community in determining the scope of eugenic policies. In a manner more akin to the German eugenicists, Apáthy correspondingly attached eugenics to racial protectionism and national traditions. Reflecting on Hungary's multiethnic society, he was also motivated by a concern over the country's national stability and ethnic composition, which he placed among the goals of practical eugenics. To reiterate, this was a wide-ranging programme of eugenic engineering that unquestionably centred on the Hungarian race. Crucially, this is where Apáthy's eugenic thinking merged seamlessly with his nationalism. He ultimately hoped that eugenics would become an effective instrument for a modern Hungarian state in search of social and biological regeneration amidst difficult historical times.

Developing this theme in greater detail, Apáthy suggested the creation of a Hungarian Eugenic Committee (Magyar Fajegészségügyi Bizottság) – instead of a section, as initially planned – whose first responsibility was to summarize the eugenic interests of various Hungarian societies. In addition to the Association for Social Sciences, three other societies attended the Eugenic Committee's opening meeting: the Medical Association of Budapest, the National Association of Public Health and the Turanic Society. Each society was asked to appoint five members to the Eugenic Committee. Other societies were similarly invited to join and contribute financially, namely, the Society of Social Sciences, the Heraldic and Genealogical Society (Heraldikai és Genealógiai Társaság) and the Geographical Society (Földrajzi Társaság) – each nominating five members.[151] Apáthy, moreover, encouraged each of the above to form "its own eugenic department", in order that a wide range of topics – from social and medical to aspects related to public health and folk psychology – could be examined in relation to eugenics.

The role of the Eugenic Committee was thus also a strategic one: to harmonize "the eugenic findings resulting from the studies of different Hungarian societies" as well as to publicize "important eugenic accomplishments to the Hungarian public".[152] This was to be a gradual process, one that Apáthy believed would bring together cultural and political concerns regarding the future of the Hungarian nation. Revealingly, the Eugenic Committee was to be financed by a mixture of state and voluntary contributions. Concluding his opening speech, Apáthy summarized the Eugenic Committee's course of action as follows:

> Our immediate tasks are: eugenic propaganda, the dissemination of eugenic ideas throughout the country, and to enlighten the public through lectures, special conferences, debates about what is *fajegészségtan* and what is *fajegészségügy*. We must win supporters, find and enlist collaborators from all over the country. We have to start a real eugenic crusade, and let the Eugenic Committee of Hungarian Societies be the general staff of this crusade.[153]

At the end of the meeting, Teleki was confirmed as Chairman of the Eugenic Committee. Other members of the Eugenic Committee included István Apáthy, József Ajtay, Jenő Gaál, Emil Grósz, Benedek Jancsó, Rezső Bálint, Sándor Korányi, Leó Liebermann, Vilmos Tauffer, Lajos Török, Zoltán Dalmady, Béla Fenyvessy, Ferenc Hutyra, Géza Lobmayer and Henrik Schuschny.

A subsequent meeting of the Eugenic Committee was held on 7 April 1914.[154] The list of Hungarian societies had been finalized. In addition to the initial four societies, the Committee now included the Royal Hungarian Natural History Society (Királyi Magyar Természettudományi Társulat), the Geographical Society, the Hungarian Economic Society (Magyar Közgazdasági Társaság), the Heraldic and Genealogical Society, the Ethnographic Society (Néprajzi Társaság), the Society for Child Study, the National League for the Protection of Children (Országos Gyermekvédelmi Liga) and, finally, the National Federation of Women's Associations (Nőegyesületek Országos Szövetsége) (see Fig. 11).[155]

One notable absence from the Eugenic Committee was the Society of Social Sciences, as well as the three eugenicists whose 1910 lectures had occasioned the first public debate on eugenics in Hungary: Lajos Dienes, Zsigmond Fülöp and József Madzsar. Considering the crucial role both the Society of Social Sciences and these individuals had played in the

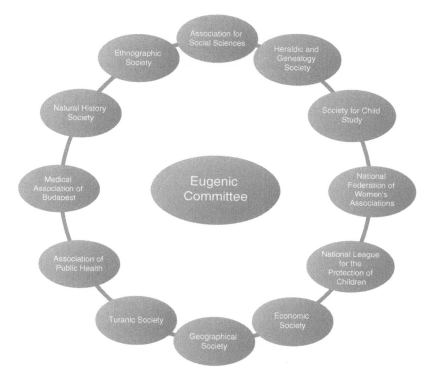

Figure 11   Hungarian Societies represented in the Eugenic Committee

development of Hungarian eugenics in the first decade of the twentieth century, such an omission was indeed surprising, if not entirely unexpected. In his letter to Apáthy on 16 January 1914, Teleki commented unfavourably on *Huszadik Század*'s political "direction".[156] He also offered Apáthy another word of warning on 13 March 1914: "I believe it would be a good idea to postpone the involvement of Jászi & co until the next meeting [7 April] and sanction it then, as I heard voices against it."[157]

The message was clear. Teleki's aim was not to bring together all Hungarian eugenicists, but to mobilize those sharing a similar political orientation.[158] This attitude pointed towards a change of leadership in the eugenic movement already noticeable for several years: from the Society of Social Sciences to the Association for Social Sciences. Eugenics was no longer a minor component of the eclectic programme for social reform and modernization envisioned by the Society of Social Sciences and publicized in the *Huszadik Század* during the first decade of the twentieth century. Under the Association for Social Sciences' careful institutional choreography, instead eugenics became a narrative of biological improvement centred upon the Hungarian nation and race. Apáthy's eugenic views were presented in precisely the same terms.

As agreed at the first meeting, the League for the Protection of Children sent the following five representatives to the Eugenic Committee: Béla Kun Szentpéteri, Béla Szentkereszty, Vilmos Neugebauer, Lipót Edelsheim-Gyulai and Mrs Gábor Vay. Additionally, the Hungarian Economic Society and the Turanic Society nominated Alajos Kovács and Alajos Paikert, respectively. As such, the Eugenic Committee was now composed of 27 members (see Table 1), nearly half of the 60 permanent members envisioned once all societies had sent delegates.[159]

Table 1   Members of the Hungarian Eugenic Committee (April 1914)

Chairman	Pál Teleki
Members and Societies	
Association for Social Sciences	István Apáthy
	József Ajtay
	Jenő Gaál
	Emil Grósz
	Benedek Jancsó
Medical Association of Budapest	Rezső Bálint
	Sándor Korányi
	Leó Liebermann
	Vilmos Tauffer
	Lajos Török
National Association of Public Health	Zoltán Dalmady
	Béla Fenyvessy
	Ferenc Hutyra
	Géza Lobmayer
	Henrik Schuschny
National League for the Protection of Children	Béla Kun Szentpéteri
	Béla Szentkereszty
	Vilmos Neugebauer
	Lipót Edelsheim-Gyulai
	Mrs. Gábor Vay
Economic Society	Alajos Kovács
Turanic Society	Alajos Paikert
Secretary	Gyula Ribiczey
Other members	Géza Hoffmann
	Pál Angyal
	Győző Alapy

Considering their scientific interest, the members of the Eugenic Committee came from various social and academic backgrounds; most also held important executive positions at universities in Budapest, Kolozsvár and Debrecen.[160] To use Mark B. Adams's terminology, these individuals formed "a personal network".[161] Nikolai Krementsov has enriched this description with the useful observation that a personal network "is more than a mere sum of individuals that form it – it is an entity with its own holistic characteristics and operating procedures".[162] Hungary's new Eugenic Committee was precisely one such "entity", aiming to bring eugenicists into dialogue with each other and engage them with a shared vision of social action and biological improvement.

Cultural societies were not the only ones, of course, invited to join the Eugenics Committee. Many public institutions and ministries – including the Municipality of Budapest, the Ministry of the Interior, the Ministry of Education, the Ministry of Justice, the Ministry of Defence, the National Statistics Office and Budapest's Statistical Office – were invited to send one representative each.[163] Regarding public engagement, two public lectures on eugenics were announced for April 1914, one to be organized by the National Society for Popularizing Science (Országos Ismeretterjesztő Társulat), the other by the Medical Association of Budapest.[164] The Eugenic Committee also undertook the education of the general public through a eugenic course consisting of 12 lectures, in advance of the national eugenic congress planned for the autumn of that year.[165] Alongside the practical eugenic programme presented by Apáthy on 24 January, this ambitious cultural and educational agenda reflected a renewed commitment to social and biological improvement in Hungary and to safeguarding the nation's racial qualities. Such was the institutional framework through which eugenicists viewed their active role in the making of a modern Hungarian state.

The institutionalization of eugenics in Hungary occurred at the point when various interpretations of eugenics found common ground, and, furthermore, when the belief was held that eugenics should be recognized as a national science that brought together various categories of scientists, politicians and public intellectuals. This was not, to reiterate, mere coincidence. The evidence cited above provides, for the first time, a clear description of how the collaboration between Hoffmann, Apáthy and Teleki in addition to the Association for Social Sciences occasioned the creation of the Hungarian Eugenic Committee in the early months of 1914.

Consolidation

Hoffmann and Apáthy played critical roles in the establishment of the Hungarian Eugenic Committee between January and April 1914 and both would continue to dominate the discussions on eugenics in the ensuing months.[166] Hoffmann, in particular, began publishing in earnest, in German and Hungarian, covering a panoply of topics. These ranged from the contemporary relevance of Plato's eugenic thinking[167] to international eugenic literature[168] and the activities of various eugenics societies around the world.[169] Hungarian eugenicists could only benefit from this wealth of information, Hoffmann reasoned, at a time when international knowledge transfer was typically difficult. Hoffmann's intention was to inform Hungarian readers about practical developments in racial hygiene, eugenics, biometrics, research on families and so on, as pursued by these societies.[170]

The process of the institutionalization of eugenics in Hungary so far described was not effortlessly smooth. It encountered opposition and generated tensions, particularly among women activists and feminists. As discussed in the previous chapters, Hungarian feminists, like Rosika (Rózsa) Schwimmer and Vilma Glücklich, took a keen interest in eugenics and championed it alongside women's social and political emancipation. Eugenicists, in turn, were often invited to speak to feminist organizations, as happened in November 1913 when René Berkovits gave a lecture on eugenics to the Feminist Society in Nagyvárad.[171] Internationally, too – as discussed in the previous chapter – there were attempts to open eugenics to wider issues associated with femininity, sexuality and gynaecology. An eclectic discipline was thus emerging, one which became known under the name of *Frauenkunde* and one whose early popularizer was the Berlin gynaecologist Max Hirsch. In 1914, Hirsch invited a number of prominent European sexual reformers, feminists and eugenicists to join the academic board of his new journal, *Archiv für Frauenkunde und Eugenetik*. Next to Havelock Ellis, Eugen Fischer, Wilhelm Schallmayer, Alfred Grotjahn, Max Marcuse, Mathilde Kelchner and Barbara Renz there were also two Hungarian gynaecologists and eugenicists: János Bársony and Ödön Tuszkai. These authors, as Paul Weindling has noted, "displaced the feminist demand for social rights and improved status, by a view of women's health centred on social biology and eugenics".[172]

Not surprisingly, then, there was discomfort about what feminists perceived as the eugenicists' attempt to reduce women to their reproductive role and to the confines of marriage and family life. As one Hungarian feminist author put it, "Not the registry office, but the idea of eugenics will bring about holy motherhood. Healthy women want to give birth to healthy children."[173] Such allegations were levied, for example, at the writer and journalist Zoltán Szász, who skilfully reworked the emotional relationship between men and women along eugenic lines in his book entitled *A szerelem* (*Love*). Szász proclaimed love to be "the most powerful agent of racial improvement"[174] and, accordingly, devoted an entire chapter to the relationship between "love and eugenics".[175] The "eugenic dream" of creating "a strong, beautiful and intelligent race"[176] was – for him – the consequence of unremitting love. But, as one reader promptly noted, Szász's vision of eugenic love rather accentuated women's passive roles as mothers and wives than advocated their much-anticipated social and political emancipation.[177]

The establishment of the Eugenic Committee did not produce the anticipated reconciliation between Hungarian feminists and eugenicists. On 5 February 1914, for example, Schwimmer's new feminist journal, *A Nő* (*The Woman*), admonished the eugenicists for recommending a programme of social and biological improvement without considering it from the women's point of view. Neither was the journal satisfied that a society like the Turanic one was invited to participate while the Feminist Association was not.[178] Furthermore, Hoffmann – named in the article as "the herald of the movement" – was criticized for his anti-Malthusian views and for neglecting the eugenic research carried out by feminist activists. Apáthy's eugenic programme and proposal to breed a "new Turanic" race were similarly received with scepticism. In short, this was "a eugenic association composed only of men".[179] Were the feminists "not loud enough?"[180]

Hungarian feminists were "loud enough" and their voice was heard. Yet the eugenicists' response was not what the feminists expected. On 7 April, at its second meeting, the Eugenic Committee preferred the more traditionalist National Federation of Women's Associations to the radical Feminist Association. On this occasion, Mrs Gábor Vay, representing the Federation, became the first woman to join the Eugenic Committee. Hoffmann, too, reacted quickly to the accusations. On 20 April 1914, he clarified his position in an article published in the same feminist journal, *A Nő*.[181] He began by explaining the central role mothers occupy in their eugenic programme and offered to correct any "erroneous assumption" concerning the relationship between eugenics and women.[182] Next, Hoffmann reiterated his support for quantitative eugenic policies, accentuating the "misleading notion" popularized by the "neo-Malthusian researchers" that high mortality was connected to the birth rate and that "by increasing the latter you also increase the former".[183]

The idea of duty to the race was a determining one, central to the survival of the "racially pure Hungarians", Hoffmann insisted. By "proclaiming the restriction of the number of children", neo-Malthusianism not only worked against the future of the race but – according to Hoffmann – also offered women a false sense of empowerment. No real "emancipation of women" could exist without motherhood, Hoffmann argued.[184] What was needed, then, were social assistance schemes and financial support for women, so that the bearing and rearing of children became less demanding on them and their families. No "trained eugenicist" would have disagreed, he believed, with economic incentives for large families. Hoffmann's closing remarks left no one in doubt about his position: the feminist and neo-Malthusian claim to emancipate women by restricting their reproductive role was not "eugenic but dysgenic".[185]

Another discussion to attract Hoffmann's attention during this period was the relationship between heredity and environment. Reviewing a book dealing with the life and future of children in rural areas in Hungary, published by the Secretary of the Society for Child Study, Lipót Nemes,[186] Hoffmann reassessed some of his previous hereditarian determinism. What he proposed now was a synthesis: those eugenicists who neglected the impact of the environment were as misguided as the social hygienists who disregarded the importance of heredity.[187] These arguments were, of course, central to debates on eugenics by the mid-1910s, in Europe and elsewhere. In Hoffmann's case, however, they were integrated into his growing interest in population policies, an interest that gained momentum after he moved to the Austro-Hungarian Embassy in Berlin.

On 11 February 1914 – just a week after the first meeting of the Eugenics Committee in Budapest – he gave a lecture to the Berlin Society for Racial Hygiene on American eugenics and became a member.[188] In this capacity, he attended most of the German Society for Racial Hygiene's meetings and congresses between 1914 and 1917.[189] Hoffmann was not a passive observant, however. Already on 7 April 1914, he wrote to the President of the Kaiser Wilhelm Society (Kaiser Wilhelm Gesellschaft), Adolf von Harnack, suggesting the creation of an "Institute for Racial Hygiene (or Racial Biology, or perhaps a Research Institute for Racial Hygiene)"[190] under the auspices of the Kaiser Wilhelm Society. "Should this plan be pursued further", Hoffmann volunteered to write a "memorandum", outlining the purpose and main activities of such an institute.[191] Harnack, in response, while acknowledging "that racial hygiene had been one of the aims of medical research since 1900",[192] nevertheless did not consent to Hoffmann's plan.

Another important eugenic event, which Hoffmann commented upon in Austrian, Hungarian and American journals, was the German Society for Racial Hygiene's congress in Jena held between 6 and 7 June 1914.[193] The main focus of the congress was on the declining birth rate, a topic chosen – according to Hoffmann – "in view of the fact that restrictive eugenics ordinarily dominate similar assemblies and the literature, thereby causing in the public the wrong impression that positive eugenics are of subordinate importance".[194] This preliminary observation expressed Hoffmann's belief – made abundantly obvious in his polemic with the Hungarian feminists – that to restrict population growth constituted a serious threat to national health and racial efficiency.

Hoffmann fully endorsed the positive eugenic measures to support large German families adopted by the congress. It was a 10-point programme, aimed at racial regeneration through social, economic and eugenic policies. Measures included: "inner colonization [back-to-the-farm movement]", whose purpose was to encourage the return to healthy living and working in the village; the "creation of family homes for large families"; and "economic assistance" for "married mothers who survive their husbands".[195] Alongside the financial support for large families, the German Society for Racial Hygiene also recommended higher taxes on "alcohol, tobacco" and other "luxuries", as well as a tax on "all who on account of their physical inability do not serve in the army". Abortion and sterilization were to be better regulated, "for instance by the provision that two physicians be consulted before an operation is executed", and increased efforts to be devoted to the "fight against all evils encroaching upon the ability to reproduce, in the first place against gonorrhoea and syphilis, tuberculosis, alcoholism" and other "industrial poisons and evils endangering the health of the working women". And there was more. "Certificates of health before marriage" were to be made obligatory.[196] Finally, national responsibility was to be cultivated not only by scientists but also by artists and educators. "Large prizes" were promised "for excellent works of art (novellas, dramas, plastic arts)" which glorified motherhood, "the family and simple life". The expectation was that, when applied, these measures would awake in the German population "a national mind ready to bring sacrifices and a sense of duty towards coming generations".[197]

The course of action adopted by the German Society for Racial Hygiene in Jena reflected the increased importance of quantitative eugenic policies. Most importantly, it signified the expansion of eugenic responsibility to include a nation-wide programme of national welfare. Accordingly, when the influential Central Office of Public Welfare (Zentralstelle für

Volkswohlfahrt) organized a conference devoted to "The Preservation and Increase of German National Strength" in Berlin between 26 and 28 October 1915,[198] the German Society for Racial Hygiene used the occasion to hold one of their regular meetings. According to Fritz Lenz, it was "unanimously accepted" that war brought population policy (*Bevölkerungspolitik*) and eugenics together.[199] Hoffmann – who also attended the meeting – noted that "the conference laid much stress upon everything which may elevate the birth rate of the best in the nation, but thought the introduction of sterilization of defectives or of marriage certificates untimely as yet".[200] Eugenic principles, adopted a year earlier at the Jena congress, were also reaffirmed.

These programmes gradually took over, placing natalist ideas of demographic growth and large families at the centre of eugenic thinking. The debate over the declining birth rate was, of course, not new or specific to Germany alone. In each European country, it assumed various overtones according to national circumstances.[201] Before 1914, Hungary's demographic trends, for instance, were not greatly dissimilar from the general Eastern European pattern of high birth rates, high mortality rates, and a low marriage age, as well as the Western European tendency to limit family size.[202] There was nevertheless a general assumption underlying the eugenicists' preoccupation with the declining birth rate, natalism and the future of the race, namely "the promulgation of a new faith in which salvation was to be found not in the sacrifice of one eugenic man, but in the propagation of many".[203]

No one, of course, could foresee how prescient these Hungarian and German eugenic initiatives were during the first half of 1914. Apáthy's practical eugenic programme and the recommendations made by the German Society for Racial Hygiene in Jena signify, however, broader changes in European eugenic thinking. At the time, like their British, French or Italian counterparts, Hungarian and German eugenicists tried to grapple with several eugenic imperatives, centred equally on the quantitative and the qualitative improvement of the population, and all geared towards a more exclusive definition of national efficiency and racial protectionism. The effects of these debates were soon to be felt throughout Europe, as both the tone and the substance of the eugenic discourse were about to change considerably. Only three weeks later, on 28 June 1914, Gavrilo Princip, a Bosnian Serb, assassinated Archduke Francis Ferdinand in Sarajevo. Within two months, Europe was engulfed in a world war: the eugenicists' worst nightmare.

5
Health Anxieties and War

In a scene from the 1916 film, *Wien im Krieg* (*Vienna in War*) a doctor gives new recruits a medical examination, or a *Musterung*. Although this was standard procedure at the time, in the context of the First World War, this *Musterung* was invested with additional eugenic significance by the film.[1] The purpose of the examination was to reassure the viewers that the soldier fighting the enemy in the trenches was the nation's finest man, the personification of physical and mental strength. The eugenic dichotomy between "fit" and "unfit" individuals was thus conveyed with artistic vision, albeit one informed by the authority of medical sciences. "Dear Fatherland, rest assured!", read the message on the screen, once a muscular and healthy man has been examined and accepted for military service.[2]

The reality was, however, starkly different to the buoyant optimism portrayed by the film.[3] Nearly 800,000 Austro-Hungarian soldiers died until the spring of 1915.[4] Although the military situation improved, following victories over the Russian army in July and August, the Austro-Hungarian casualties continued to rise, adding another half a million men by the end of 1915.[5] The subsequent Russian offensive of June 1916 further lost the Austro-Hungarian Monarchy another 1.5 million men (including 400,000 taken prisoner), making it painfully clear that the nation's best were being heroically sacrificed in the trenches. Moreover, when Hungary's hitherto neutral neighbours Italy and Romania entered the war on the side of the Entente in 1915 and 1916 respectively, it became increasingly apparent that the war was fought less to avenge the death of an Austrian prince and more to ensure territorial and national aggrandizement. Not surprisingly, then, two years into the war, the Austro-Hungarian governments were forced to accept the conscription of reservists and militarily untrained civilians, a practice which, in turn, contributed to the lowering of physical standards of those men drafted into the army, the elimination of several categories of exemption, and the extension of military obligation from the ages of 21–42 to 18–50. Fashioned by warfare and the spectre of racial degeneration, physical fitness had become the very foundation of the nation's military prowess.

As will be discussed in this chapter, after the outbreak of the war, concerns with the deterioration of the nation's health dominated the government's social and medical agenda. With every passing month of the war, public health and medical experts and doctors deplored not only the loss of human life inflicted by the war, but also its devastating consequences for the combatant nations and their civilians. These experts laboured widely to translate their knowledge into practical policies, whether it was the control of epidemics, campaigns against the spread of venereal diseases, the protection of mothers and infants or the social reintegration of injured soldiers and veterans. Statistics of military and civilian health were indeed demoralizing. According to the Austrian bacteriologist Clemens Pirquet, infectious diseases and other health problems caused by the war affected approximately 3.2 million people in the Austro-Hungarian Monarchy alone. For example, over a million were infected with syphilis and other venereal diseases, 430,000 from tuberculosis and 330,000 from malaria.[6] Records kept by the Clinic of Dermatology and Venereology affiliated to the Royal Hungarian University of Science in Budapest indicated that 9 per cent of the city's total population was infected with venereal diseases in 1915. This number increased to 16 per cent in 1916 and 26 per cent in 1918.[7]

Engaging with these problems, venereologists like Zsigmond Somogyi, Tamás Marschalkó and Dezső Hahn, and the pathologist Béla Entz, among others, outlined the complicated relationship between the spread of pathogens, living conditions and the environment. These experts warned the military authorities and the government that official campaigns against sexually transmitted diseases and epidemics were not only needed in order to protect civilian health but they were also essential to the post-war reconstruction process.[8] Moreover, as the realities of war began impacting on all spheres of domestic life and work in Hungary, new medical and social concerns arose.[9] The physiologist Géza Farkas and the neurologist Károly Schaffer, for instance, were equally concerned about nutrition during the war or with side-effects such as neurasthenia.[10] The pharmacist Lajos Száhlender, on the other hand, acrimoniously described the war as a lethal biological experiment, condemning the usage of chemical weapons, especially the use of poison gas.[11]

The outbreak of a typhus epidemic in Serbia in November 1914 further accentuated these anxieties about health and disease.[12] Almost half a million cases were reported by mid-1915, with devastating effects for Serbia's population and its military performance in the war.[13] Due to its geographical proximity, Hungary was directly exposed to the typhus epidemic. In response, disease control programmes and epidemiological monitoring were immediately introduced, including arrangements for the transportation of infected soldiers. Special military observation stations were also created by the Ministry of National Defence (Honvédelmi Minisztérium) in several Hungarian towns (for example, Szatmárnémeti, Miskolc, Kassa, Munkács) under the leadership of a civilian health commissioner entrusted with preventing the spread of epidemics and attending to those already contaminated, along with a military officer tasked with enforcing discipline and supervising the proper functioning of quarantines.[14] Cases of contagious diseases were to be isolated and treatments administered. From a medical point of view, this preventive strategy was critical to determining the type of treatment necessary.[15] From a military perspective, however, it was suggested that the evacuation of patients undermined the army's fighting strength. It was, therefore, essential that the medical services had the facilities to enable them to reach accurate, early diagnoses, and that the expertise provided by the medical doctors was effectively utilized.

There were two practical dimensions to this issue. First, there was the assessment of hospital facilities, nursing homes, and specialized medical departments. Physicians and public health activists questioned whether the government was providing adequate medical assistance and maternal care in addition to proper social assistance and public health programmes. Second, there was a growing concern about a decline in birth rates and high infant mortality in Hungary, largely due in equal measure to military conscriptions, the spread of contagious and venereal diseases, as well as declining hygienic conditions more generally. In this respect, the involvement of the Hungarian Royal Defence Force (Magyar Királyi Honvédség), and especially that of its own medical officers, added supplementary expertise and increased the chances of protecting civilians from epidemics.[16] Establishing effective sanitary control was supported by the creation of military health stations in the combat zone, both permanent and temporary, as well as in the transit sector and in the unaffected regions.[17]

Public Health and Social Hygiene

Reflecting these health concerns and the increased political interest in medical management, an exhibition devoted to the relationship between public health and war was organized in Budapest between April and May 1915 (Fig. 12), under the illustrious patronage of Archduke Joseph August, Prince of Hungary and Bohemia, and his wife Augusta.[18] The main organizers were the Hungarian Red Cross and the National Ambulance Service, together with the Municipality of Budapest and the Social Museum. Numerous charity and medical associations were also involved,[19] alongside hospitals, nursing homes and adoption centres. At a time when the country was gradually coming to grips with human and material loss, this exhibition on public health and war, and the publicity it generated, seemed timely and necessary.

The main source of inspiration was – as noted by one of the organizers, György Lukács, Director of the Royal Prince Joseph Sanatorium Association in Budapest (József Király herceg Szanatórium Egyesület) – the Exhibition on the Care of Sick and Wounded Soldiers (Ausstellung für Verwundeten und Krankenfürsorge im Kriege) organized at the Reichstag, in Berlin, between December 1914 and January 1915.[20] Similar to the German exhibition, the one organized in Hungary combined medical expertise with war propaganda, through photographs, films, charts and models. It introduced the public to the latest medical equipment and how it was used in the war, while simultaneously highlighting the importance of preventive medicine and the control of epidemics for the civilian population.[21]

According to the secretary of the exhibition, Ernő Tomor – medical superintendent at the same Royal Prince Joseph Sanatorium Association in Budapest – the War and Public Health Exhibition was a "very special and extraordinary event", as it happened "during the time of great historical events, such as a world war".[22] A military conflict of such magnitude, Tomor continued, "will not only decisively affect the destiny of nations, but will also change the ways we think about our lives. Above all, the war will radically change our conceptions of health care". Hungarian public health services were compared to those of Germany, as these "were the most perfect and exemplary among all belligerent countries".[23]

With respect to their own public, the organizers in Budapest pursued two main goals. First, the Hungarian exhibition aimed to inform the general public about the destructive consequences of infectious and venereal diseases; second, it purported to convince Hungarian political

Figure 12 Poster for the War and Public Health Exhibition (reproduced from Szántó and Tomor, eds, *Had- és Népegészségügyi kiállítás katalógusa* (Budapest, 1915), frontispiece)

and medical elites to act promptly in order to protect the health of the population amidst serious social and economic changes caused by war, famine and displacement.[24] The first goal was, in fact, achieved without difficulty. Using its vast collection of materials and the knowledge from previous exhibitions on public health, the Social Museum organized presentations on "the human body and its vital functions", "child health", "alcoholism", "venereal diseases" and "tuberculosis".[25] Within the first week 100,000 people had visited the War and Public Health Exhibition, approximately 10 per cent of Budapest's entire population.[26]

Accomplishing the exhibition's second goal – that is political and government support for preventive medicine and public health – was rather more difficult, but the organizers were determined to convey a sense of urgency otherwise rarely portrayed in the Hungarian public sphere through printed materials and visual representations of disease and infection. Such public exposure, moreover, highlighted the fundamental importance of medicine in protecting the society and the nation. In effect, by addressing specific public health problems caused by war, the exhibition successfully publicized the crucial responsibility of medical experts. Perhaps most importantly, it allowed them to appeal across the political divide, addressing wide-ranging issues concerning the health of the nation. Progressively more and more medical experts were catapulted to the forefront of national politics, assisting government officials and lobbying for the implementation of their eugenic ideas. Their political activism reflected the growing commitment to social hygiene, public health, social assistance, preventive medicine and eugenics. Considering how central topics such as venereal diseases, tuberculosis and alcoholism became in the eugenic discourse in Hungary and elsewhere during the war, the eugenicists' involvement was only to be expected. Not surprisingly then, some of the medical experts involved in the War and Public Health Exhibition – like Rezső Bálint, Leó Liebermann and Géza Lobmayer – were also members of the Eugenic Committee.

Tomor was also the author of *A socialis egészségtan biológiai alapjai* (*The Biological Basis of Social Hygiene*) published in 1915.[27] Following Alfred Grotjahn and Alfons Fischer, Tomor presented eugenics as an integral part of social hygiene committed to the understanding of degenerative factors in society and race.[28] The practical goals of social hygiene and of "social heredity" – as he called the aspect of social hygiene devoted to eugenic research – were not dissimilar to those grouped by Galton under "national eugenics" or to those identified by Ploetz under "Rassenhygiene". There was, Tomor insisted, a conceptual symbiosis between eugenics and sociology, one that conveyed a normative and prescriptive view of reality and one which Galton in fact summarized in his 1904 definition of eugenics. Yet, "to improve the race", Tomor argued, "required more theoretical knowledge still, particularly in the field of 'experimental heredity'".[29] Until this theoretical knowledge became available, social hygiene would fulfil the eugenic role afforded to "social heredity" in attempting to understand the relationship between heredity and degeneration.[30] Engaging with the latter issue, Tomor further discussed various theories in natural sciences, applied biology and eugenics, from Darwin to Weismann. Mendel's theories of inheritance received additional scrutiny. Tomor was also familiar with the eugenic literature in both Europe and the USA, as revealed in his repeated references to Galton, Ploetz, Gruber, Schallmayer and Davenport.

It was towards the end of the book, however, that Tomor redefined Grotjahn's theories of social hygiene and their applicability to the current war conditions. It was difficult, he remarked, to evaluate in Hungary "the physical condition of the population" based on statistics about the "increase or decline in the number of those with various mental and physical disabilities".[31] One would only be able to describe the extent of racial degeneration in Hungary in accurate terms if the entire Hungarian population were subjected to "a comprehensive anthropometric investigation" – similar to the one attempted in England after the Boer War by the Inter-Departmental Committee on Physical Degeneration. As such research was prohibitively expensive and time-consuming, the only reliable sources remained official statistics on military conscription. While it was not possible to establish whether "Central European nations were experiencing widespread physical degeneration", Tomor nevertheless agreed with Grotjahn that only when those of "inferior" racial quality (*kisebb értékű*; *minderwertig*) were accurately identified could the state adopt the much-needed preventive "medical and social policies".[32] There was no denying that war, Tomor concluded, "transformed the way in which questions about degeneration and eugenics" were asked. "The war", he continued,

> changed the very concepts we used when discussing the physical and psychical energies of people, providing an immense material that needs now to be processed and researched. There is no moment of reprieve or relaxation for medical sciences; on the contrary, this is a period of intense activity, which without doubt will contribute greatly to the clarification of the basic questions of social hygiene.[33]

Books such as Tomor's provide a sense of how diverse the debates on eugenics were during this period in Hungary, while many practical endeavours, such as the first Hungarian congress of public welfare planned for 1915, did not get beyond their planning stage. The Association of General Public Charity (Általános Közjótékonysági Egyesület) was the driving force behind this congress, which is yet another example of how eugenic ideas permeated public debates and government initiatives on infectious and venereal diseases, as well as alcoholism and social assistance.[34] The list of themes prepared for the congress, for instance, included "practical eugenics", alongside tuberculosis, alcoholism and child welfare, as well as the protection of mothers and children.[35] The organizing committee had hoped that the congress would bring together public health experts and eugenicists with the ambitious task of addressing the country's social pathologies and medical problems. For reasons unknown, the congress did not take place, but the initiative reveals the broad engagement with eugenics during the first months of the war.

As health reformers in Hungary sought greater access to state funds, many advanced the eugenic agendas of social and biological control. As the war escalated, eugenicists too demanded a more direct involvement of the state in the management of public health policies and social assistance. Ensuring political support for its programme was, after all, one of the aims of the Eugenics Committee. At first, however, it seemed that the war dealt a severe blow to the Committee's attempts to establish a fully operational eugenic movement. In addition, the leadership rarely found an opportunity to meet in wartime conditions. When not visiting the front, Teleki was mostly in Budapest, while Apáthy spent most of his time in Kolozsvár. Hoffmann, on the other hand, was in Berlin. While it was difficult for the Eugenic Committee to synchronize a new meeting of its affiliate societies – dispersed as their own individual members were across the combat zone and behind the front lines – eugenic work was not altogether absent. Tangible practical eugenic achievements were perhaps missing, but eugenic thinking had permeated many sectors of society by 1914. It was during the war, in fact, that Hungarian eugenicists succeeded in having their much-desired political impact.

At the beginning of the war, eugenic activism was noticeable in two areas in particular: the protection of mothers and infants, and the fight against venereal diseases. If, in some respects, eugenicists merely restated ideas and practices in large measure already established in the fields of public health and preventive medicine, there was a new sense of emergency to their activism, insisting that the future of the Hungarian race was compromised unless immediate action was taken by the government. As will be discussed in the remaining sections of this chapter, arguments of this nature were intended to underline the indispensable necessity of eugenics as a means of social and racial control.

Motherhood and War

As hereditarian language increasingly shaped war debates on fertility, healthy mothers and children as well as a means for their social protection, it also underpinned new eugenic discourses that equated large families with a strong race and national efficiency. Given such concerns, eugenicists viewed the protection of future generations both in social and racial terms.[36] It justified not only immediate measures against birth control but also eugenic ideas of national rebirth.[37] Central to this language was the ideal of the family whose eugenic management was now demanded in the name of the Hungarian nation and its state. To be sure, the protection of the family was always at the heart of many eugenic narratives of social and biological improvement, but during the war it assumed increased prominence.[38]

Facing dramatic human losses, eugenicists believed that the nation's demographic potential and racial strength would be seriously undermined unless the state provided coherent policies towards the family combined with the protection of mothers and infants. The relationship between the future of the nation and reproduction was already an established eugenic trope, but the war transformed it into a national obsession. The birth of healthy children was increasingly viewed less as an exclusively private matter and more as a matter of major concern to the state. Demographic growth was thus increasingly seen as the proper starting point of any practical eugenic policy.[39]

Reflecting on this situation in a 1915 article – written first in Hungarian for the medical journal *Orvosi Hetilap*, and then expanded in German for the *Archiv für Frauenkunde und Eugenetik* – the above-mentioned Hungarian gynaecologist, János Bársony (Fig. 13), reflected on the long-term consequences of war and the crucial importance of eugenics in safeguarding the demographic future of the nation.[40] "Racial selection and improvement" was the responsibility of every community – Bársony postulated at the outset of his article – as the war ruthlessly damaged "our human material".[41] Fears of demographic decline and racial degeneration were seemingly justified by statistical evidence about the increasing number of "inferior individuals" in the population. Eugenics, in turn, needed to respond efficiently to wartime challenges and individual traumas and thus prepare the world for the post-war reconstruction. "After the war",

Figure 13 János Bársony (courtesy of the Semmelweis Museum, Library and Archives of the History of Medicine, Budapest)

Bársony believed, "eugenics – the doctrine of racial improvement – will step into the foreground with its full strength".[42]

Bársony discussed "the present from a eugenic point of view" and envisioned plans for the future, highlighting the unique combination between demographic growth and racial health. What was needed, he argued, was a strategy to ensure the "improvement of the race" (*Hebung der Rasse*) caused by war. The first suggested course of action was to increase the birth rate. Bársony criticized certain fertility traditions existing in Hungary, especially the one-child system, and bemoaned that "the family with six children is regarded as a rarity of the past, and there are entire regions in which the 'one-child system' dominates".[43] Family limitation, together with birth prevention and abortion, "were the reasons why the Hungarian race was stagnating".[44] At the Gynaecological Clinic of the University of Budapest, where he worked, Bársony experienced such practices first-hand. He thus woefully noted that of the 3625 registered pregnancies at his clinic in 1913, 935 were aborted. Moreover, out of the 640,000 children that were born on average in Hungary, 140,000 died within their first year.[45] Reflecting on these statistics, Bársony stressed the importance of a eugenic education for prospective mothers and prenatal care in reducing the high rates of abortion and infant mortality. Furthermore, he deployed a nationalist rhetoric of racial responsibility and eugenic health to defend his ideas of motherhood. "To be very active", in the interests of mothers and children, "was the very first step towards racial growth and improvement".[46] The eugenicist's task was to investigate the causes of the nation's declining health and to propose adequate remedies.

Given these premises, eugenicists facing a nation-wide devastation caused by war needed to do more than merely diagnose Hungary's health and medical problems. On the one hand, Bársony encouraged physicians to prepare young women for maternity and alert them and their families to the eugenic importance of their offspring; on the other, he suggested that the state should expand its charitable activities and increase its social welfare and social assistance expenditure. Bársony's second method to guarantee the "improvement of the race" underlined precisely this point: "The new generation should not only be large, numerically speaking, but also primarily healthy. The health of the parents is the first condition for the recovery of the race to happen."[47] More generally, the reappraisal of the eugenic role of the woman as a mother resulted in a nuanced evaluation of the relationship between eugenics, race and maternity. Bársony did not hesitate to declare that the woman's body was "created to be exclusively in the service of racial reproduction".[48] Women's reproductive capabilities, therefore, needed to be controlled and managed scientifically. There was thus a convergence of interests between the future of the nation, the control of reproduction and the protection of mothers. In order to raise the racial quality of future generations, Bársony advised the Hungarian government to "begin by protecting women", for "the protection of our race begins at conception".[49] The priority of the existing political elite should hence be to use eugenic propaganda to create a sense of social responsibility towards the biological future of the nation.

Stefánia Association for the Protection of Mothers and Infants

The eugenic preoccupation with the health of future generations, speculations about the declining birth rate, infant mortality and the protection of mothers and infants was most powerfully expressed in the activities of a new association, the National Stefánia Association for the Protection of Mothers and Infants (Országos Stefánia Szövetség az Anyák és Csecsemők Védelmére). It was established 13 June 1915, with administrative support from the Mayor of Budapest, István Bárczy, and under the patronage of Princess Stéphanie of Belgium, formerly the widow of Rudolf, Crown Prince of Austria – at the time married to her second husband, the Hungarian count, Elemér Lónyay. Albert Apponyi, the chairman of the Lower House of Parliament and former Minister of Religion and Education (1906–10), was elected the Association's president. Other founding members included József Madzsar, who became the Stefánia Association's managing director, and Ottokár Prohászka, the Bishop of Székesfehérvár. A number of renowned Hungarian paediatricians and child reformers were involved as well, including Pál Heim, János Bókay, Jr, Miklós Berend and Pál Ruffy.[50] The statistician Alajos Kovács and Vilmos Tauffer, both members of the Eugenic Committee, completed the group.[51]

As a charity and private initiative, the establishment of the Stefánia Association mirrored developments elsewhere in Central Europe, as Tara Zahra has demonstrated in her study of child welfare in the Bohemian lands.[52] By relying on paediatricians and eugenicists, and by endorsing their ideas of infant hygiene, the Stefánia Association also resembled the German Association for Infant Protection (Deutscher Vereinigung für Säuglingsschutz) and, especially, the Empress Auguste-Victoria House for Combating Infant Mortality (Kaiserin Auguste-Viktoria Haus or KAVH), established in Berlin-Charlottenburg in 1906.[53] Both institutions were directly involved with the protection of mothers and infants, as well as child welfare and the popularization of modern ideas of motherhood (Fig. 14). The aim was to place infant hygiene – now seen "as part of an increasingly centralized, rationalized, and medicalized effort at population control"[54] – in service to the state and nation.

During the war, the Stefánia Association contributed to the expansion of child welfare services, as exemplified by the training for professional district nurses, a practice that was introduced in Hungary in 1916 and in Germany in 1917.[55] Not surprisingly, perhaps, in his 1914 report to the American Association for Study and Prevention of Infant Mortality, Henry J. Gerstenberger – Medical Director of Babies' Dispensary in Cleveland, Ohio – praised the state treatment of infants in Hungary and Germany as unrivalled anywhere in Europe and North America. "Hungary", he further noted,

> is the only country that cares for its dependent infants and children in a national manner. The underlying principle of this organization is the following: every child which cannot be properly cared for by its relatives has claim upon care and help from the State, without the loss of any of its personal or family rights.[56]

Figure 14 The Child Welfare Centre of the Stefánia Association in Gödöllő (courtesy of the Rockefeller Archive Center)

Reflecting recommendations for a centralized system of child welfare, amidst growing nationalist concerns, Gerstenberger's report revealed, in both Germany and Hungary, a growing sense of responsibility towards future generations.[57] Child reformers in both countries firmly believed in the protective gaze of the state.[58]

A tendency to shift the eugenic responsibility of mother and infant welfare from the family to the state had already emerged in Hungary before the war. Thus, a National Association for the Protection of Mothers and Infants (Országos Anya- és Csecsemővédő Egyesület) was created in 1908, devoted to the social assistance of pregnant women with low income, and to infant care. The war, however, highlighted the need for a centralized network of infant and child welfare, which – together with the recurrent eugenic interest in the protection of mothers and family – became the Stefánia Association's defining features. While there clearly was a synergy between various private and public agencies, and the support of Budapest's municipal authorities, it was József Madzsar's dedication and enthusiasm that secured the initial achievements of the Stefánia Association.

As discussed in previous chapters, Madzsar was one of the main supporters of eugenics in Hungary during the first decade of the twentieth century.

Together with Lajos Dienes and Zsigmond Fülöp, Madzsar was especially instrumental in preparing the 1911 debate on eugenics organized by the Society of Social Sciences. Yet, he was not invited to join the Eugenics Committee – no doubt due to his socialist sympathies and connections with other radical thinkers and societies, like the Hungarian Association of Free Thinking (Szabadgondolkozás Magyarországi Egyesülete) and the Galileo Circle (Galilei Kör).[59] In terms of their eugenic ideas about the moral education of the population as well as the instruments of its social and biological improvement, however, there were no substantial differences between Madzsar and prominent members of the Eugenic Committee like Apáthy, Teleki and Hoffmann.[60] Furthermore, as the meetings of the Stefánia Association indicate, Madzsar did not seem to have any difficulty in working with conservative-nationalist and religious leaders like Apponyi and Prohászka.[61]

When the war broke out, Madzsar was the executive vice president of the Section on the Protection of Mothers within the Budapest Central Relief Committee (Budapesti Központi Segítő Bizottság). From this position he argued in favour of eugenic natalist and protectionist policies for mothers and infants.[62] He then articulated and strengthened the relationship between eugenics and the protection of mothers and infants in two programmatic texts, both serving as the Stefánia Association's practical programme.[63] According to Madzsar, the Association's purpose was twofold: on the one hand, to stimulate the activity of the state in the management of child welfare services; on the other hand, to increase public awareness about the eugenic role of motherhood. In practical terms, the Stefánia Association endeavoured to help Hungarian mothers in the following ways: (1) to appreciate the biological value of their families for the national economy; (2) to give birth to healthy children in optimal conditions (hospitals, clinics and so on); (3) to encourage breastfeeding; and, finally, (4) to fulfil their duties as mothers.[64] In order to accomplish these goals, five sections were created: social-political, legal, administrative, medical and propaganda.[65] While Madzsar emphasized the role of the state in safeguarding the interests of the nation, he also envisioned the role of the Stefánia Association as central to maternity and child services in Hungary during the war and afterwards.

To illustrate his arguments, Madzsar compared birth rates and infant mortality in various European countries. Based on statistical reports he argued that between 1901 and 1905, for instance, Italy had the highest birth rate, 48.1 per cent, and France the lowest, 21.2 per cent, while Hungary occupied a median position with 37.2 per cent. "Hungary", he noted, "might have a high fertility, but she also has high mortality; numerous births but also numerous deaths".[66] Madzsar's principles were as much medical as they were economic and social, as they insisted both on the importance of biological protection of mothers during birth and protective measures afterwards.[67] Practical solutions were urgently needed, and Madzsar provided some in a public lecture suggestively entitled "A jövő nemzedék védelme és a háború" ("The Protection of Future Generations and the War"), delivered at the Society of Social Sciences in three successive parts on 27 November and 4 and 11 December 1915. Similar to many of his previous publications, this lecture was motivated by Madzsar's preoccupation with lower fertility rates, infant mortality, the control of reproduction and social management.[68] Despite the existence of an attitude favourable to eugenics among the Hungarian public and political elites, Madzsar felt that the government had been too reluctant to engage in a systematic policy of public health and eugenics.

The future of the national community was disheartening, Madzsar continued, and it was not only due to internal losses but also to the relative demographic potency of neighbouring nations, like Romania and Serbia. Madzsar sketched an apocalyptic vision of a society facing biological extinction: "If we look into the future, a threatening vision appears in front of us: if we will not be able to recover soon from the losses suffered, we might be defeated by those races and nations who were only spectators of this immense devastation."[69] The government may not be able to predict the number of men lost in war, but by introducing protectionist legislation to reduce infant mortality it may ward off anxieties about depopulation and speculations about the nation's diminished racial vitality.

Building on the statistical analyses of mortality and birth rates for 1914 and 1915 that illustrated how Hungary experienced a decline in birth rates, Madzsar hoped to change perceptions of social hygiene and of those pathological factors that contributed to the health of a population. Between October 1914 and October 1915, for example, the number of births was reduced by almost half.[70] Contraception, moreover, was widespread as families were more concerned about their material survival than guided by the "future interests of the race". To counteract these problems, Madzsar suggested the adoption of intensive eugenic policies, both qualitative and quantitative. "The qualitative side of population policies", which in times of peace prevented "those burdened with genetic diseases, the mentally ill and recurrent criminals [to] inundate the society with great masses of descendants",[71] should be continued after the war. "Undoubtedly", he warned, "the quality of the race will fall to lower levels after the war, than before."[72]

Madzsar thus insisted that eugenics should govern the selection of those fit to reproduce, especially as Hungary had lost many of those "enthusiastic, courageous and ready for sacrifice", and many with "healthy bodies" during the war. As a first measure, however, towards the recovery of national vitality, Madzsar recommended a quantitative demographic policy "at least until the deficiency in numbers" was corrected.[73] The reduction of infant mortality was also deemed essential. In short, Madzsar described the interplay between qualitative and quantitative eugenics as the most essential component of any future-oriented public health and social hygiene policy.[74]

This concern with the deterioration in health of future generations caused by war, coupled with a decline of birth rates and the increase of infant mortality, remained central to the eugenic debates during the first two years of the war. At the end of 1916, Madzsar completed his empirical demographic analysis of some of the causes contributing to the decline of birth rates and limitation of fertility in the Hungarian capital. The book entitled *A meddő Budapest* (*The Sterile Budapest*) largely followed the economic analysis of demographic trends in Berlin published in 1913 by the demographer Felix Aaron Theilhaber under the title *Das sterile Berlin*.[75] Like Berlin, Budapest was in need of biological regeneration. Immigration from the countryside into the cities in order to replenish the human losses caused by war was encouraged, alongside the proper involvement of state institutions in public health and welfare. Madzsar's pessimist conclusions on war-related causes of social and biological degeneration complemented Bársony's demands for the state control of reproduction through eugenic policies. But it was due to assessments of the nation's health such as these that the Hungarian political elite mobilized in the name of eugenics.

Political Mobilization

A year after the Stefánia Association was established, its president, Albert Apponyi, called for the direct involvement of the state in social and biological welfare, in his report to the Lower House of Parliament, occasioned by the debate on the fifth government report on the usage of emergency powers in times of war.[76] As one of the most respected Hungarian politicians of the time, Apponyi was a passionate advocate of measures to increase the number of racially healthy Hungarian families. "The question of the protection of mothers and infants", he asserted, should become "an important part of the future social policy" in Hungary.[77] With the birth rate declining, Apponyi believed that "the question of demographic policy, which is of paramount importance following the terrible loss of blood we have suffered during the war, has now gained even more importance than before".[78] Apponyi's preoccupation with the biological quality of the population and the protection of the future generations resonated with the eugenic movement. "From the point of view of eugenics", he noted, "it is important for the nation not only that many children are born but mainly that they are healthy and strong".[79] Voicing his criticism of neo-Malthusian ideas and practices, Apponyi reasserted the importance of racial reproduction and requested that state funds be distributed to large families. His eugenic argument for the protection of mothers and infants also carried a strongly moralistic and nationalistic message.

The Minister of the Interior, János Sándor, provided the reply to Apponyi's report. Sándor assured members of the Lower House and the general public alike that the government understood "the importance of this matter". His confidence acquired a new significance considering the national war effort. "Hungary", he agreed, "needs a new generation brought up in good health, one able to sustain this country for another millennium".[80] The eugenic focus on motherhood informed a political vision of the family and children as the nation's future. In this respect, the state and the eugenicists strove simultaneously to augment social protection for mothers and infants. The priority of the state – Sándor made clear – was to sustain the patriotic morale of the population during war, while at the same time to awaken a sense of responsibility towards future generations.

This exchange of ideas in the Hungarian Parliament between two important public officials illustrates how widespread eugenics was by this stage; equally important, there was a good deal of convergence between the eugenic activities of various public associations like the Stefánia Association and the practical programme promoted by the Eugenic Committee. Political differences among Hungarian eugenicists continued to exist, but a consensus gradually emerged with respect to the role of the state in protecting the population during the war, as well as in shaping the future of the Hungarian nation after the war. Stressing the importance of immediate action, the Eugenic Committee followed the Stefánia Association's example and petitioned the government for support. To this end, a third meeting of the Eugenic Committee was planned by two member societies, the Association for Social Sciences and the League for the Protection of Children, for 28 December 1915.

The meeting was to be devoted to war and venereal diseases, with Lajos Nékám, a professor of dermatology at the University of Budapest, as the main speaker.[81] It is unclear whether the meeting took place or not, as a letter from Teleki dated 18 December 1915 announced to the members of the Eugenic Committee that, due to the war, all "future meetings were postponed until further notice".[82] What is clear, however, is that members of the Eugenic Committee did work together with Nékám on preparing a memorandum about the spread of venereal diseases among the Hungarian army and civilian population, as well as its devastating consequences for the body of the nation, which they then submitted to the government.

On 4 January 1916, a small delegation consisting of two members of the Eugenic Committee, Jenő Gaál and Lajos Török, together with Lajos Nékám and László Széchenyi – who represented the League for the Protection of Children – met Prime Minister István Tisza, the Minister of Interior János Sándor, and the Minister of Defence Samu Hazai. The representatives of the Hungarian government were all "familiar with the extraordinary urgency" of the problems outlined by the eugenicists and commended the Eugenic Committee on their initiative.[83] It was recognized that in order "to surmount these exceedingly great difficulties, exceptional measures were needed" and "that the effective engagement of various governmental agents were indispensable and urgent".[84] Gaál also graciously thanked Nékám on behalf of the Eugenic Committee for his "precious service to the cause" and "his great expertise and valuable experience".[85]

The memorandum on venereal diseases was the Eugenic Committee's only practical achievement, as Hoffmann noted in 1916. Having existed for only two years, the Eugenic Committee struggled nevertheless to achieve as much as possible from the practical programme outlined by Apáthy in January 1914. Although it benefitted from initial support from the government, "when the war broke out", further activities "stopped".[86] Hoffmann further recognized that war made it difficult for the Eugenic Committee to pursue its activities, but this predicament did not affect the Hungarian eugenic movement in general. On the contrary, war prompted other eugenicists to work together with politicians, health reformers and population policy advisers in order to articulate common programmes of national and racial protectionism.[87]

The pressure put on the Hungarian government to intervene in regulating the nation's health grew in intensity. In May 1916, Mrs Sándor Szegvári, for example, on behalf of the Feminist Association submitted a petition to the government in which she requested the introduction of medical certification before marriage.[88] This became apparent once more in August 1916 when the Lower House of Parliament discussed the new income tax law. The opposition, under the guidance of Albert Apponyi, requested tax reductions for families with many children and a tighter control of childless families.[89] The proposal did not make it in the final version of the law, but the House "unanimously adopted a declaration addressed to the government, insisting that the new income tax law be introduced after the peace settlement and should promote families with many children".[90] In addition, György Lukács presented a memorandum to the Ministry of Justice, requesting that one important change was made to the new income tax law: "A marriage could only happen between individuals who could prove with an official medical certificate that they were not suffering from a contagious venereal disease." This measure – Lukács assured the government – would not reduce the birth rates, since it was not "a marriage injunction for life", but one intended to protect the health of the population instead, as "sexually transmitted diseases caused fertility problems and in some cases even infertility".[91]

How were these new developments commensurate with the eugenic tradition represented by the Eugenic Committee? As mentioned, by early 1916, the Eugenic Committee ceased to function, but not before it provided Lukács and Nékám with an opportunity to articulate their new eugenic programme of social and national protection based on the control of venereal diseases.[92]

The Association of National Protection against Venereal Diseases

The memorandum on venereal diseases submitted by the Eugenic Committee to István Tisza and other members of his government in January 1916 consolidated Lajos Nékám's eugenic reputation. Nékám (Fig. 15) had been associated with the eugenic movement since his participation in the conference on public health organized by the Association for Social Sciences in 1912. He was also one of Hungary's foremost dermatologists.[93] Not surprisingly, then, when the Royal Hungarian University of Science in Budapest opened its new, state-of-the-art Clinic of Dermatology and Venereology, Nékám was appointed its Director.[94]

On 7 July 1916, Nékám together with György Lukács established the Association of National Protection against Venereal Diseases (Nemzetvédő Szövetség a Nemibajok Ellen). As its name suggests, the Association of

Figure 15   Lajos Nékám (from *Dolgozatok Nékám Lajos Prof. negyedszázados tanári működésének évfordulójára* (Budapest, 1924), frontispiece)

National Protection's main goal was to protect the Hungarian nation from the supposed social and racial damage caused by venereal diseases. It also endeavoured to assist the authorities with medical expertise, as well as to encourage the population to lead a "healthy and virtuous life".[95] As the language of preventive medicine was shaping the debates on the national body and the particular means to protect it, so it underpinned new eugenic representations of society, and with it of those belonging to the nation's racial body. Well-established narratives of health and disease were now embroiled with eugenic descriptions of "racial poisons" (venereal diseases, tuberculosis, alcoholism and so on) and their undermining effect on the nation and race.

The protection of the nation depended on the careful monitoring of the population. This drive towards human management, in turn, created – to use Alison Bashford's inspired description – "eugenic *cordons sanitaires*".[96] The crusade against venereal diseases, in Hungary and elsewhere, was one such eugenic *cordon sanitaire* aimed at isolating and containing the current generation's pathologies. Between 12 April and 25 May 1917, the Association of National Protection organized its first major activity: a general study commissioned by the Ministry of the Interior on the subject of venereal diseases, pornography and prostitution. Various social, charitable, welfare and religious organizations in Hungary were asked to suggest "legal, regulatory and social measures in order to prevent the spread of venereal diseases and thus the further deterioration of the race".[97] Some of those who contributed with expertise were also members of the Eugenic Committee: Lajos Török, Géza Hoffmann, Pál Angyal and Vilmos Tauffer.[98] This government initiative should therefore be seen within the context of a new biomedical concern in Hungary with the racial body, which – as suggested here – brought together eugenic categorization of civic and national responsibility with much broader interpretations of public health governance.[99]

As the Hungarian government grew increasingly worried about the impact of the war on the health of the civilian population, it also became more open to eugenic arguments concerning social and biological degeneration. By mid-1916, the two areas in which government involvement was immediately required, namely the protection of mothers and infants and the protection against venereal diseases, benefitted from the creation of two new societies: the Stefánia Association and the Association of National Protection. These achievements contributed to the consolidation of the eugenic movement in Hungary and simultaneously also offered its much-needed reinforcement through the work of József Madzsar, Albert Apponyi, György Lukács and Lajos Nékám. These eugenicists became actively involved not only with their respective organizations, but also contributed directly to the general dissemination of eugenics in official political circles, devising guidance and drafting instructions. In this sense, their strategy was successful. Their strong political ties to the Hungarian government – especially with regards to Apponyi and Lukács – revealed the growing political importance of eugenics. Moreover, the presence of leftist eugenicists like Madzsar in this affiliation indicates that attempts were finally made to engage with eugenic questions of national importance across the political spectrum. Although the degree of actual involvement with practical politics varied from one individual to the other, the emergency of the war provided the overarching intellectual context that could unite disparate trends within the Hungarian eugenic movement.

Such initiatives, as will be discussed in the next chapter, had far-reaching consequences. Indeed, national health was increasingly seen as the Hungarian government's immediate responsibility. This, in turn, affected the impact of eugenics on political decisions. Debates on the nation's health and its future were not only expressions of eugenic concern with the social and biological effects of the war. In addition to occasioning the introduction of social and medical policies dealing with the protection of mothers and infants, and with the spread of contagious and venereal diseases, eugenics also generated a resurgence of nationalist concerns about the deterioration of the Hungarian nation's racial qualities.

The official acceptance of eugenics by prominent politicians like Pál Teleki, Albert Apponyi, György Lukács and others paved the way for an officially sanctioned form of eugenics. As the next chapter will reveal, the most notable outcome of this transformation of the political elite's attitude towards eugenics was that the state became involved with ideas of social and biological protection. Two years into the war, the question with which eugenicists had struggled with since the beginning of the century, namely whether the Hungarian government was capable of adopting eugenic measures geared towards the protection of its population was finally answered. Far from succumbing to the difficulties of war, the eugenicists waged an organized campaign to promote their ideas and put them into practice. Indeed, it is possible to speak of Hungary as the only country in Europe where eugenics ultimately triumphed during the First World War.

6
Eugenics Triumphant

During the First World War, Hungarian social scientists, physicians, demographers and anthropologists began to construct a specifically national eugenic philosophy – one based upon the fusion of various academic disciplines, among which sociology and medicine were the most prominent. That eugenics was placed into critical convergence with many narratives of social and political change was not incidental. This chapter therefore offers first a discussion of some of the most important eugenic theories of war proposed in Hungary after 1914. This overview is necessary in order to prepare the ground for a more detailed analysis of the practical measures adopted by the Hungarian eugenicists in the last years of the war in the wake of their success in establishing a new eugenic society in 1917. As they undertook the complex task of developing structures to put their ideas into practice, the eugenicists emphasized that improving the health of the nation would, in turn, stimulate the state's capacity to withstand the social and biological damage caused by the war.

Darwinism and War

Establishing the effects of war on race and nation was a central component of eugenic theories proposed after 1914.[1] Eugenicists generally condemned war for nurturing a number of social and biological problems, including prostitution, the spread of contagious and venereal diseases, and declining birth rates. Some authors highlighted, in particular, the detrimental effects of war on the family unit as a whole, especially on the mothers and infants. All of them nevertheless agreed that due to these problems, together with the growing number of human casualties, the nation's racial vitality was seriously under threat.[2] War, it was further suggested, led to the elimination of the physically and psycho-logically healthy, allowing the "weak" to increase in number and thus undermine the nation's future.

Alfred Ploetz, for instance, had pointed to the dysgenic effects of war as early as 1895 and proposed that the "worst individuals" be drafted into military service in order for the "healthy individuals to be saved".[3] Mihály Lenhossék (Fig. 16) offered a similar view in his 1914 lecture to a public debate on war organized by the Royal University of Science in Budapest (Budapesti Királyi Magyar Tudományegyetem).[4] Subscribing to the Darwinist idea of natural selection, and to the Spencerian notion of "the survival of the fittest",[5] Lenhossék accepted that war could enhance social and national cohesion, and stimulate emerging patriotic sentiments of sacrifice and devotion to the national community: "War awakens, develops, and accentuates the noble qualities and virtues of the human soul. War brings hearts closer in self-sacrifice, patriotism and mutual support."[6] Consistent to his anthropological expertise, Lenhossék defined "race" in Darwinist terms; that is, as a group of individuals characterized by constitutional and psychological traits

Figure 16 Mihály Lenhossék (courtesy of the Semmelweis Museum, Library and Archives of the History of Medicine, Budapest)

inherited through the generations. He believed in the generic unity of the "human race", but accepted that, within it, there was racial diversity. However, this physical heterogeneity should not, he cautioned, create the impression that some races were "superior" to other races. More importantly, by postulating war as the product of "civilization" and "culture" rather than heredity, Lenhossék distanced himself from biological determinism. He even hoped that civilization would eventually generate its natural immunity to war, "just like a human organism produces both self-hazardous poisons and their antidotes".[7] Although Lenhossék recognized that endemic warfare among nations and races had been an important selective force in human history, he did not view war in terms of instinctive pugnacity. In another study dealing with the relationship between war and eugenics, he further warned that modern military conflicts were dysgenic since they wasted the "best blood" of the nation on the battlefield.[8]

Such views found supporters among progressive religious leaders who, like Sándor Giesswein, had long defended modern theories of evolution but nevertheless warned against the appropriation of Darwinism by pugnacious theorists of war and nationalists.[9] There were others, however, like the Catholic Bishop of Székesfehérvár, Ottokár Prohászka who saw in war the possibility of redemption for a depraved humanity.[10] Finally, some argued that war perfectly illustrated processes of natural selection in social and national spheres, and expressed the superiority of one race over another in the perpetual struggle for survival.[11] War, according to these authors, was a Malthusian mechanism through which to regulate over-population and the demand for economic resources. The Director of the Department of Zoology at the Hungarian National Museum (Nemzeti Múzeum Állattani Osztálya), Lajos Méhely, represented this view. Méhely believed that war was effective in mobilizing the racial abilities of the nation, while simultaneously counteracting physical degeneration and racial miscegenation.

Méhely was one of the few eugenicists in Hungary to engage publicly with what he perceived as naïve interpretations of war, not only at the level of practical politics, but also eugenically and racially. In his *A háború biológiája* (*The Biology of War*), a crude analysis of war as seen through the prism of the racial struggle for existence, published in 1915, Méhely forcefully argued against those who failed to understand the aggressive tendencies in human nature that were driven by an instinctive drive for survival and an endemic fear of extinction.[12] Méhely proposed that "similar to animals, human species, races and nations live in constant competition and strife",[13] and explained that "as soon as a species or population finds itself in the threshold of being or non-existence, war sets in with utter cruelty".[14] An avid supporter of biological militarism, Méhely thus offered a theory of historical conflict based on genetic determinants and biological laws. According to Méhely, those who denied that "the struggle for life in its Darwinist form" characterized human societies or that "excessive reproduction" was controllable were plainly refuted by "twenty million heavily armed individuals getting ready to kill each other in all corners of the Earth".[15]

Méhely invested warfare with its own laws allegedly derived from the natural world: "*war is not a human invention but a regular aspect of organic life*".[16] Appropriating Darwin's view that "if man wishes to reach further, then the intensity of his struggle must not dwindle, or else he will sink into inert indifference and helplessness",[17] Méhely gave his own interpretation of the struggle for existence, and its role in shaping modern warfare: "War will exist as long as there is life and a living human being on Earth." It ensued that war was humanity's factor of progress: "*The human race continues to evolve and war is nothing else but the stepping-stone of human evolution.*"[18] Playing an indispensable role in fostering evolution, war also served as the natural ally of "*various races and sub-races*" in history. To deny the importance of war was, in fact, to deny the importance of evolution, and thus to misconstrue human history. "Human peace", Méhely further claimed, "together with the moral and material blessings of this peace, including the creations of the human spirit, are all in fact groundwork for war."[19] Méhely's supplementary emphasis on biological morality opposed so-called "humanist theories of war", namely those which assumed that

> the noble actions of the human spirit and the wonderful creations of the human mind be self-sufficient; nevertheless from the point of view of biology *these values are mere resources*, although ones which decide the survival of peoples and nations in war – the greatest test in the struggle for life.[20]

By visualizing the Hungarian nation as a fighting organism amid hostile neighbours, Méhely was in fact encoding the classical Hungarian theme of the "death of the nation" (*nemzethalál*) anew in biological terms.[21] As a result, he praised war for the biological rejuvenation and eugenic selection it entailed. Patriotism, Méhely believed, was in retreat in a society overwhelmed by intellectualism and contemptuous of discipline and self-sacrifice. A new generation of Hungarian nationalists was being formed in the trenches, one that was racially healthy and willing to sacrifice itself in the service of a greater cause. War, in other words, was intrinsically bound to the human condition; it was an opportunity for both a spiritual and physical renewal, seen as the antidote to degeneration and the chance to increase the biological qualities of the nation. As Méhely insisted, natural selection prompted by war would eliminate "the pale, the weak, the nervous, and, from a military point of view, a simply worthless generation – despite its sophisticated spiritual life".[22] The result will then be "the breeding of patriotic, strong-willed, disciplined and physically strong generations of citizens".[23]

But who were "the pale, the weak, [and] the nervous" individuals inviting Méhely's opprobrium? Méhely did not point to any particular ethnic group or social milieu, though his contemporaries knew only too well to whom his nationalist diatribes were ascribed. As Méhely's nationalist vision indicates, the Jew came to symbolize the degenerate intellectual whose inability to completely identify with Hungarian patriotism translated into various anti-Semitic accusations, including that a great number of Jews had avoided military services in favour of "exploiting the labour of the civilian population".[24] The Jews thus became, as so frequently perceived, directly responsible for the people's dire living conditions and for undermining the war effort.[25]

Having established this distinction between "strong" and "weak" Hungarians, Méhely additionally implied that a new nationalist morality based on the primacy of race would emerge during times of war, one that would offer the Hungarians the prominent role among the nations of Europe they so rightly deserved. The following passage clearly encapsulates Méhely's ultimate conviction that war gave the Hungarians a chance to prove their racial credibility:

> From a national point of view our war has already brought us a valuable result – namely that Hungarians are greatly respected by the peoples of the monarchy and by our German ally as well – which is a token of a better future. Hungarian faith and bravery have gained us the friendship of all those who were in this respect out of reach; they have ameliorated problems that have existed between us and other nationalities in the monarchy and they have more or less erased the darkness which had eclipsed our relationship with those nationalities, and thus, in exchange for our great sacrifices we have created the conditions for the consolidation of our state. What is left now is to make use of these conditions to our best interest.[26]

Méhely stretched the potential of Darwinism to the limit in order to argue for a distinctively eugenic renewal of the nation's biological foundations. "We think of our past and future sacrifices with a bleeding heart", he acknowledged, "yet we are proud to look forward because we feel that our nation will emerge purged from the flames of war, consolidated and enriched with great moral values."[27] Unapologetic in his insistence that war was beneficial to the Hungarian nation, Méhely concluded passionately that although "most of the heroes rest in peace in foreign ground now, the light of their spirit illuminates our path and will lead us in the sublime work of our new conquest".[28] Ultimately, war was both an instrument of collective selection and the test of the nation's social fitness and morality.

Méhely's straightforward and unequivocal Darwinist assertion that the Hungarian race could only benefit from war may appear to repudiate Fülöp and Lenhossék's accounts but, in fact, these authors did not differ fundamentally in their treatment of the eugenic problems affecting the Hungarian nation during the war. Other Hungarian eugenicists – while not sharing Méhely's glorification of war as an expression of nature's survival of the fittest – nevertheless accepted that Europe had become an arena of ethnic competition for economic resources and natural space. Moreover, as a multi-national state, the Austro-Hungarian Monarchy was particularly exposed to internal fissures and tensions arising between the politically dominant nations – the Hungarians and the Germans – and other less privileged ethnic groups demanding more political power and recognition. Hoffmann, for example, was among those who documented Hungary's increased racial antagonisms and who, at the same time, passionately defended the role of eugenics in protecting the Hungarian nation from social degeneration and demographic decline. Where Méhely glorified war's beneficial influence on strengthening Hungarian racial qualities, Hoffmann discussed the need for eugenic policies to preserve such qualities in the first place. Prior to examining Hoffmann's views on war in more detail, it is worth pausing for a moment to reflect on an article dealing with the theoretical and practical difference between racial hygiene and eugenics that he published in July 1916.

Racial Hygiene and Eugenics

As discussed in the previous chapters, Apáthy preferred to use the term *fajegészségtan* to *eugenika* (although both of these words translate as eugenics) in an attempt to formulate a Hungarian vocabulary of social and biological improvement. As a term *fajegészségtan* entailed far more than a linguistic subterfuge – nor is it sufficient to explain its frequency in Hungarian public and scientific discourses simply in terms of Apáthy's rejection of foreign models of eugenics such as the British and the German. *Fajegészségtan* (eugenics), moreover, shared a set of common meanings and a vision with another term: *fajegészségügy* (practical eugenics). Apáthy used the former term to express the universal science of eugenics and the latter to express the practical application of eugenic ideas for social and biological improvement. Apáthy held a nascent eugenic worldview tailored towards Hungarian realities that involved a specific style of thinking concerning the Hungarian race and nation, one defined in Chapter 1 as the biologization of national belonging.

To some extent, Hoffmann cultivated Apáthy's terminology and the same worldview, albeit insisting that *eugenika* as a term was too widespread among Hungarian social scientists and the general public to be completely eliminated from the vernacular. It may thus appear there was conceptual synchronicity between Apáthy's and Hoffmann's interpretation of eugenics. In fact, the two interpretations were quite different. In an article published in the *Természettudományi Közlöny* in July 1916, Hoffmann returned to these theoretical ruminations on the meaning of eugenics and racial hygiene. Entitled "Fajegészségtan és eugenika" (which according to Hoffmann should be translated as "Racial Hygiene and Eugenics"), this important article was both a response to Apáthy's article "Fajegészségügy és fajegészségtan" ("Practical Eugenics and Eugenics") and a reassessment of Hoffmann's own article "Eugenika" ("Eugenics"), both published in 1914.

First, Hoffmann explained Galton's theory of eugenics and his faith in controlling heredity in order "to improve human material and to eliminate degeneration from one generation to another"; in other words, race improvement through selective breeding.[29] In contrast, German theories of eugenics were more "abstract". Instead of "focusing on the individual and the family", German eugenicists accentuated the importance of "an entire group of people throughout time".[30] It is this group of people that Ploetz called a "race". According to Hoffmann,

> Ploetz chose the word *race* because as a word it expresses the biological essence of a community of individuals better than *people* (which rather relates to those who live together), *nation* (which expresses a political decision), or *society* (which focuses on human relations). Race, in this context, does not refer to either a blood relation or the descent of an individual – because there are no pure races in this part of the world, as all civilized people have emerged from the combination of several different races. Moreover, *the word race employed by*

> *Ploetz does not stand for race as a species* or for the German, Turanic
> or Jewish race; it signifies instead *a unit of life consisting of succeeding*
> *generations of people living together.*[31]

Different descriptive conventions surrounding the word race (*Rasse* in
German and *faj* in Hungarian) notwithstanding, Hoffmann proposed –
following Apáthy – that the general science of race improvement be
referred to as *fajegészségtan*, yet he did not translate it as eugenics but
as racial hygiene, similar to Ploetz's term *Rassenhygiene*. Hoffmann too
offered a binary terminology: *fajegészségtan* (racial hygiene) and *fajegészségügy* (practical racial hygiene): "*fajegészségügy* is *fajegészségtan*'s
practical application – *praktische Rassenhygiene* or *Rassenpflege*, in
German". Thus viewed, "the objective of practical racial hygiene" was
"the prosperity of the race, meaning that it was not sufficient to elevate,
according to the possibilities, only the racial value or racial average of
individuals; there was also the need for the dormant values of the race
to assert themselves in life and for the race alone to cope with life".[32]
Part of the adoption of this new terminology, moreover, was Hoffmann's
concern that eugenics, in its English usage (and British and American
eugenics more generally), was too limited, as it focused solely on the
control of reproduction.[33] Racial hygiene, on the other hand, was concerned with more than "breeding the most valuable offspring". Its task,
in short, was "to evaluate and amend each measure for the best interest
of the race, and here we are not talking about momentary or economic
interests".[34] Racial hygiene and eugenics overlapped, Hoffmann admitted, but were not identical (see Fig. 17).

   In addition to Germany, this broad definition of race improvement – as
racial hygiene or *fajegészségtan* – also found supporters in the Scandinavian
countries, Austria and Hungary. According to Hoffmann, the time had
come in Hungary for eugenicists to transcend the debate over the theoretical application of eugenics and begin considering the practical importance
of racial hygiene:

> Both from a theoretical and a practical point of view, racial hygiene
> offers a more wide-ranging possibility. Taking into account several other
> circumstances of life, it manages to avoid one-sidedness while it func-
> tions alongside the other social endeavours and not against them, as it
> tries to bring all of our institutions together for the benefit of the race.[35]

Furthermore, Hoffmann asserted the importance of heredity and biology
in the understanding of racial development, as the field had diversified

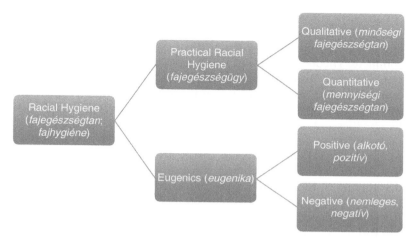

Figure 17   Hoffmann's theory of racial hygiene

to include not only eugenics, but also racial genetics and racial biology. Conceptual clarification was, he insisted, essential. This was, to reiterate, a period of eugenic experimentation, and not only in Hungary. Hoffmann was completely aware that by expanding the meaning of eugenics to include all aspects of race improvement – hence *fajegészségtan* (racial hygiene) – there was no boundary to its sphere of activity. In Hungary, for instance, various "institutions and activities (infant welfare, the fight against tuberculosis, alcohol-related illnesses or venereal diseases, morality, philosophies of life, youth education, foreign affairs and so on)" were already pursued in the name of racial hygiene.[36] While these efforts were commendable, Hoffmann nevertheless believed that much more attention was given to social and medical protection rather than to eugenics. Conceptual clarity was required on this occasion, Hoffmann insisted. By highlighting the differences between various interpretations of eugenics, Hoffmann had wanted "to avoid unproductive disputes". On a theoretical level, comments such as these reflect Hoffmann's dissatisfaction with the applicability of the terms *racial hygiene* and *eugenics* in Hungary. "It is essential", he concluded, "to clarify the concepts employed. Instead of having each writer, each orator, rename them over and over again; the only way to prevent the amalgamation of various views is to consequently employ the justified and accepted version of each clearly determined concept."[37] Hoffmann's allegations were not entirely unfounded.

This was Hoffmann's first article in Hungarian since the war began and as such it represents a departure from his earlier work on American eugenics. His embrace of German theories of racial hygiene was now complete. Theoretical without being repetitive, this clarification of his views on racial hygiene and eugenics is noteworthy for its more comprehensive aims. Hoffmann felt that the eugenic movement in Hungary was experiencing a critical moment: the Eugenic Committee had ceased its activities and new organizations – like the Stefánia Association – were taking its place. War had brought the protection of mothers and infants to the mainstream of the eugenic debate in Hungary, as well as the fight against contagious and venereal diseases. Hoffmann welcomed these developments, though in a more restrained way. For he now believed that it fell to the eugenicists to take centre stage and engage directly with the damage caused by the war to the Hungarian national community.

It cannot be denied, however, that Hoffmann offered a critique of the prevalent eugenic theories in Hungary, including Apáthy's, in the name of a nationalist vision of unrestrained support for the development of the Hungarian race. It is clear from the preceding discussion that his adoption of the language and methodology of racial hygiene had strong ideological overtones. It was also another means of defining a specific Hungarian eugenics, one according to which the protection of the race depended on practical measures, and it is these that Hoffmann turned to in his next book, *Krieg und Rassenhygiene* (*War and Racial Hygiene*).

The Dysgenic Effects of War

By 1916, Hoffmann had become a pivotal figure of the Hungarian eugenic movement. He endeavoured to provide this movement with its own national agenda and an intellectual programme. Like other Hungarian eugenicists at the time, though, he struggled to bridge the alleged cultural divide between theories of eugenics developed in Britain, the USA and Germany and their practical application to Hungarian realities. For Hoffmann, racial hygiene was a wide-ranging philosophy of national protection, inseparable from the dynamics of the political, social and cultural life of the country. It was also the most suitable strategy to heal the injuries caused by the war to the body of the nation and thus restore the Hungarian race to its health. It was not only a new interpretation of eugenics as racial hygiene that Hoffmann adopted from his experience in Germany as a member of the German Society for Racial Hygiene, but also an increased interest in population policy (*Bevölkerungspolitik*) – specifically the proper balance between qualitative and quantitative measures for the improvement and protection of the race.[38]

These arguments resurfaced in Hoffmann's 1916 publication *Krieg und Rassenhygiene*.[39] In the first part of the book, Hoffmann dealt with the mechanisms of "national extinction" experienced during war, namely racial degeneration and demographic decline. War, Hoffmann argued, amounted to the "total destruction of nations", yet it protected the "inferior" and the "weak" members of the race.[40] How would these dramatic transformations impact the future of the race?, Hoffmann asked. And further,

> To imagine that the indescribably great effort of this war and, in general, of this era could go to waste, and that everything we cherish could fall prey to foreign races – only because earlier or later generations of descendants, either in terms of their numbers or of their racial quality, were or will be less worthy than us – is just frightening.[41]

Hoffmann considered racial determinism to be the "basis of historical evolution", explaining that it was "only natural" that the superior nation triumphed over the inferior ones. However, his was no biological militarism. Hoffmann portrayed war as having "a damaging effect" on the nations involved in conflict, since "the strongest, healthiest and bravest individuals fall victims to war while the sick, weak and far less heroic ones survive and produce offspring".[42] Racial hygiene was capable of mobilizing national forces as well as state institutions towards the development of the most essential of all means of national survival: racial protection.[43]

After describing the counter-selective consequences of warfare, Hoffmann went on to discuss the influence of war and culture on racial development throughout history. War always obstructed the development of strong racial unity between various tribal and ethnic communities. According to Hoffmann, "races of Germanic descent", inhabiting most of Western and Northern Europe, did not manage to forge a long-lasting "racial solidarity", as allegedly existed between races of Turanic descent, including Hungarians, Bulgarians and Turks living in Southern and Southeastern Europe. Turanism, Hoffmann believed, was one form of expressing such racial solidarity between different ethnic groups not united by economic or political interests.[44]

Of equal importance was that the war also acted as an agent for biological selection among those individuals belonging to the same ethnic group.[45] Hoffmann lamented that the war had exposed the participating nations to various forms of biological and racial extinction, obstructing

population growth and producing a dysgenic effect on the hereditary
constitution of the main European races. Hoffmann then proceeded to
assess the "racial burden" and "degeneration" that the war had on the
nation's "genetic standards", arguing for the protection of marriage, the
introduction of prophylactic measures against venereal diseases and,
especially, the discouragement of "inferior people" from reproduction.[46]
As far as the survival of the race was concerned, Hoffmann linked it
equally to morality; in other words, to one's devotion to the national
community. Another focal point for evaluating the relationship between
the war and racial hygiene was the domain of culture. As it attempted to
regulate the struggle for existence among various nations, culture was –
for Hoffmann – "essentially an obstacle to the development of the race".
Culture protected "inferior individuals" by allowing them to reproduce.
The "welfare of the race" necessitated an "unrelenting struggle for existence" within the same national community, as it offered the means to
protect the race from the destructive consequences of war.[47]

This form of race protectionism required that racial hygiene be given
the leading role in the process of national recovery, seconded by an
equally well-articulated population policy. And, as the damages caused by
the war increased, Hoffmann insisted that the more extensive segments
of the political elite and the population be made aware of the importance
of these policies. Without a significant expansion of public awareness, the
negative effects of the war would, in turn, result in a profound national
crisis. Hoffmann's wish to inform the opinion the public held, so that it
acquired "a racial hygienic way of thinking", was fully exploited in the
last section of the book, dealing with the methodology of racial hygiene.[48]
Considering that Europe faced a difficult present and an uncertain future,
Hoffmann presented measures of racial hygiene that he deemed necessary
to guaranteeing "the revival of the race" after the war. As his confronta-
tional, provocative language attests, he thought immediate action was
necessary, especially with respect to the stability of the family and the
"devotion to future generations".[49]

After outlining his methodology, Hoffmann then discussed diverse
practical racial hygienic strategies, which he assembled into "quantitative
and qualitative" categories, with the latter further divided into "positive
and negative" forms. Positive racial hygiene promoted the "breeding
of those superior", while negative eugenics "impeded the reproduction of
those inferior".[50] Reflecting his eugenic focus on reproduction, Hoffmann
thus centred his argument on interventionist state policies to assure racial
efficiency. Instituting a coherent domestic policy of family planning
needed to be complemented by an equally efficient demographic policy.

Two countries served as prominent examples, the USA and Germany. If the former was the herald of "qualitative racial hygiene", having already introduced measures to "prevent the reproduction of the inferior", the latter was largely preoccupied with "quantitative racial hygiene". Only recently, Hoffmann remarked critically, had German eugenicists turned towards "qualitative racial policies". To illustrate this point, he detailed the programme adopted by the German Society for Racial Hygiene at its 1914 congress in Jena, including its "furthering of internal colonization with privileges of succession in favour of large families"; the "abolition, as far as possible, of certain impediments to marriage"; the "legal regulation of procedures in cases necessitating abortion or sterilization"; and, finally, "awakening a national mind ready to make sacrifices, and have a sense of duty towards the coming generations".[51]

An armoury of "racial hygienic requirements" – which also included "better housing, the promotion of large families, the fight against sexually transmitted diseases, alcoholism and tuberculosis" – enforced the persuasive power of Hoffmann's arguments. Yet, for these measures to be adopted in Hungary, eugenics had to be institutionalized. The architects of this process – enthusiastic eugenicists like Hoffmann – emphasized that the purification of society and its biological continuity depended on the transmission of new racial codes and mentalities to the general public. Here, the examples provided by North American eugenics reigned supreme.[52] For Hoffmann, it was only in the USA that "the highest earthly goal: the future welfare of the race" was fully integrated into social and population policies.[53]

Reviewing the book in the *Archiv für Rassen- und Gesellschaftsbiologie*, Fritz Lenz described Hoffmann as "the most active champion of practical racial hygiene in Europe" at the time.[54] Indeed, when one considers Hoffmann's publications, such a portrayal seems accurate. In the same year *Krieg und Rassenhygiene* was published in Germany, Hoffmann published a slightly modified version in Hungarian under the title "Háború és fajhygiéne" ("War and Racial Hygiene").[55] He then published a follow-up article entitled "Fajegészségtan és népesedési politika" ("Racial Hygiene and Population Policy"), in which he discussed in more detail some of the themes simply outlined in his book.[56] Here again, Hoffmann integrated the population policy with the domain of racial hygiene and articulated a terminology to reflect this symbiosis. He reaffirmed his commitment to both qualitative (*minőségi*) and quantitative racial hygiene (*mennyiségi fajegészségtan*).[57]

Towards the conclusion of the article, he described how quantitative and qualitative racial hygiene could be applied to Hungary. Drawing upon rising fears of racial degeneration and the declining health of the population, Hoffmann offered provocative evaluations of the consequences resulting from drafting the community's healthy members. Hoffmann then put forward two population policies based on race protectionism: the control of emigration (*vándorlás*) and rural settlement (*telepítés*). With respect to the former, "the aim was neither complete interruption of emigration nor resettlement, because we need to differentiate between individuals".[58] Hoffmann suggested that only those individuals considered "desirable as citizens of the state" should be asked to remain in the country, while those "who from an individual, racial or political point of view are harmful to the community" should be encouraged to emigrate. It would be ideal, he pointed out, if *"the smallest possible number of superior individuals would emigrate and the greatest possible number of those superior individuals who had emigrated would return; as well as the opposite, if the greatest possible number of undesirable individuals would emigrate and never return"*.[59]

This distinction between "desirable" and "undesirable" individuals is equally apparent in Hoffmann's proposal of rural settlement. The eugenic responsibility towards the peasants – seen "as the core of the Hungarian race"[60] – was an innovative element in Hoffmann's eugenic discourse. He valued the peasantry as an embodiment of racial fertility and national strength.[61] "Rural settlement", he believed, "served this objective".[62] According to Hoffmann, only the healthy and fertile peasants were to be chosen as prospective settlers. "To colonize with sickly, undesired families" would be counterproductive for the national economy. "To colonize with individuals", Hoffmann continued, "who are incapable of proper reproduction or who simply do not want to procreate would be a disproportionately costly charity with absolute disregard for the future."[63] Both the social and economic transformations of the nation were to be driven through projects of eugenic renewal. As Hoffmann emphasized, "only racial hygiene measures which consider the next generations can be of lasting value".[64]

Hoffmann was not alone in invoking the spectre of war in order to attract political attention to the importance of eugenics, racial hygiene and population welfare. The Austrian social hygienist and eugenicist Julius Tandler, for instance, while rejecting claims that this war was "racial or national"[65] in a 1916 article, similarly accepted that from a eugenic point of view the impact of the war on the population was disastrous. Like Hoffmann, he too identified the "quantitative and qualitative damage to the nation's body".[66] More importantly for our discussion here, Tandler's article prompted Sándor Szana, Director of the Royal State Child Asylum

in Budapest (Budapesti Királyi Állami Gyermekmenhely), to publish a similarly titled article in the next issue of the *Wiener Klinische Wochenschrift*.[67]
According to Szana, Tandler had addressed one of the most important
issues of social hygiene and eugenics pertaining to the war: "the regeneration of the injured national strength".[68] The debate concerning the war
and eugenics, Szana believed, brought to the surface a growing concern
with the nation's social and biological vitality.

The political content of some of the eugenic narratives of national
protection may have changed after 1914, Szana noted, but some eugenic
concerns predated the war. Szana's own expertise was in child welfare,
an area in which Hungary had made significant progress since 1900.
"Hungary", Szana informed his Austrian readership, was "the only state
in which the Ministry of the Interior has its own section of child protection". The Hungarian government could thus draw on existing statistics
and a number of "social hygienic studies in determining infant mortality and declining birth rates".[69] Facing such extensive changes in the
structure of the population, particularly the loss of those individuals
who were valuable "from a biological and racial hygienic point of view",
Szana reaffirmed the possibility of biological regeneration and the need
for a stronger sense of national purpose. It was of the utmost importance, ultimately, that the "nation's regenerative strength" was not lost
after the war.[70] In practical terms, Szana considered the "racial-biological
value of war orphans" and that of "emigrants". Only with state support,
Szana insisted, could war orphans – most of them sons and daughters
of valuable members of the race – contribute to the regeneration of the
race; otherwise, they will become destitute and prone to degenerative
issues like alcoholism, as well as contagious and sexually transmitted
diseases. As for emigrants, the return to their country of origin was, for
Szana, "the fastest, most valuable, and most powerful instrument of
regeneration".[71]

Like Hoffmann, Szana seemed fully integrated into the European
debates on racial hygiene, eugenics and population policy. In another
book, also from 1916, he compiled a set of guidelines on the application
of Hungarian population policy seen through the prism of child protection, his own area of expertise.[72] The time for "new ideas", he noted,
has long gone. What was needed, according to Szana, were practical
measures and the direct involvement of state institutions. The field
of "social hygiene" needed to be reconceptualized too, to reflect these
major changes in thinking about population welfare. His work on child
protection was based on the principle of racial improvement, but he
noted that the war made certain "concepts" – like "population policy"

and the "preservation and increase of national strength" – very popular, particularly in German academia and official discourse.[73]

Szana assembled Hungarian ideas about "the preservation and increase of national strength, and racial improvement" into a number of inter-related areas of expertise. The first of these was concerned with "infant and child mortality". Szana identified the most "vulnerable communi-ties" in Hungary, particularly in the rural areas, and suggested the crea-tion of a "Child Protection Council" (Gyermekvédelmi Tanács), within the Ministry of the Interior, to coordinate nation-wide programmes of child welfare and offer expert advice to the government.[74] Another area in which population policy exerted significant influence was in the "protection of motherhood". Here, Szana argued in favour of health insurance schemes and social assistance for mothers, alongside improvements in facilities of maternal care and financial incentives for midwifes. From a eugenic point of view, Szana accepted that abortion and contraceptives were often necessary, though made the public aware of their impact on the health of the mother and on the growth of the population in a more general sense.[75]

The continued expansion of the protection of mothers and infants described above led Szana to assume an even greater state interven-tion in social welfare across public institutions, from education in schools and universities to special courses for police officials.[76] Of equal importance was the increased state and medical responsibility that impacted directly on "race protection and race improvement". In this respect, Szana advocated marriage restrictions for those suffering from contagious and venereal diseases, and the introduction of the Wasserman test before marriage to detect syphilis.[77] A more efficient way to improve racial health, Szana believed, was the introduction of a health certificate, which from birth onwards would record all of the medical complications of an individual that may be passed from a parent to their child. A person's family medical history would thus be recorded and monitored. Another area deemed crucial to racial improvement was physical and hygiene education in schools, and Szana expected the Ministry of Religion and Education to engage with such programmes.[78]

Together with the fight against contagious diseases like tuberculosis and alcoholism, the linkage of child welfare and physical education solidified the role that eugenics played in defining population policy and as a result the health of the nation.[79] In the wake of the dra-matic demographic changes brought about by the war, Szana – like Hoffmann – also deplored the loss of valuable racial members of the nation due to emigration.[80] Other topics reviewed by Szana from the point of view of population policy and eugenics included: the relationship between marriage and illegitimate children,[81] the importance of nutrition and the distribution of arable land[82] and unemployment.[83] All of these guidelines and suggestions for practical measures in the field of eugenics and population policy were supposed to assure Hungary's much-needed demographic growth. Szana stated it clearly and simply: "Only families with more than three children can maintain the race."[84]

These suggestions were compelling, not because they appealed to either the winner or the losers of the war, but because they offered hope after the military carnage was finally over. At the time, eugenicists across Europe argued in similar terms, although they were in military opposition. Eugenic theories of war proposed in Hungary discussed so far illustrate different views, ranging from Lenhossék's culturalism and Méhely's crude Darwinism to Szana's child protectionism and Hoffmann's synthesis of racial hygiene and population policy, but they all share one crucial concern, namely the future of the nation. As seen in the previous chapter, other eugenicists, who were active in campaigning for motherhood and childcare, social hygiene, and the protection against venereal and contagious diseases, also shared this concern for the nation's welfare. As Hoffmann put it in a somewhat unusual article entitled "Nyelvében él a nemzet" ("The Nation Lives in Its Language") – also published in 1916 – "the nation was in a life-and-death battle".[85]

The various eugenic strategies outlined so far were essentially attempts to take the nation's social and biological decline seriously and to suggest ways in which the Hungarian race might not only survive the war but also thrive. This twofold eugenic project conveyed biological as well as social strategies. From the protection of mothers and infants to the control of emigration and rural settlement, the proposals discussed above thus reveal the complex relationship between eugenics, public health, social hygiene and population policy. At the same time, eugenic narratives – like the one offered by Apáthy in his article "Háború és fajegészségügy" ("War and Practical Eugenics") – possessed far more than social and biological significance. The war, Apáthy argued, had provided the eugenicists with a new ideological source of legitimacy and with it a new political vocabulary and imagery. The bond between the members of the national community was inevitably weakened during the war, Apáthy continued, but a sense of collective identity had triumphed.[86] As will be discussed in the next section, specific eugenic allegiances emerged during this period, often in unexpected ways. The result was direct government intervention and a state-sponsored eugenic welfare programme, culminating in the establishment of the Society for Racial Hygiene and Population Policy.

National Military Welfare Office

A first step in this direction was taken in 1916, when the Hungarian government created the National Military Welfare Office (Országos Hadigondozó Hivatal) and entrusted Teleki with its leadership.[87] The Military Welfare Office evolved from the the Welfare Committee for Disabled and Injured Soldiers (Csonkított és Béna Katonákat Gondozó Bizottság) – created at the initiative of the Prime Minister, István Tisza, in 1915 – and its successor, the Welfare Office for War Invalids (Rokkantügyi Hivatal).[88] The existence of a military network was crucial to the Military Welfare Office's activities, which were divided into three main groups: disabled soldiers, war widows and war orphans.[89] Additionally, the Military Welfare Office aimed to educate the servicemen and military officials:

> the chief aim at the beginning was to spread sound eugenical [*sic!*] ideas. Lectures were held in all institutions where the mutilated and other victims of the war were treated or taught, pictures were posted everywhere, and leaflets distributed in great numbers. The army commandants also distributed these leaflets to the soldiers.[90]

Speakers enlisted for these occasions included Géza Hoffmann, who lectured on racial hygiene, and the physician Mihály Pekár, who discussed medical issues.[91] Organized by the Military Welfare Office, with financial support from the Ministry of War, a nation-wide eugenic campaign thus became possible for the first time in Hungary. It was also clear that eugenicists like Teleki and Hoffmann shared certain eugenic values with other officials in the government and the public administration, values that were shaped not only by the war but in large measure by a common belief in the future of the Hungarian race.

Of equal importance was Teleki's appointment as president of the Military Welfare Office. This decision illustrates the Hungarian government's growing involvement in social protection, eugenics and population policy.[92] In turn, Teleki integrated his eugenic vision of national health within the overall framework of Hungary's social and racial improvement through state assistance given to disabled soldiers, war widows and war orphans.[93] Such a vision was clearly expressed in Teleki's speech to the Lower House of Parliament on 2 March 1917.[94]

Hungary was facing a demographic catastrophe, Teleki announced gravely. Citing Madzsar's statistics from his 1916 *A meddő Budapest*, Teleki highlighted the fact that the birth rate had dropped by 50 per cent since 1914.[95] It was imperative that public health, eugenic and population policy measures were introduced immediately. It was, he said, "clear that the war has ruined us not only quantitatively but also qualitatively".[96] Moreover, he noted the absence of an official policy on a number of critical issues such as the social reintegration of wounded soldiers, or the repatriation of those refused immigration status in the USA. These issues directly affected the nation, but – Teleki complained – they were incorporated into various social policies pursued by the government without guidance from the eugenicists, and without sufficient material resources. The military conflict had imposed a new set of eugenic priorities, Teleki continued. Whereas before the war the concern with racial hygiene had been largely theoretical, the war had made practical eugenic policies supervised by the state paramount.

Another issue addressed by Teleki concerned the repatriation of those Hungarians refused entry in the country by the American immigration offices. According to Teleki, their "repatriation to Hungary must be accomplished based on racial and health considerations".[97] It was particularly important that the Ministry of War be given information on the subsequent progress of these individuals, so that the effectiveness of various repatriation schemes and eugenic proposals could be properly evaluated. At this point, Teleki quoted from a document prepared for the Eugenic Committee, in which Hoffmann remarked that "those who – due to feebleness, disease or other reasons – could not thrive" in the USA were sent back to Hungary.[98]

Furthermore, Teleki endorsed rural settlement schemes (*telepítés*), also advocated by Hoffmann and prominent Austrian eugenicists like Max von Gruber. As expected, "only a desirable individual or family was to be encouraged to settle" in the countryside. These families needed not only to be healthy but also fertile as the allotted property was not inheritable unless the family had a sufficient number of children.[99] Yet, Teleki warned the government that political determination alone would not suffice. Disparate eugenic activism by a number of individuals and public associations, like the League for the Protection of Children and the recently established Stefánia Association, needed to be channelled towards accomplishing common objectives. To this end, Teleki informed the Lower House that a meeting took place on 6 February at the Social Museum in Budapest, attended by "20–30 leading public personalities". At this meeting, the creation of a Centre for

Population Policy (Népesedéspolitikai Központ) was proposed, whose purpose would be to "influence the public opinion, the scientific community, the legislative bodies, and the government". The prospective Centre would be divided into "numerous departments", but one in particular was singled out: "a legislative committee" dealing with eugenics. Teleki highlighted the Centre's "collaboration with the government" and requested that a "parliamentary commission" be established "that would be in permanent contact with the Centre". Teleki considered the latter idea to be the "most necessary", as he believed that Parliament should become "the most important institution of population policy, the first and most important centre of social welfare in this country".[100] Parliament was at the centre of political life in Hungary, and it was there that Teleki hoped to convince political parties of the importance of racial hygiene policies.

Teleki concluded his speech on a solemn note: "I remind you and ask all members of Parliament – especially those who are members of the government – not only to begin this social-political work, but to finish it as well." Even though he sought to avoid direct references to the war's potential outcome, he made it clear that Hungary would be one of the most severely affected among the belligerent powers. He provided both a lucid diagnosis and a warning: without a nation-wide programme of racial protection, the end of the war "will have catastrophic consequences" for Hungary.[101] Hence, he insisted, "every law we issue and every measure we take must have a sociopolitical objective".[102]

There was no doubt that Hungary's demographic structure was about to change dramatically as a result of war. As a politician and chair president of a number of societies and organizations, Teleki highlighted the eugenic dimension of the social disintegration caused by the war. In his new capacity as the president of the Military Welfare Office, he was particularly concerned with the social reintegration of disabled soldiers and invalids after the war.[103] Of primary importance in this regard was a "circular letter on eugenics" that Teleki sent to offices and employees of the Military Welfare Office in the autumn of 1917.[104] Two salient principles were outlined in this letter, as Teleki explained at the outset:

> The provision for injured servicemen is not only to alleviate this generation's plight, but has the task of remedying the quantitative and qualitative damage inflicted upon the whole of our national body by the war. As our nation's best are either dead or injured, those living – and among these the very best – must be taught that raising a maximum number of children is a national duty. Making allowances for a

new generation – one suitable in both quality and quantity – is the hallmark of a forward-looking Military Welfare Office.[105]

By encouraging the marriage and reproductive activity of war invalids, Teleki hoped to reconcile the pessimistic descriptions of Hungary's demographic future, due to the loss of racially valuable members of the nation, with eugenic ideas of national efficiency. In reference to the former, Teleki suggested to "integrate population policy and racial hygiene within the framework of military welfare in order to strengthen the Hungarian nation's health". According to Teleki, it was necessary that the state and welfare organizations "emphasize, at every opportunity, the requirements of racial hygiene".[106] This "circular letter" served exactly this purpose.

Following on from this, Teleki referred to some practical measures to reassure disabled soldiers and other war invalids of their social and racial value. First, "the movement of the peasant population to the large cities" was to be discouraged, as it was "proven that urban families tended to limit the number of their children".[107] Urbanization and industrialization were blamed as much for contributing to negative fertility trends as for corrupting the nation's youth. Teleki recommended instead the "love for the countryside", so that disabled soldiers could appreciate "the advantages of rural life and the disadvantage of living in large cities (poverty, lack of proper housing, temporary employment, weakening of family ties and so on)".[108] He equally described the crucial role of the family in fostering family values and the protection of the nation's racial qualities. No future renewal of the nation was possible, Teleki believed, without strengthening the Hungarian family.

Teleki also considered modern cultural trends, like urban intellectualism, and "the longing for intellectual activities and similar careers" to be equally detrimental to the eugenic renewal of the nation: "The intellectual proletariat proves the reproduction of this social class to be dangerous. From a population policy point of view this trend is particularly troublesome".[109] The existing shortages of manual labour in agriculture and industry only strengthened Teleki's emphasis on the putative importance of physical work for the renewal of the nation's weakened economic resources. Eugenicists, he believed, should not encourage "the desire for intellectual careers". On the contrary, "just as in the case of migration towards the large cities, they should stress – first of all – that manual labour, especially skilled manual labour or an agricultural occupation, even though it may appear otherwise, offers a better and safer living than an intellectual one".[110]

This portrait of rural life and its blessings was, of course, highly roman-
ticized, but eugenicists in Hungary and elsewhere at the time increasingly
viewed the uncorrupted peasantry as a source of racial revival. Contrary
to urban dwellers, the peasants were devout and rarely challenged the
precepts of race protection. In these respects, Teleki's eugenic arguments
were all established; as discussed earlier, Hoffmann too considered the
peasantry to represent "the core of the Hungarian race", eulogizing traditional Christian values about family and the community. Church and
state, religion and eugenics, all conveniently converged.

There was another, closely related theme that Teleki highlighted in
his letter: the importance of the family. As heroes who shed their blood
for the country, disabled servicemen and war invalids needed to feel
that "their family conditions" were "considered when they were advised
on career opportunities, sent on holidays or transferred to another
workplace". Preferential treatment was especially suggested for those
with "large families, since these are considerations that would make
these people firmly believe in family life and nurture in them the belief
that by raising children they will achieve public recognition". A fulfilling family life was paramount to elevating eugenic responsibility
towards the nation. This realization was a natural corollary to that other
ubiquitous eugenic theme: the danger represented by "sexually transmitted diseases and the abuse of alcohol" and the need to "preserve the
society's desired moral purity".[111]

Finally, it is worth mentioning Teleki's unfailing hereditarianism. If one
believed that acquired characteristics were inherited, then the physical
damage caused by the war was unavoidably transmitted to future generations. Teleki criticized this Lamarckian interpretation, declaring that "the
sporadically encountered, but erroneous view that the physical wounds
endured by soldiers can negatively influence the quality of their offspring
must be invalidated; an acquired characteristic, such as disablement, is
not hereditary".[112] While those with "inherited defective illnesses" were
not encouraged to marry – or if they did they should not have children –
"single disabled people" were encouraged to marry and have children.
Teleki's eugenic decisiveness related to his conviction that any post-war
programme of national health should not marginalize those disabled by
military action; on the contrary, as long as their biological capital was
deemed worthy these individuals were to be reclaimed by the nation.

The eugenic arguments expressed in this letter were hardly new – and
Hoffmann's influence on Teleki is easily discernible here – there was,
nevertheless, an additional fervour of purpose and political legitimacy
in the way in which Teleki presented them. Eugenic propaganda was

needed in order to introduce the Hungarian political establishment and the general public alike to new ideas on health and hygiene, seen as requirements not only for the post-war regeneration of the national community, but also for ensuring that future generations benefitted from the appropriate biological environment to prosper and grow.

There were also reasons to rejoice. For Hoffmann, Teleki's "intention of introducing practical eugenics" made "Hungary the first country on the European continent to accept eugenics as a government measure".[113] He was called from his diplomatic post in Berlin to assist the eugenic work initiated by the Military Welfare Office. Disabled servicemen and war invalids, Hoffmann explained, were ideal "from a hereditary point of view" to serve as the first practical eugenic experiment in Hungary. They were "best adapted for the first trial measures, which could later be extended to the population as a whole".[114] According to Hoffmann, the most important of these practical eugenic measures "was the distribution of land to the mutilated in such a way that the best individuals in the hereditary sense of the word received sufficient land to support a family and that the stipulations of the contract encouraged the rearing of children".[115] Another important decision was taken with respect to returning soldiers, who were "redirected from the large and overcrowded cities to the country, the latter being better adapted to a healthy family life". Finding work after demobilization was also part of this eugenic arrangement:

> If we had two positions to fill, e.g. that of a janitor who probably could not rear more than two or three children, or would rather stay single, and that of a manager on a country farm, then we sent a man whose propagation seemed not advisable to fill the janitor's post, and sent the healthy and otherwise desirable man to the farm.[116]

The Military Welfare Office also invested in the nation's eugenic reproduction. Thus, the soldiers deemed racially valuable were given "advice as to the duty of the healthy to rear many children", while those deemed morally, medically and racially "defective" were encouraged "to terminate his bad stock in his own interest". Such procedures, Hoffmann claimed, "were taken only after thorough investigation and medical examination". The promotion of marriage followed. Race-specific counselling was available to those concerned. "Much interest", it seemed, "was shown by men as to the advisability of their marriage and propagation. Efforts were made to convince the sick and the mutilated that their defects were not hereditary". It was rewarding, Hoffmann continued, "to see how the advice given enlightened these poor victims of the war". Such direct intervention into the private life of Hungary's citizens was to be devised with support from state and public administration: "All local authorities and the different government offices were asked to assist these efforts of race regeneration, to spread sound eugenical [*sic!*] ideas among the population, and to act accordingly when fulfilling their official duties."[117]

Another government measure that the eugenicists also endorsed was the financial assistance to "public officials and state employees after the birth of their children". Equally important was Apponyi's proposal – as discussed in the previous chapter – to modify the income tax law. This was finally accepted in 1917. As a result, childless families had to pay 15 per cent more than those with children while those with just one child had to pay 10 per cent more than others.[118] By encouraging reproduction, these measures reflected the government's increasingly eugenic pronatalism.[119] They also reflected a successful eugenic politicization and a higher degree of public receptivity to eugenic policies, particularly in the capital, Budapest. The administrative and financial support provided by the Mayor of Budapest, István Bárczy, was essential in this respect. As seen, Bárczy played a crucial role in the establishment of the Stefánia Association for the Protection of Mothers and Infants in 1915; he was also instrumental in the creation of Hungary's first Welfare Centre (Népjóléti Központ) in Budapest in May 1917.[120]

The Welfare Centre was divided into 18 sections, including Disability (President Gyula Dollinger), Public Health (president Sándor Korányi), Protection of Mothers and Infants (President József Szterényi), Child Protection (President Mrs Gábor Vay) and Adult Education (President Kuno Klebelsberg). The Sections on the Protection of Mothers and Infants, and on Public Health attracted some of the most active eugenicists and social hygienists: József Szterényi, József Madzsar, Miklós Berend, János Bókay, Ernő Deutsch, Sándor Szana, Vilmos Tauffer, Albert Apponyi, Gyula Donáth, György Lukács, Dezső Hahn, Imre Dóczi and Mihály Pekár.[121] An Institute for the Protection of Mothers and Infants, associated with the Welfare Centre, was established in July 1917 and divided into three departments: Protection of Mothers (Director Vilmos Tauffer), Child Protection (Director Miklós Berend) and Social Hygiene (Director József Madzsar).[122] Here we encounter yet again the familiar reconfiguration of eugenic networking observed in the case of the Stefánia Association and the Military Welfare Office.

The Budapest Welfare Centre's accomplishments in the area of eugenics, social welfare, public hygiene and health care were quickly noticed on a national level. Entrusted with health decisions in the capital, the above-mentioned eugenicists aimed to extend their expertise to the rest of the country. Health was proclaimed above politics and in the public interest. To strengthen this view, eugenicists promoted the idea that the medical and social problems of the Hungarian nation during and after the war could only be remedied through adequate eugenic and population policies as well as a nation-wide welfare programme. Their vision of the Hungarian state, much like the practical proposals they made, was based on the acceptance – on both the left and the right of the political spectrum – of a eugenic paradigm of social and biological improvement. Protecting the health of the population had finally become a national priority, and this was aptly illustrated by the national congress on public health, which convened in Budapest between 25 and 28 October 1917.

National Congress on Public Health

This congress was a remarkable achievement as evidence for the emergence of a eugenic consensus in Hungary during the war. The most important Hungarian sociologists, demographers, social hygienists, feminist activists and eugenicists attended it.[123] It was also the only time when the eugenicists discussed in this book – from all Hungarian cultural, medical and welfare societies – socialized as a collective. United by a general concern with the widespread social and biological damage to the Hungarian population caused by the war, they had managed to put aside their political and personal differences, thus hoping to influence the imminent debates in Parliament on the future of the Hungarian state.

The inaugural session, on 25 October, was chaired by Kálmán Müller, the vice president of the National Council of Public Health (Országos Közegészségügyi Tanács) and Albert Apponyi, the president of the Stefánia Association. Leó Liebermann delivered the keynote lecture on "state responsibilities in the field of public health".[124] It was important, Liebermann noted, that the state accepted the premise that public health reform was linked to the general welfare of the nation and that it acted accordingly. Physicians, social reformers, health activists and eugenicists had all been pressuring the Hungarian state and various governments to improve the nation's biological welfare. To simply "centralize all work on public health" was no longer sufficient, Liebermann argued. What was needed was the creation of a Ministry of Public Health and Welfare (Közegészségi és Népjóléti Minisztérium).[125]

The first day of the congress, held on 26 October, was devoted to the following topics, each with its own panel of specialists: "population policy from an economic point of view", followed by "population policy from a racial point of view" and "the protection of mothers and infants". The day concluded with a session on "child protection". Kálmán Müller chaired the first panel on "population policy from an economic point of view". The Minister of Commerce, József Szterényi, was the main speaker and Hoffmann the discussant. Other participants included Oszkár Jászi, Gyula Donáth, Miklós Berend, Béla Schmidt and Dezső Buday.

One of the important issues discussed by Szterényi in his lecture was the relationship between economic development and demographic growth, a topic embodying general concerns over the decline of birth rates and fertility in Hungary. Szterényi proposed a rationalization of both personal and family life, and highlighted the importance of large families as a prerequisite for the nation to reach its biological potential.[126] In his comments, Hoffmann reacted positively to Szterényi's insistence on quantitative demography, but he warned against differential fertility and its possible outcome: disproportionate population growth. Notably, he remarked that the government's current financial support for families largely neglected the importance afforded to the "quality of reproduction". These schemes, Hoffmann noted, "motivated, in fact, inferior layers of the population to reproduce, whereas those families, which stand out above the average, were not sufficiently stimulated by these allowances to encourage them to increase their reproductive rate and create larger families".[127] If those deemed racially valuable did not reproduce, or reproduced in small numbers, "the average value of the nation will decrease from generation to generation", and Hungarians – similar to other great cultures in history – will eventually experience their "downfall". Hoffmann's recommendation, to avoid this path, encompassed

> economic measures that reflect both the quality and quantity of the successors. Such measures are primarily those mentioned by the speaker: the change of inheritance law, which is perhaps one of the most radical measures from the point of view of racial hygiene, and the considerable tax to be levied on families with less than four children.[128]

This was precisely the type of eugenic regulation that was expected from the state. According to Oszkár Jászi, it was also important that the promoters of the new thinking about population in terms of national efficiency, reproduction and eugenic planning in Hungary provided the state with much-needed solutions. To illustrate this point, Gyula Donáth, in his remarks, reminded the participants of a previous meeting organized at the Social Museum in February that year (also mentioned in Teleki's March speech to the Lower House of Parliament) and the planned Centre for Population Policy (Népesedéspolitikai Központ), which – if created – could bring these academic networks, public associations and political interests within an institutionalized framework. By comparison, the Austrian government, Donáth continued, agreed to establish a new Ministry of Health and Social Welfare. In Hungary, in addition to the recently created Ministry of Welfare (Népjóléti Minisztérium), the establishment of the Centre for Population Policy would serve as an example of the Hungarian government's commitment to the nation's welfare, while strengthening the collaboration between eugenicists and the state.[129]

The panel debating "population policy from a racial point of view" was chaired by Pál Teleki and featured József Madzsar as the main speaker. Hoffmann again took up the role of discussant. Such an arrangement was in itself remarkable considering how different Teleki and Hoffmann's ideas on eugenics were from Madzsar's. Other participants included Ernő Tomor, Ernő Deutsch, Zsigmond Engel, Sándor Bródy and Ödön Tuszkai. In his short welcoming address, Teleki sought to dismiss the concerns expressed during the conference that the methods of racial hygiene advocated by eugenicists "seem cruel, and are indeed cruel against individuals". His support for these measures posited an overlap of interests between individuals, the state and society, interests which, Teleki expected, should be at the centre of any campaign of national eugenics. According to Teleki,

> These measures and endeavours are based on real social ideas and are necessary if we wish to establish a healthy and strong race, and raise a healthy and strong generation. I believe that we very much need to familiarize ourselves with these ideals and carry out these racial hygienic measures when the war is over. Without delay, we need to prepare ourselves for what comes after this world war.[130]

It was the institutional framework provided by the Eugenics Education Society in Britain and the Society for Racial Hygiene in Germany that Teleki celebrated, as it was these societies that served as models for a Hungarian eugenic society. Such a society, Teleki continued, would be able to "promote eugenic endeavours and encourage the public to appreciate the need for a healthy race and raising a healthy future generation".[131]

Madzsar's examination of population policies from a eugenic point of view focused on social and biological degeneration. He remained a follower of the British eugenic tradition and did not hesitate to use it to contextualize his own ideas of population selection and control. He agreed, however, with Teleki's view that eugenics had evolved from a philosophy of selective breeding and control of reproduction to an all-encompassing narrative of racial protection.[132] Certain conceptual differences were obvious, though Teleki and Madzsar agreed in principle on the need for eugenic strategies to improve the racial quality of the population and, at the same time, to protect it from further degeneration.

This was an important component of the eugenic consensus that this congress aimed to promote. Hoffmann was honest when he told the audience that he "could not find anything" in Madzsar's lecture that he could criticize. "I myself", he added, "am rather supportive of the thoughts expressed here." Yet, when he appraised the lecture from the point of view of racial hygiene, Hoffmann noted that it should have been called "population policy from the eugenic point of view". Madzsar did not, according to Hoffmann, "discuss racial hygiene, but rather one aspect of it, called eugenics".[133] Insisting on the difference between racial hygiene and eugenics was by now a constant in Hoffmann's thinking on the subject.[134] He felt that this congress was an important occasion to draw attention to it. "Racial hygiene and eugenics are not the same", Hoffmann noted. "Eugenics looks only at one individual from the point of view of heredity. Racial hygiene, by contrast, examines entire human communities and asks what means can be used to promote their racial development." Hoffmann also believed that Madzsar's explicit focus on "the negative aspects of eugenics", as well as his overt preference for "the elimination of the worthless degenerate elements" in society was justified, but this came with a warning. "The current knowledge of genetics is rather limited", Hoffmann noted, and did not allow for a complete evaluation of the "inferior genetic" traits in society.[135]

Other participants were critical of Teleki's and Hoffmann's imitation of German racial hygiene and population policy. Ernő Tomor, for instance, questioned the suggestion that Hungarian eugenicists and population policy advisers should follow the example set by the German Society for Racial Hygiene and disagreed with other speakers who believed that when it came to eugenics Hungary was "lagging behind other countries".[136] This was to Hungary's advantage, Tomor argued. Hungarian eugenicists could thus "avoid making the same mistakes as the German eugenicists". Moreover, they should not shy away from "criticizing other countries" or from refusing to "follow one, be that even our ally". What Tomor recommended instead was that Hungarian eugenicists "choose from among all intellectual models" the ones they deemed suitable for the Hungarian context. Only after these models were transformed and adorned with "Hungarian features" should they be applied "to our circumstances and made into a Hungarian science".[137]

Whether German racial hygiene was suitable to Hungary or not also encouraged another participant, Ernő Deutsch, to intervene. "German racial hygiene", he remarked, "had not only a biological foundation but a national one as well". Hinting at Hoffmann's comments, Deutsch also argued that German racial hygiene was not simply about "improving the quality and quantity of any community of people, but of the German one in particular". What was needed, according to Deutsch, was a "theory and practice of racial hygiene specific to the Hungarian nation".[138] To this and previous comments on his lecture, Madzsar replied that he did not endorse a racial interpretation of eugenics. Moreover, he did not distinguish between "a German, an English or a Romanian racial hygiene, as German racial theorists and racial politicians" did. "Biology", Madzsar insisted, "has nothing to do with the German version of racial hygiene".[139]

Connected to these debates on eugenics, discussions of racial hygiene and population policy also featured prominently in the next two sessions: one on "the protection of mothers and infants", the other on "child protection". Albert Apponyi chaired the first, with Vilmos Tauffer as the main speaker. Dénes Szabó was the discussant and Sándor Doktor, Mrs Oszkár Szirmai and Aladár Fáy were some of the participants. In his introduction Apponyi concurred with the eugenicists that the "protection of mothers and infants" was "one of the most important branches of public health and social policy".[140] It was also, according to Apponyi, an integral part of the emerging Hungarian population policy:

> The war and the terrible losses suffered by the nation highlighted the importance of population policy. Evidently, population policy influences all measures concerning public health, which deal with the tasks of prolonging the life of an individual and enhancing their vitality. However, the future lies with the babies still to be born and those already born. By saying "babies to be born", I mean prenatal

protection. The protection of those women who prepare to give birth to a child is an indispensable precondition of the very protection of infants, which, in turn, is one of the vital components of the population policy.[141]

For Apponyi, population policy – defined as the science dealing with "the maintenance and the practical reproduction of the nation's human capital" – depended on two factors: increasing the birth rate and the appropriate welfare measures to ensure the survival of mothers and infants. Accepting that the "one-child" system had caused "a decline in vigour and numbers of the best of the Hungarian race", Apponyi encouraged the introduction of "measures such as tax cuts and other benefits provided to large families and increased taxes levied on those who refrain from having a family, thus avoiding contribution to the growth of the nation's human capital".[142] What is more, Apponyi asserted that any such policy had to be grounded in public support. It was critically important to inform and shape "the moral judgement of the nation, because no economic policy and public health policy can be carried out without a healthy moral judgement". Ultimately, this congress had also been organized to develop and advance "direct measures in the field of public health" in order to offer women the most propitious conditions to give birth "to healthy descendants" and protect their own physical condition. Therefore, Apponyi concluded, the protection of infants is inseparable from the protection of mothers: "The two concepts are closely connected and in no way may one of them be considered without the other."[143]

Apponyi's precise introduction was followed by Vilmos Tauffer's detailed presentation of the Stefánia Association's aims and achievements as well as other strategies to protect mothers and infants in Hungary. Ennobled by its civic responsibility to the Hungarian nation, the Stefánia Association provided a network of support across the country. The eugenic and welfare philosophy behind many of its activities, as outlined by Apponyi, Madzsar and others, represented a multifaceted attempt to address Hungary's complex problems with lower fertility and declining birth rates, while at the same time providing a modern concept of motherhood and childhood.[144] The last session on the first day of the congress was organized in a similar fashion to the previous ones. János Zichy, a member of the Upper House of Parliament, chaired the meeting, while Sándor Szana lectured on "child protection".[145]

This host of social and medical issues affected not only individual families but society as a whole. To this end, the next morning, on

27 October, sessions were devoted to "administrative reforms in public health"[146] and "urban and rural health",[147] followed by the afternoon sessions on "health and insurance"[148] and "the hospital".[149] The school, the orphanage, the hospitals and other public institutions all provided opportunities for the health experts to assess the scope and needs of medical reform. Some of these experts, including József Katona, Sándor Szabó, Gyula Filep, Pál Szende, Pál Ruffy, Dezső Hahn, Zoltán Rónai, Gyula Dollinger, László Epstein and László Jakab, prioritized medical involvement in their papers and comments. They emphasized the importance of hygienic care and sanitary management as essential to safeguarding the health of the entire population.

On the last day of the congress, held on 28 October, participants discussed the "importance of food" and "housing reform" from the point of view of public health, hygiene and sanitation.[150] The congress concluded with two sessions: one "on tuberculosis" based on a lecture delivered by Sándor Korányi,[151] the other "on alcoholism", which focused on Lajos Dienes's lecture on the subject.[152] In between these two sessions, Lajos Nékám spoke on venereal diseases.[153] Again, a number of prominent Hungarian physicians and public figures participated in the discussions, including István Bárczy, György Lukács, József Hollós and Mrs Sándor Szegvári.[154] Public health, preventive medicine and social hygiene became intertwined, sharing a distinctive concern with national welfare. At the end of the congress, all participants, based on their own area of expertise, agreed on the introduction of welfare measures in order to strengthen the nation's social and biological foundations.

The scope and depth of this congress on public health illustrates the prominence and public relevance this topic had acquired in Hungary at the time.[155] With the war entering its last year, more and more Hungarians realized that victory was no longer within their grasp. It followed naturally from this conviction that to allow the further deterioration of the Hungarian population was completely irresponsible, resulting in a possible catastrophe. This congress provided, as the organizers clearly believed, the most reliable evaluation of the nation's health during the war. In terms of eugenics, moreover, it revealed the remarkable convergence between various narratives of social and biological improvement that had been formulated and reformulated in Hungary since 1900.

During the last two years of the war, however, there were a number of new, interlocking elements that hinted at the imminent finale, whose disastrous consequences the eugenicists were desperately trying to placate. Most importantly, by stressing the importance of the nation's racial health, the eugenicists – together with public health experts and social hygienists – unhesitatingly called for the official recognition of their vision of social and biological improvement, one they deemed coterminous with the future of the Hungarian state. As will be discussed in the next section, this vision – and the political consensus that it generated – successfully provided the essential conditions for the establishment of a new eugenic society in Hungary.

The Hungarian Society for Racial Hygiene and Population Policy

The congress on public health brought together most Hungarian eugenicists. Among them, there were former members of the Eugenic Committee like Pál Teleki, Géza Hoffmann, Leó Liebermann, Emil Grósz, Vilmos Tauffer and Sándor Korányi. They were accompanied by some of those – like József Madzsar, Zsigmond Engel, Oszkár Jászi, Sándor Doktor, Dezső Hahn, Zoltán Rónai, László Detre, Lajos Dienes, Ödön Tuszkai and Ernő Deutsch – who also participated in the debate on eugenics organized by the Society of Social Sciences in 1911 and the debate on public health organized by the Association for Social Sciences in 1912. There were others, moreover, – like Gyula Donáth and Dezső Buday – who had been regular contributors to debates on social and biological improvement in Hungary since the early twentieth century. Finally, there were participants – like Albert Apponyi, Sándor Szana, Ernő Tomor, György Lukács and Lajos Nékám – who became supportive of eugenics through their involvement with broader programmes of social and national protection during the war. These individuals belonged to a wide variety of professional, political and cultural associations, each reflecting its own identity. On this occasion, however, they chose to adopt a common eugenic language to articulate a critique of the war and its devastating consequences for Hungary.

The congress also facilitated crucial meetings between Teleki, Hoffmann and other eugenicists with respect to their future plans regarding the eugenic movement in Hungary. A week later, on 4 November 1917, Teleki informed Apáthy of their intention to transform the Eugenics Committee into a Hungarian Society for Racial Hygiene and Population Policy (Magyar Fajegészségtani és Népesedéspolitikai Társaság) devoted to the "regeneration of the Hungarian nation's racial and numerical importance". The inaugural meeting was scheduled for 24 November at the Hungarian Academy of Sciences (Magyar Tudományos Akadémia) in Budapest under the patronage of the Academy's president, Albert Berzeviczy.[156]

Apáthy replied on 9 November, sending Teleki a number of suggestions concerning the rationale and functioning of the new society. According to Hoffmann's letter to Apáthy, dated 17 November,[157] Teleki and other eugenicists welcomed Apáthy's involvement. "We took on board as many of your suggestions and criticisms as possible", Hoffmann explained to Apáthy, but "the draft of the statutes" remained largely unchanged "even if in theory it may not be perfect".[158] Hoffmann accepted, however, Apáthy's suggestion that there was as much need for a new eugenic society in the provinces as in Budapest. More importantly, he also agreed with Apáthy that adding "population policy" to the society's name was "unnecessary" and "even harmful". It was, however, "the only way to stop the establishment of a rival Society of Population Policy" in Hungary.[159]

Here Hoffmann alluded to continuous tension that existed between the German Society for Population Policy (Deutsche Gesellschaft für Bevölkerungspolitik)[160] and the German Society for Racial Hygiene. In Austria, for instance, the Society for Population Policy (Österreichische Gesellschaft für Bevölkerungspolitik), also established in 1917, had supplanted the need for a distinct society for eugenics.[161] Hoffmann hoped that by establishing a eugenic society devoted to both racial hygiene and population policy, "the double movement which divided the efforts of race regeneration in Germany was united in Hungary from the beginning. Students of social science and biology worked together in the greatest harmony."[162]

Compared to the more elaborate programme devised by Apáthy for the Eugenics Committee in 1914, the general aim of the Society for Racial Hygiene and Population Policy, as outlined in the draft of the statutes, was simply "to study the Hungarian nation's racial hygiene and population policy". Its work was defined as "first and foremost scientific. Party politics, religious and sectarian issues were excluded."[163] In terms of objectives, the following were proposed:

a. The exploration and diagnosis, together with the scientific examination of the dangers that threaten the body of the Hungarian people – primarily in connection with the decline in the number of births.

b. The use of those tools and methods of research, discussion, thematization, and expert knowledge with which to improve the quality of a population depleted and degenerated by the war and increase its numbers at the same time.

c. The support of such objectives that are aimed at the protection of the existing human material; for example, by decreasing the death rate.[164]

Objectives such as these required a "guiding ethos", defined as "racial self-consciousness", so that the preoccupation with "future generations and with healthy and large families" was articulated in "every aspect of the social, economic, political and moral life". The education of the population concerning the importance of racial hygiene was both theoretical and practical. In addition to providing "expert opinion and knowledge", the Society for Racial Hygiene and Population Policy endeavoured to organize "field trips" and "discussions with business associations", as well as prepare "questionnaires and public debates of the more important issues". A library containing the literature on racial hygiene and population policy was also envisaged, together with the creation of a new journal, the publication of books and the organization of exhibitions.[165] The Social Museum in Budapest, for instance, offered to lend its collection of photographs and materials on eugenics for a national exhibition on war and welfare to be organized together with the Military Welfare Office.[166]

In terms of its internal structure, the Society for Racial Hygiene and Population Policy had an honorary president ("always the then current Minister of Welfare"), a president, vice presidents, a presidential council, an Executive Committee and members who formed the General Assembly. Ten departments were also envisioned, each with its own specific activity and research (see Fig. 18). "Any individual whose character was beyond reproach and believed in the leading principles of the Society" could become a member.[167] In terms of membership fees, contributions and donations the following was stipulated:

> Supporting members are those private individuals and public institutions, who apply for membership and promise to pay for it for at least three years. Private individuals shall pay at least 20 koronas a year, while public institutions should pay 100 koronas membership fee; college students and secondary, higher elementary, and vocational school pupils should pay 2 korona. In justified cases, the Presidential Council can grant exemption from the charge of membership fees.
>
> Founding members are those who donate at least 1,000 korona to the Society. For public institutions this donation should be at least 2,000 korona.[168]

Honorary membership was restricted to foreign scholars, nominated by the Executive Committee but appointed only after approval from the Ministry of Interior. As for life members, those were to be elected from "the general members who achieve special merits in furthering the field

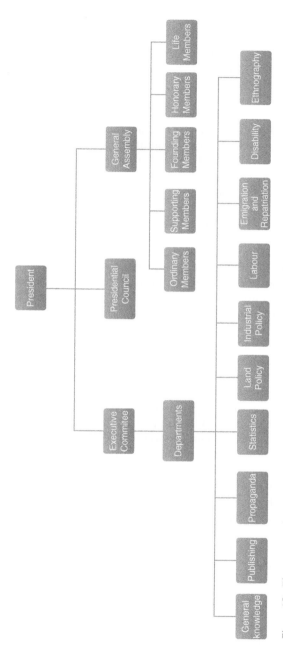

Figure 18 The proposed organization of the Hungarian Society of Racial Hygiene and Population Policy

of racial hygiene and population policy".[169] The General Assembly was to meet once a year, but an "extraordinary meeting could be summoned if the Presidential Council, a tenth of the members, or at least a hundred of them, request it". In terms of responsibilities, the General Assembly was in charge of the following:

> a) Changes in the Society's statutes; b) approve new membership; c) elect the President and the Presidential Council for a six-year tenure; d) approve the budget; e) audit the annual accounts; f) discuss and approve the annual report; g) review the appeals against the decisions of the Presidential Council; and finally, h) decide on the dissolution of the Society.[170]

The inaugural meeting of the Society for Racial Hygiene and Population Policy took place, as planned, on 24 November 1917 at the Hungarian Academy of Sciences in Budapest. Teleki delivered the opening speech, in which he highlighted the crucial importance of racial hygiene and population policy. The war, he noted, was the nation's sobering moment. The country's economic loss was only exceeded by the total number and the racial worth of those Hungarians who died in the war.[171] It was their responsibility to the future generation – Teleki told the audience – to act immediately. To this end, this Society was established as a means by which specialists and government officials could work together towards the survival of the Hungarian race.

The next speaker was Hoffmann, who offered a brief history of racial hygiene, eugenics and population policy. He distinguished between "Gobineau's school of racial anthropology", which was interested in the physical characteristics of human races and "Galton's school of eugenics", which was preoccupied with the improvement of the hereditary qualities of future generations. British and American eugenics conspicuously focused on the individual, while the German racial hygiene concentrated on the community. Hoffmann also reiterated the complementarities between population policy (qualitative and quantitative) and racial hygiene (positive and negative). The new eugenic society, he explained, combined practical population policies with the scientific methods of racial hygiene.

The third speaker was Lajos Nékám, who linked national protection with the prevention against venereal diseases and racial hygiene. Adding to the demographic anxieties about the future of the Hungarian race expressed by Teleki and Hoffmann, Nékám endorsed a scientific management of society. He too highlighted the regenerative potential of eugenics and the medical experts' role as custodians of the Hungarian

Table 2   Members of the Hungarian Society for Racial Hygiene and Population Policy (November 1917–February 1918)

President	Pál Teleki
Vice presidents	Géza Hoffmann
	György Lukács
Presidential Council	István Apáthy
	János Bársony
	László Buday
	Emil Dessewffy
	Gyula Donáth
	Emil Grósz
	Ernő Jendrassik
	Sándor Korányi
	Mihály Lenhossék
	Sándor Szana
Executive Committee	József Szterényi
	Lajos Nékám
Department of Ethnography*	Zsigmond Bátky
	István Györffy
	Kálmán Lambrecht
	Tibold Schmidt
	Gyula Sebestyén
Secretary General	Dezső Laky

* This department was established in February 1918, as discussed in the next section.

race. Finally, the Secretary General, Dezső Laky, presented the Society's objectives.[172] The General Assembly – consisting of 130 personalities of Hungarian scientific, cultural and political communities[173] – then approved the Society's statutes and its leadership (see Table 2).

In addition to Teleki and Hoffmann, only three members of the Eugenic Committee – Apáthy, Korányi and Grósz – remained on the board of the new eugenic society. Also, from the eleven Hungarian societies represented on the Eugenic Committee, only two remained connected: the Turanic Society and the League for the Protection of Children. The new institutional partners were now the Association of National Protection and the Military Welfare Office. Collaborative work was essential, and the Society for Racial Hygiene and Population Policy encouraged a broad collaboration with other societies and institutions in Hungary sharing the same anxiety over the health of the Hungarian nation.

The first meeting of the Presidential Council and the Executive Committee was scheduled for 4 December 1917 and was devoted to the creation of a department dedicated to the promotion of eugenic marriages. An outline was prepared on this occasion by Árpád Bródy, the Secretary of the National Workers' Insurance Fund (Országos Munkásbiztosító Pénztár). Marriage was – as repeatedly discussed in this book – essential to eugenic programmes of social and biological improvement. For Bródy, eugenic marriage dovetailed with population policy, in that both encouraged Hungarian men and women to consider the racial importance of their union. Specialists working in such a department would advise the population against "unhealthy marriages" and promote socially and racially responsible ones. "The nation will slowly die out", Bródy concluded, "if the best principles of eugenics are not applied for its salvation".[174]

When put into practice, the idea of eugenic marriages translated into "dating meetings" between "women who wanted to marry and had no suitable partner" and wounded soldiers. First, "a questionnaire was filled out, and, accompanied by a picture of the man without his name, it was shown" to the woman who had "applied for such information in the notary offices or in the state institutions where the wounded were treated and the soldiers taught".[175] Second, "medical advice was given to couples wishing to marry". The Hungarian Society for Racial Hygiene and Population Policy hoped that a law would soon be introduced to establish the introduction of medical certificates before marriage. "Medical offices to be used for this purpose" were prepared "in different clinics and other public institutions." "The medical rules to be followed by the examiners were worked out in detail by the most competent physicians."[176]

Another immediate activity was to provide Hungarian libraries with the classics of world eugenic literature,[177] and a special request for the provision of books on heredity and eugenics was sent to German publishers. "A range of bookstores", moreover, had agreed to purchase and store books on eugenics, as well as to "circulate to their customers the bibliography on eugenics" prepared by Hoffmann and other members of the Society for Racial Hygiene and Population Policy.[178] Not surprisingly, "a library" designed to satisfy the need for "racial hygiene texts" was also planned – to be located in the society's building on Heltai Ferenc street (Fig. 19).[179]

The Society for Racial Hygiene and Population Policy also attempted to popularize eugenics among professionals and government civil servants. To this effect, Hoffmann designed a six-week course on racial hygiene and population policy for the officers of the Military Welfare Office, and started teaching it on 7 January 1918.[180] In addition, "members of Parliament, the clergy, as well as various relevant professions

Figure 19 The building of the Central Statistical Office, where the Hungarian Society for Racial Hygiene and Population Policy was located between 1917 and 1919 (courtesy of the Library of Hungarian Central Statistical Office, Budapest)

and state departments have all been asked to consider racial hygiene in their line of work, and to assist in its promotion".[181] And yet by themselves, the dissemination of eugenic tracts, the attempts to enlighten officials on the negative effects of the war, along with the introduction of eugenic legislation were not going to solve the difficulty of racial degeneration alone – the population at large had to be persuaded to support these eugenic programmes.

The synthesis between racial hygiene and population policy indicated the seriousness with which the eugenicists treated the impending demographic crisis resulting from the war. It was an all-encompassing agenda. Public institutions such as schools and universities were targeted as ideal locations to disseminate eugenic information about family values and the protection of the nation's racial qualities. Courses were thus organized at the University of Budapest to be taught by Mihály Lenhossék, Lajos Nékám, Sándor Gorka and József Madzsar. The courses dealt with "the history and theory of racial hygiene, its biological and anthropological foundations, the declining birth-rate, and positive and negative racial-hygiene". As a testimony of the increased official interest in these topics, "civil servants have been officially encouraged to attend these lectures". In similar fashion, a number of public lectures and conferences were planned, concerning "the insurance system, the tax system, questions surrounding immigration and emigration, settlement policy, and the one-child practice currently spreading in Hungary's rural areas".[182] Benefitting from the support of the Ministry of War, the Society for Racial Hygiene and Population Policy aimed at expanding its activities throughout the country, beginning with a series of lectures on topics related to eugenics to be organized in the major Hungarian provincial cities.

Although the Society for Racial Hygiene and Population Policy was defined "first and foremost" as a scientific society, it did not claim to be apolitical. For example, one topic that caused a passionate debate was the inclusion of women's right to vote into the government's proposed suffrage legislation. The government wanted to extend the right to "those women who had successfully completed eight years at a state school", to "war widows with at least one legitimate child", and to "those successful women who were active members of a scientific or cultural society".[183] A first meeting to discuss women's suffrage was arranged on 18 January 1918. Members were asked to consider the proposal from the point of racial hygiene and population policy and to communicate their arguments in favour or against it. The Presidential Council was entrusted with presenting the Society's motion in the Lower House of Parliament.[184] As no consensus emerged, a second meeting followed on 30 January.[185]

Eventually, the majority of the participants agreed to endorse the women's suffrage. As Hoffmann noted, "in Hungary, educated elites, particularly in the capital, hold extremely modern views, and to be considered a reactionary is a disgrace". He did, however, oppose the motion, considering that

> the fundamental difference between hereditary value and culture was rarely recognized, which is why the majority of those defending women's political rights employed purely cultural arguments at the meeting – that women's suffrage would result in better education, better child care, and a well-developed social law system – while biological disadvantages were either overlooked or underestimated.[186]

As can be seen from the description of these activities and meetings, the Society for Racial Hygiene and Population Policy was active from the outset. Some of its objectives were never achieved, however. For example, there is no record of how many departments outlined in the draft were, in fact, established. On the other hand, on 15 February 1918 a new, ethnographic section was established, with the following members: Zsigmond Bátky, István Györffy, Kálmán Lambrecht, György Lukács, Lajos Nékám, Tibold Schmidt, Gyula Sebestyén, József Szterényi and Pál Teleki.[187]

Concurrent with this practical programme was the ambition to establish a proper eugenic journal, following the example set by *The Eugenics Review* and the *Archiv für Rassen- und Gesellschaftsbiologie*. In a letter sent to Apáthy on 22 January 1918, Hoffmann announced the decision taken by the Society for Racial Hygiene and Population Policy, the Military Welfare Office, the League for the Protection of Children, the Stefánia Association and the Association of National Protection to establish a new journal named *Nemzetvédelem* (*The Protection of the Nation*).[188] Clearly, Hoffmann had hoped to secure a contribution from Apáthy for the new journal, but more importantly he had wanted – as expressed in another letter from 9 April 1918 – to convince Apáthy to join the editorial board. It was important, he told Apáthy, that the issues of racial hygiene and eugenics received appropriate treatment in the new journal.[189]

The establishment of the Society for Racial Hygiene and Population Policy in 1917 completed the long and convoluted process of nationalization that eugenic ideas of social and biological improvement had experienced since the beginning of the twentieth century in Hungary. By securing political support, the Society also subsumed its practical eugenic agenda within the parameters of a regenerative narrative of nation and state. As Hoffmann triumphantly noted in the *Archiv für Rassen- und Gesellschaftsbiologie* in 1918, "In Hungary, racial hygiene is a mission acknowledged and promoted by the state", and that with the exception of Sweden, "no other European country is so determined to put racial hygienic demands into practice as Hungary is."[190] This achievement, of course, is precisely what makes the Hungarian case so important in the history of international eugenics and modern politics.

As demonstrated in this chapter, during the war eugenics became a useful instrument for supporters of state welfare and those who urged the introduction of a population policy to protect the nation's racial quality and contribute to its demographic growth. Like in other European countries, eugenicists in Hungary developed an interrelated strategy, which promoted racial values and a new nationalist morality, articulated in a medical and social scientific language. With growing scientific and political authority, they also recommended the introduction of schemes in preventive medicine, public health and social hygiene, promoting both eugenic diagnosis and social therapy. As a result, the Hungarian Society for Racial Hygiene and Population Policy – together with other societies and government agencies created after 1914 – proposed innovative schemes for protecting the nation's racial health, seen as the essential component in the future reconstruction of Europe after the war. As will be discussed in the next and final chapter, by the autumn of 1918, a defeated political regime together with a crumbling imperial state, struggled to retain power in Hungary in the face of extensive social and national unrest. In conjunction with other political currents, eugenics generated defensive mechanisms that were simultaneously forming in response to an increasingly unstable domestic and international political life.

7
The Fall of the Race

By 1918, governments across Europe were confronted with the prospect of significantly weakened nations, population displacement, dismemberment of families, as well as social and political unrest. Through numerous publications and public activities, eugenicists had attempted to raise awareness of the severe medical, biological and demographic problems affecting the belligerent countries, while insisting on the need to protect the nation's racial future through immediate population policies. Such interpretations of the nation's social and biological improvement were no longer viewed as marginal, and eugenics had meanwhile acquired a clear and ambitious programme: regenerating the race.[1]

One important rationale for eugenics' increased social, cultural and political influence during the war was its capacity for adapting its precepts to harmonize with each country's specific conditions. In Hungary – as discussed in the previous chapter – this capacity encompassed renewed correspondence between eugenics and the adjacent fields of public health, social hygiene and preventive medicine. The result was an entangled narrative about the protection of the nation, one that assembled the various biomedical ideas and practices, ultimately providing solid foundations for the intense eugenic activity during the last years of the war. These theoretical debates and practical initiatives were carefully orchestrated, because what the eugenicists hoped to achieve ultimately was the political codification of their vision of a healthy national body.

In 1914, few eugenicists in Hungary believed that the Central Powers would lose the war. After two years of conflict, however, it became clear that the biological and social destruction inflicted by the war were far more damaging to Hungary than the positive national fervour it also generated. Hungary's finest individuals were lost during the war, they maintained, in addition to a long inventory of medical, social and economic problems. In this context, the establishment of the Hungarian Society for Racial Hygiene and Population Policy at the end of 1917 represented the culmination of a growing number of public and official initiatives devoted to the nation's social and biological protection. It was also the celebrated moment of eugenics' long-standing yearning for official recognition.

This chapter charts the last episode in the history of early twentieth-century Hungarian eugenics, covering the last year of the war. It is important to appreciate the innovative and laborious eugenic work initiated by the Hungarian Society for Racial Hygiene and Population Policy, both domestically and internationally. From the outset, prominent members like Teleki and Hoffmann proposed a distinct reformulation of eugenic responsibility towards the nation, alongside a reconfiguration of its intellectual sources. If until 1914 Galtonian eugenics was prevalent among Hungarian cultural and social elites, during the war a more racialized version of eugenics, inspired by German racial hygiene, concerned less with the control of reproduction and more with the nation's racial protection, became dominant. The embracing of this eugenic tradition also contributed to a number of international conferences that brought German, Austrian and Hungarian eugenicists together, culminating in the first German–Austrian–Hungarian conference on racial hygiene and population policy organized in September 1918.

An Improbable Supporter

Before discussing the activities of the Society for Racial Hygiene and Population Policy in 1918, it is important to pause briefly and reflect again on the relationship between Apáthy and Hoffmann. Distinguishing between various interpretations of eugenics developed in Hungary was one intellectual objective pursued in this book. Of these, two were particularly important: Apáthy's definition of eugenics and Hoffmann's theory of racial hygiene. As seen, the latter interpretation – based solidly on the German model, but with certain features inspired by the American experience – was also adopted by the Society for Racial Hygiene and Population Policy. This reorientation in the style and content of Hungarian eugenics did not go unchallenged[2] and Apáthy, in particular, reacted immediately.

The announcement of a new eugenic society in Hungary took Apáthy slightly by surprise. To start with, he emphatically opposed the fusion of eugenics with population policy, notwithstanding Hoffmann's assurance that such a decision was strategically important for the unity of the eugenic movement in Hungary. Second, he openly criticized certain aspects of Hoffmann's theory of eugenics, particularly his infatuation with German racial hygiene. In many respects, Apáthy regarded his critical intellectual posture as a vindication of what he believed to be the "proper" definition of eugenics, one which he had elaborated in 1914 and was not replaced by Hoffmann's synthesis of racial hygiene and demographic policy.

In his last eugenic text entitled "A fajegészségtan köre és feladatai" ("The Field and Tasks of Eugenics") published in the early months of 1918 in *Természettudományi Közlöny*, Apáthy chastised Teleki, Hoffmann and other members of the newly established Hungarian Society for Racial Hygiene and Population Policy for their imitation of the German Society for Racial Hygiene. Yet, this profession of loyalty to the indigenous Eugenic Committee was not simply a reflection of Apáthy's exaggerated Hungarian patriotism. Through his own academic notoriety, Apáthy did more than most to promote a eugenic vision of social and biological improvement based on the science of evolutionary biology. He felt the new eugenic society had ignored the rigorous eugenic methodology he had been elaborating for almost a decade. It is important, therefore, to appreciate Apáthy's hesitant support of the Hungarian Society for Racial Hygiene and Population Policy in this context.

Apáthy divided the article into two parts: one dealing with Hoffmann's article "Fajegészségtan és eugenika", the other with his book *Krieg und Rassenhygiene* – both published in 1916.[3] From the outset, Apáthy complained about Hoffmann's superficial treatment of Hungarian contributions to eugenics, including his 1912 book *A fejlődés törvényei és a társadalom* (*The Laws of Evolution and Society*), and the gratuitous eulogizing of German literature on the subject as well. Furthermore, although the Eugenic Committee of Hungarian Societies created in 1914 had not had the most auspicious launch due to the outbreak of the war, Apáthy did not consider it defunct. The Society for Racial Hygiene and Population Policy was, nevertheless, established in 1917, reflecting "the immediate need for eugenic measures". Apáthy thus hoped "that this new society will function successfully and will succeed in motivating the Hungarian government and civil society towards their practical eugenic responsibility".[4] Yet when invitations were sent to the prospective members for the inaugural meeting, they were asked to read Hermann Siemens's 1917 book *Die biologischen Grundlagen der Rassenhygiene und Bevölkerungspolitik* rather than a Hungarian book on the subject, such as his. Affronted by this gesture, Apáthy remarked sternly, "I am convinced that the Hungarian reader would have found a more clear biological terminology in my above-mentioned little book [*A fejlődés törvényei és a társadalom*], but of course less eugenic quackery."

In this article, by way of correction, he offered to provide Hungarian readers clarifications about "eugenic questions and the tasks of practical eugenics".[5] Again, central to this endeavour was Apáthy's distinction between eugenics (*fajegészségtan*) and practical eugenics (*fajegészségügy*):

> *If eugenics is the sum of those scientific findings that demonstrate how it would be possible to replace this generation with one better in body and spirit; then practical eugenics is the sum of those practical measures that indeed could achieve this goal, that is, a generation better in body and spirit than that of today.*[6]

These eugenic hopes for a better future were dashed by the war, which in its wake left the Hungarian nation depleted of its valuable racial members. The majority of those "mentally and physically superior, of whom we could have expected excellent offspring" were killed, Apáthy noted, while "inferior individuals, who will only weaken the physical and psychological average of the coming generations" were left to procreate. With respect to the application of eugenics after the war was over, Apáthy was nothing short of a visionary: "How may we dare to begin nurturing the good crests of a new forest from remaining good scions, when perhaps two or three decades later a new and more devastating war will come and destroy the sturdiest trunks once again?"[7]

These postulates were accompanied by an elaborate biological rationalization, which Apáthy always employed very successfully when discussing eugenics from the point of view of the evolution of living organisms. In this way, he demarcated himself clearly from the view – which he attributed to, among others, Hoffmann – that eugenics was largely a subdivision of racial hygiene. Discussing Hoffmann's *Krieg und Rassenhygiene*, Apáthy maintained that neither Hoffmann nor even Galton offered "an exact definition of eugenics *as science*". He offered one, together with its corresponding taxonomy (in five clusters), repeating in fact exactly what he wrote in his article on "Fajegészségügy és fajegészségtan", published in *Magyar Társadalomtudományi Szemle* in 1914.[8] In this respect, Apáthy's definition of eugenics remained unchanged.

As noted earlier, Apáthy firmly believed that eugenics was essential to civilization and national progress, although he did not hide his disillusionment with the current situation in Hungary. Essentially, he maintained the view that eugenics was not only a biological strategy of racial improvement, but also a moral doctrine whose purpose was the happiness of the whole of society and not just that of some sections of it (or certain groups). Apáthy circumscribed eugenic intervention in society within this social morality: "Practical eugenic measures often require a strong heart and hard faith in human evolution, and this is nothing but self-sacrifice."[9]

This sentiment was reinforced by Apáthy's conviction that a nation-wide programme of eugenic selection would ultimately prove ineffectual. There was, he believed, a dilemma faced by any Hungarian eugenicist preoccupied with "the future of our nation after the war". Eugenicists, population policy experts and politicians were all determined to find the necessary means to encourage population growth. Apáthy, like others, accepted that "the necessary demographic growth required the birth of many children; that is to say, it required many fathers and mothers". He was, however, apprehensive about encouraging any Hungarian man to father children. On the one hand, he believed that "from a eugenic point of view the war had left many prospective fathers unfit for reproduction". On the other hand, he recommended a selection of those deemed unsuitable. And here was the eugenic dilemma: by selecting prospective fathers one reduced their number and thus the number of offspring. By not selecting them one would, though, run the risk of increasing the number of undesired offspring. "How could we break this vicious circle?", Apáthy asked. His final suggestion was exceptionally non-interventionist. No eugenic methods were officially adopted in Hungary in order "to prevent those who were unsuitable to become mothers and fathers. In my opinion, we are obliged to push for such methods; but until then, we must instruct and educate, we must heal and cure, physically and spiritually alike." War and eugenics were ultimately incompatible and, as Apáthy suggested, "no step forward can be made in the field of eugenics until humankind is rid of the present war and the spectre of another war".[10]

Hoffmann was genuinely surprised by Apáthy's article and comments. He sent a letter to Apáthy on 16 April 1918, expressing his distress over the "disagreement" that occurred between them.[11] Contrary to what Apáthy intimated, Hoffmann assured him that he had read his *A fejlődés törvényei és a társadalom* "with very much interest and enthusiasm". Neither was Apáthy's description of Hoffmann as less of a patriot warranted. "I am infatuated with us, the Hungarians", he told Apáthy, but it was indeed true that his Hungarian nationalism did "not prevent" him from "also loving the German race". Connected to this pro-German attitude was Hoffmann's admiration for "the Schallmayer–Gruber–Ploetz–Lenz–Siemens school of racial hygiene". It was this admiration that attracted Apáthy's opprobrium. To Hoffmann's disappointment, Apáthy not only followed a different methodology, but did not even mention the German eugenic tradition. To add insult to injury, Apáthy also called Schallmayer – an author whom Hoffmann "together with the entire German literature on racial hygiene" considered one of the founders of modern eugenics – "a dilettante".[12]

Hoffmann did, however, provide reasons why he did not discuss Apáthy's 1912 book at greater length: first was his lack of expertise in the fields of biology and physiology. Second, he did not agree with Apáthy's definitions of eugenics. In a similar way, Hoffmann believed that the eugenic debate organized by the *Társadalomtudományi Társaság* in 1911 and Tomor's 1915 book on *A sociális egészségtan biológiai alapjai* – although both had "merits" not least of their informative presentation of Galtonian eugenics – offered the "uninitiated, uninformed readers a distorted image of racial hygiene".[13]

This explanation was intended to rectify Apáthy's opinion about Hoffmann's intentions. "It is rather disturbing for me", Hoffmann confessed, "to see that Your Excellency considers me an obnoxious dilettante, who is unwilling to pay attention to Your Excellency's work". He assured Apáthy of his respect and expressed a somewhat belated regret that he did not choose him as his "most beloved teacher instead of Schallmayer and Ploetz".[14] Thus, Hoffmann commended Apáthy's most recent article published in *Természettudományi Közlöny* as expressing ideas "entirely in accordance with my own",[15] even though it portrayed him in such unfavourable terms.

Apáthy's remarks were, of course, excessive. Hoffmann's unfailing popularization of Hungarian research on eugenics abroad was, in fact, unmatched at the time. He was without doubt the most recognizable Hungarian eugenicist. In a comprehensive article entitled "Rassenhygiene in Ungarn", published in the *Archiv für Rassen- und Gesellschaftsbiologie* in 1918, Hoffmann portrayed eugenic developments in Hungary as innovative and comparable with those in countries like Germany, Britain and the USA. This article was less an exposition of Hoffmann's own theories of eugenics and more a synopsis of the concerted efforts by individuals and institutions in Hungary towards the popularization and practical application of eugenics.

The Hungarians may have been "a small nation", wedged uncomfortably between "the East and the West", but – according to Hoffmann – they had "a healthy sense for everything that served the preservation of their race".[16] The great Hungarian dramatist of the nineteenth century, Imre Madách was, for example, depicted as a precursor of modern eugenics, before Francis Galton. According to Hoffmann, the idea of human improvement dominated Madách's 1861 masterpiece, *Az ember tragédiája*

(*The Tragedy of Man*). It was, however, at the beginning of the twentieth century that eugenics began to be discussed scientifically and publicly in Hungary. Examples provided by Hoffmann included the eugenic debates of 1910 and 1911, as well as the writings of Hungarian eugenicists like István Apáthy, József Madzsar, Lajos Dienes and Zsigmond Fülöp, in addition to others – such as Ernő Tomor, János Bársony, Lajos Nékám, Zoltán Dalmady, Dezső Buday, Ödön Tuszkai, Jenő Vámos and Károly Balás – who were equally interested in public health, social hygiene and population policies.[17]

One particular area where eugenics in Hungary developed substantively concerned the protection of mothers and infants. In this context, Hoffmann mentioned the Stefánia Association and its series of lectures on eugenics, organized by József Madzsar and Sándor Gorka. Particularly influential, Hoffmann believed, was Madzsar, who also lectured on social hygiene at the University of Budapest.[18] Finally, issues related to the racial character of the Hungarian nation raised by eugenicists intersected with the activities of other cultural societies, such as the Turanic Society, under the leadership of Pál Teleki and Alajos Paikert. Hungarian eugenicists, in short, were capable of producing valuable scholarship and experiencing the practical application of their ideas. Hoffmann exuded confidence: "On racial hygiene and related areas in Hungary, one could present new reports almost daily."[19]

Another member of the Society for Racial Hygiene and Population Policy's Presidential Council, the renowned medical anthropologist Mihály Lenhossék (see Fig. 16), certainly agreed with Hoffmann.[20] Returning to eugenics in a lengthy article entitled "A népfajok és az eugenika" ("Races and Eugenics") published in *Természettudományi Közlöny* in April 1918, Lenhossék similarly tried to separate eugenics from racial determinism.[21] On the one hand, he placed eugenics and nationalism in distinct categories. "Eugenics is not the same as the love of one's race or patriotism", Lenhossék maintained.[22] On the other hand, he detached eugenics from ideas of racial superiority: "one cannot talk of eugenics when it is suggested that one race, based on its historical rights, should survive or be above other races".[23] Lenhossék similarly condemned all forms of "racial arrogance", indicating that the German, French, British and Yugoslav nationalists had unjustifiably claimed "racial supremacy for their own races, an extreme racial consciousness which the war amplified by inspiring reciprocal hate and antipathy".[24]

In response to these forms of racism, Lenhossék regarded eugenics as an applied discipline capable of making a valuable contribution to national betterment without resorting to racist arguments. "Our national eugenics", he continued, "can have only one goal: to strengthen the nation, increase its numbers, heal its wounds; to support [it] in its struggle, and to favour the unity of this brave Hungarian nation".[25] Lenhossék's interpretation of the complex interaction between policies of racial and social improvement was enhanced by his concluding comment: "The middle class, the vertebral column of the nation must be rejuvenated from its infinite resources and enriched from the treasury of the Hungarian nation and of the pure-blood Hungarian peasantry."[26] Lenhossék thus revived Hungarian social solidarity amidst moments of crisis by deeming eugenic measures necessary to strengthen national vitality and resurrect the national body. It was this conviction that Lenhossék shared with Hoffmann and his programmes of racial hygiene and national protectionism.

It seemed that Hoffmann's direct approach achieved its purpose and an amiable relationship soon resumed between him and other Hungarian eugenicists. Suggestively, Apáthy's letter to Hoffmann from 20 April 1918 was "kind and flattering", and there was an expectation that Apáthy could attend the next meeting of the Society for Racial Hygiene and Population Policy on 29 April.[27] While Apáthy and Hoffmann remained devoted to their respective interpretations of eugenics and racial hygiene, they ultimately shared a dedication to the future of the Hungarian nation and state. In terms of the leadership of the eugenic movement, however, it was Hoffmann who emerged triumphant. His interpretation of racial hygiene and demographic policy were deemed fully compatible with the direction chosen by other prominent members of the Society for Racial Hygiene and Population Policy. Just as important as this theoretical role was Hoffmann's responsibility as the acting vice president. It was this official capacity, together with his eugenic expertise and international contacts, that contributed decisively to the success of the eugenic movement in Hungary in the last year of the war. Due to Hoffmann's consistent efforts, by the spring of 1918, the Society for Racial Hygiene and Population Policy was recognized as Hungary's foremost eugenic organization, providing an all-encompassing narrative of social and biological protection.[28]

Idealism and Realism

As mentioned above, the next meeting of the Society for Racial Hygiene and Population Policy was scheduled for 29 April 1918. The topics to be discussed were: first, "emigration, immigration and repatriation from the point of view of racial hygiene", and second, "rural re-settlement from the point of view of racial hygiene". In his capacity as managing vice president of the Society for Racial Hygiene and Population Policy, Hoffmann prepared the necessary written materials for the meeting.[29]

That the Society for Racial Hygiene and Population Policy decided to pursue these topics further was not entirely surprising. Already in his 1911 book, *Csonka munkásosztály: az amerikai magyarság*, Hoffmann was preoccupied with the racial and eugenic impact of immigration and repatriation on both the local community and the host country. He also prepared a similar document for the Eugenic Committee, one that Teleki had used in his 1917 parliamentary speech on racial hygiene and demographic policy. For eugenicists like Hoffmann and Teleki the ambition to redefine the nation's biological quality lay as much in the political realm as it did in the society at large. They desired to appeal to those racially valuable Hungarians who had left the country to return, while encouraging those deemed racially worthless to emigrate and never return.

Two alternative methods were envisioned: the American immigration system, based on eugenic inspections and racial classifications,[30] or a generalized anti-immigration policy like the one introduced in Germany, according to which those "of different races (like the Slavs, for example), criminals and politically dangerous individuals" were not allowed to settle in the country.[31] In Hungary, the following biological and social criteria were considered applicable: "1) a physical or mental disability; 2) a criminal record; and 3) poverty". Individuals falling within any of these categories were "not wanted". By contrast, those who were "a) intelligent and healthy; b) with impeccable character; and c) wealthy and hard-working" were considered beneficial to the state and the race. Eugenic and racial considerations should also be applied to those who requested Hungarian citizenship: "It is not the foreigner who honours us by becoming a Hungarian citizen, but us who honour him by accepting him in our Hungarian state."[32]

The last four decades of continuous emigration and the current world war were seen as having a devastating impact on the demographic fabric of the Hungarian nation. It was thus hoped that by regulating emigration, immigration and repatriation according to eugenic principles, the contribution to the urgency of Hungary's social and biological regeneration would be achieved. The eugenic control of the borders was not merely an aggressive policy based on a negative and systematic classification of internal "Others" and foreigners, but an aggregate of collective social and biological norms, spelled out by Hoffmann as one of the instruments of the post-war reconstruction.

A similar methodology and set of guidelines animated Hoffmann's second proposal: "rural re-settlement from the point of view of racial hygiene". Again, this was a topic already discussed in 1917 at the Military Welfare Office and raised in Parliament by Teleki and other politicians. What Hoffmann reaffirmed in 1918 was the anxiety over the racial future of the nation caused by the extensive loss of individuals in the war. The assessment of the war as dysgenic was typically expressed in this direct connection between current and future generations. Such language was common at the time, in Hungary and elsewhere in Europe. Hoffmann like other eugenicists believed that valiant Hungarians, unsurprisingly the bravest and the noblest members of the nation, had volunteered first and, as a result, had incurred most of the war casualties. The nation's racial body and its racial vitality were exceptionally weakened as a result. Most importantly, these deaths were bound to have a significant impact on future generations, and "for the future of the race, this was the most painful loss".[33]

Hoffmann thus looked more to the future than to the past, marshalling arguments in defence of rural resettlement as the "most efficient instrument of population policy". The village was envisioned as the medium through which to reinvigorate Hungary's racial strength. From the point of view of racial hygiene, the objectives were: "1. To resettle in the country those possessing valuable physical attributes and psychological qualities that were ideal to pass on to future generations"; and "2. To encourage those who had been resettled to have a large family."[34] These were often, Hoffmann hastened to add, frequently "private matters" – especially reproduction — yet, they could not be left unsupervised by the state. Direct intervention with the nation as a whole by encouraging reproduction was therefore required.

Hoffmann also explained how ruinous the rural resettlement policy would be if it was accomplished with "families regarded as unhealthy and degenerate". Following naturally from eugenic assumptions such as these was the corollary assertion that "the unworthy members of the race" were a "problem" for society, only contributing further to the deterioration of the nation. This aggressive rhetoric was something more than mere regurgitation of the language of degeneration of old. Prefiguring post-war exclusionary eugenic discourses, Hoffmann meticulously described how costly and burdensome the prolonged care of its "degenerate" citizens would be for the state. Eugenic selection and a greater appreciation of children were hence both central to these rural resettlement schemes:

> Only an individual or family can be re-settled that: a) belongs physically, spiritually and linguistically to the Hungarian race; b) does not have a criminal record; c) whose health can be verified by a physician and d) has the required number of children. If both spouses are young having a certain number of children is not a requirement, for it is the purpose of the resettlement policy to encourage them to have as many children as possible.

Hoffmann also suggested economic incentives for child-bearing when he stated that "taxes and other financial obligations" would be "reduced proportionally to the number of children in the family". However, this was ultimately a eugenic contract between the state and its citizens, Hoffmann insisted. Therefore, if the number of five children agreed upon were not conceived, "the resettlement and its benefits would be invalidated".[35]

The main aim was to resettle the Hungarian countryside with racially healthy individuals. The scheme sought to manage the conditions of rural life in order to maintain the large peasant family as the racial resource for the nation. Prospective applicants, therefore, were carefully examined against clearly defined, desirable, characteristics, demanding that the individual and his or her family be, first and foremost, healthy and loyal to the Hungarian race. Those with criminal records were excluded and so were those with hereditary diseases. Childless couples married for more than four years were also discouraged from applying. Preference was given to soldiers who distinguished themselves in the war, widows, orphans and young married couples determined to have children. Considering the pressing concern with the racial qualities of soldiers and veterans, material incentives were devised to encourage them to create families and have children. Taxes, for instance, would be gradually reduced in proportion to the number of healthy children they produced.

In return for property and arable land, families selected for rural resettlement had certain financial obligations, all designed to encourage procreation and demographic growth. Thus, "the mortgage was reduced if after two years a healthy child was born, and continued to be reduced if a child was born every two years, for a duration of ten years". Moreover, and in a move to oppose the one-child practice, a family's fecundity not only assured the sustainability of their contract with the state, but also that the land could be bequeathed to the children. Conversely, the mortgage could increase "with 25 per cent in the third and fourth year if there were no children and with 25 per cent in the fifth and sixth year if there was only one child" and so on. The increase in repayment was calculated according to a child born every two years. These schemes of rural resettlement were more than a means to remedy the demographic decline of the Hungarian population, however. It was also assumed that such policies of population control might correct deficiencies in differential fertility in certain multi-ethnic regions in Hungary.[36]

Militant and interventionist, the eugenic vision emerging from these documents was profoundly radical for its time.[37] By proposing these solutions to a country devastated by the war and troubled by social and ethnic unrest, the Society for Racial Hygiene and Population Policy and the Military Welfare Office were, in fact, proposing a vast programme of social and biological improvement, one aimed at strengthening the Hungarian national community and offering a comprehensive eugenic morality based on a broad notion of the good and productive life for the benefit of the race and the nation.[38] None of these eugenic suggestions were adopted by the Hungarian government at the time, however.[39] Rather, their lasting importance lay elsewhere, as expressions of something more reflective and whose wide significance would become apparent once the war was over: a concept of the Hungarian nation grounded in the regenerative potential of rural life and the peasantry. Eugenicists now looked to the Hungarian villages as repositories of racial authenticity and vitality that had been lost in the cities. To resuscitate a falling race one had to return it to its original matrix and recolonize the space that had once produced it.

The Protection of the Nation

The creation of the journal *Nemzetvédelem* (*The Protection of the Nation*) was the culmination of a series of attempts by the Society for Racial Hygiene and Population Policy to popularize its views to a wider audience, both domestically and internationally (Fig. 20).

Announced already in January 1918 as the common venture of five societies – the National Military Welfare Office, the Child Protection League, the Stefánia Association, the Association of National Protection against Venereal Diseases and, of course, the Society for Racial Hygiene and Population Policy – *Nemzetvédelem* was launched in the summer of 1918 (albeit without the Stefánia Association on the cover). The President of the Association of National Protection, György Lukács was appointed

Figure 20   The first issue of *Nemzetvédelem* (1918)

editor-in-chief, assisted by an editorial board consisting of Elemér Fischer (military welfare), Géza Hoffmann (racial hygiene and population policy), Béla Kollarits (venereal diseases) and Fülöp Rottenbiller (child protection).

According to the editors, the journal "was years in the making. The Hungarian Eugenic Committee first proposed it in 1914."[40] It was, however, the war that brought racial hygiene and population policy into the mainstream of public and political debate: "Our race became self-aware of its own survival."[41] The journal aimed to cover a wide range of topics, from eugenics and racial hygiene to child and maternal welfare, from population policy to protection against venereal diseases. No methodology was favoured in particular; rather, contributions were encouraged from a number of disciplines, including medicine, social sciences, biology or economics, as long as the article addressed the eugenic aspects of national improvement.[42]

The first two issues, which were published together, amassed an impressive 100 pages. As expected, the creation of the Society for Racial Hygiene and Population Policy on 24 November 1917 was fully documented from the outset. The lectures delivered on that occasion by Hoffmann,[43] Nékám[44] and Teleki[45] were published in their entirety, together with a complete description of the aims and the activities of the Society for Racial Hygiene and Population Policy.[46] Reflecting the close collaboration with German racial hygienic and population movements, it comes as no surprise that German eugenicists were invited to write in the journal. Some of them, like Max Christian and Wilhelm Schallmayer, were prominent eugenicists;[47] others, like Ulrich Patz, were largely unknown.[48]

In addition to foreign expertise, *Nemzetvédelem* encouraged a greater appreciation of the Hungarian literature on eugenics.[49] In addition to Hoffmann and Nékám, well-known authors like Dezső Buday, Zoltán Dalmady and Gyula Donáth also contributed. Buday, for instance, wrote about medical examination before marriage,[50] Dalmady discussed the importance of physical education in racial hygiene,[51] while Donáth examined alcoholism through the prism of racial hygiene and population policy.[52] In this respect, the Military Welfare Office remained essential to this process of spreading eugenic awareness among the population and political elites, as well as in terms of introducing practical eugenic policies in Hungarian society. If, for instance, war invalids who were otherwise racially valuable Hungarians were to be saved for future generations, then "practical work" was immediately needed.[53] Another example of practical work was "rural resettlement" and "emigration, immigration and repatriation", both proposals put forward by Hoffmann at the meeting of the Society for Racial Hygiene and Population Policy in April 1918, which were now published in *Nemzetvédelem* as articles.[54]

Moreover, the journal aimed to inform the Hungarian public about the latest publications on eugenics, both in Hungary and especially abroad,[55] as well as about the activities of other societies for eugenics, racial hygiene and population policies in Europe and the USA.[56] Of particular importance was, however, the section on "News" and "Events". It was here, for example, that information about the debates on racial hygiene and population policy carried out in the Hungarian Parliament was provided. One such debate occurred on 16 May 1918, for example, when the respected Jewish politician Ernő Bródy proposed certain improvements in child welfare from the point of view of population policies.[57] Another opportunity arose when the vice president of the Upper House, Albert Radvánszky, challenged his colleagues to consider the importance of the "country's future population policy, after four devastating years of war". To those MPs interested in more specialized discussions about the role of population policy in dealing with the nation's social and biological recovery after the war, Radvánszky recommended the activities pursued to this effect by the Society of Racial Hygiene and Population Policy.[58] That the government was directly involved in supporting population policies was clearly in evidence through two Ordinances issued by the office of the Minister of the Interior, Sándor Wekerle, on 23 June 1918 to the public administration and the clergy, highlighting the magnitude of the social and biological problems caused by the war, from the increase in infant mortality to the spread of venereal and contagious diseases.[59]

That racial hygiene and population policy were discussed regularly in Parliament unmistakably demonstrates that, by 1918, Hungarian eugenicists had succeeded in having their opinion taken seriously by the country's political elites and the government.[60] How successful, however, was the Society for Racial Hygiene and Population Policy in reaching out to ordinary Hungarians? Were they passive, apathetic and indifferent to what the eugenicists and their government were preparing for them, or were they active participants in the eugenic dream of racial rejuvenation? This is exceptionally difficult to gauge from the existing historical records, but if the church and its relationship with its parishioners is any indication of the general population's way of thinking, the dissemination of certain ideas (at that time the church – Catholic and Protestant – was the institution that ordinary Hungarians respected and followed most), this single example is particularly illuminating.

On 14 February 1918, the first Bishop of the Greek Catholic Diocese of Hajdúdorog (established in 1912), István Miklóssy, sent an Ordinance on eugenics to his parishioners, informing them of the Society for Racial Hygiene and Population Policy's activities towards "protecting the future generations of our people from threats and illnesses". Miklóssy believed that the church must assist the state in its attempt to save the "nation from its sins". Essentially, the educational programmes about "race preservation" offered by the Society for Racial Hygiene and Population Policy dovetailed with the broader teachings of the church about moral purity, chastity and marital fidelity.[61] Thus the eugenic transformation of society was decidedly both spiritual and physical.

As it was intensely portrayed in the pages of *Nemzetvédelem*, the eugenicists infused their theories of racial regeneration with images of an ailing nation, threatened by diseases and demographic decline. They provided, however, as discussed above, not only a eugenic diagnosis but also eugenic solutions. The importance of *Nemzetvédelem* for the history of eugenics in Hungary thus cannot be underestimated. It offered the general public a synthesis of the theory and practice of eugenics in Hungary. Among the Hungarian eugenicists associated with the journal, Hoffmann was certainly the most involved. He felt that it was his duty to solidify the official eugenic discourse in Hungary, to systematize its achievements and to establish its intellectual sources, thus illustrating its national and international relevance.[62] It is this internationalization of Hungarian eugenics that will be explored in the next section.

The Internationalization of Hungarian Eugenics

The creation of the eugenic societies in Prague, Vienna and Budapest in 1913 and 1914 did not prompt the emergence of a regional network among Central European eugenicists. It was only during the last year of the war that such initiatives materialized. Considering Hoffmann's membership in the Berlin Society for Racial Hygiene and his extensive contacts with German eugenicists, it comes as no surprise that attempts were made to intensify these relations, both on a personal and on an institutional level. One such example is the popular handbook on abortion and sterilization for physicians and population policy advisers edited by the Berlin psychiatrist Siegfried Placzek in 1918. Hoffmann contributed a chapter on sterilization practices in the USA,[63] next to important German eugenicists, physicians, psychiatrists and public officials like Wilhelm Schallmayer, Friedrich Martius, Wilhelm Strohmayer and Otto Krohne.

More importantly, Hoffmann tried to use his contacts with the German eugenicists in order to initiate a possible collaboration between the Hungarian Society for Racial Hygiene and Population Policy and the German Society for Racial Hygiene. A meeting was planned between several Hungarian eugenicists and Ploetz for the end of January 1918 but did not take place.[64] Yet, the possibility of a collaboration between German and Hungarian eugenicists was not abandoned. As Paul Weindling noted, "the internationalism of the pre-war eugenics movement was replaced by a triple alliance between German, Austrian and Hungarian eugenicists".[65] This was certainly the rationale behind the Austro-German meeting on public welfare held in 1916, for example.[66]

Considering Teleki's connections with the Ministry of War, it was, in fact, deemed more resourceful to use the already existent network that had formed between the German, Austrian and the Hungarian Societies of Fraternal Military Association. The German society (Reichsdeutsche Waffenbrüderliche Vereinigung) was created at the end of 1915, followed by its Hungarian counterpart (Magyarországi Bajtársi Szövetség) on 11 June 1916.[67] Important Hungarian politicians like Gyula Andrássy (who was also the president), Albert Apponyi, Károly Khuen-Héderváry, Sándor Wekerle, Aladár Zichy, István Bárczy and József Szterényi were members of the Hungarian Fraternal Military Association, together with public intellectuals and academics like Mihály Lenhossék, Ottokár Prohászka, Rusztem Vámbéry and Elemér Simontsits. Following the German model, various sections were created in different fields, including medicine, art, sport, science and so on. Leó Liebermann was elected president of the Medical Section.

In addition to reinforcing political and military links between the Central Powers, the Fraternal Military Association also aimed at bringing specialists from these countries together in an attempt to strengthen the exchange of scientific knowledge and cultural experience.[68] The first major conference of the medical sections of the Fraternal Military Association was organized in Berlin between 23 and 26 January 1918. The theme was the ever-pressing "reconstruction of the nation's strength after the war". The Hungarian delegation was quite impressive, consisting of over 50 delegates. The Ministry of the Interior and the Ministry of War were both represented, next to the Red Cross, the Stefánia Association and the Military Welfare Office. The following members of the Society for Racial Hygiene and Population Policy also attended: Sándor Szana, Emil Grósz, Ernő Jendrassik and Sándor Korányi.[69]

The conference was organized around three main themes: "reproduction and maintenance of offspring", "the protection and reinvigoration of the young population", and finally, "the reduction of mortality through the decisive fight of contagious diseases". Otto Krohne and Julius Tandler set the tone of the discussion in their papers on population policy. They both recommended that the state adopt a eugenic management of the population, arguing for an integrative vision of public health and family welfare.[70] Complementing this view, the gynaecologist Vilmos Tauffer discussed the imperative to control infant mortality in Hungary. Although he commended the achievements of the Stefánia Association, he insisted nevertheless that the protection of mothers and infants was, after all, a state responsibility.[71] In view of the pronatalist goals of population policy, the ongoing and concerted eugenic effort to enable the protection and the reinvigoration of the youth was the focus of Gyula Dollinger's paper.[72] According to another Hungarian participant, Imre Dóczi, the eugenic discourse on "the nation's strength" and "the nation's future" thus emerged as a site of state intervention, one that he recommended after the war as well.[73] In keeping with the importance of the nation's health, another set of medical and preventive measures were directed at contagious diseases, as Ernő Jendrassik demonstrated in his paper.[74]

With the end of the war on the horizon, this conference made it clear how "the nation's health" was politically dictated. Post-war reconstruction was the organizing principle around which population policy and eugenics were conceptualized; yet disagreement persisted in terms of their practical implementation. Hoffmann's reflection after the conference is illustrative: "the shadow of the great antagonism between conservative and radical thought which later led to the revolution already disturbed the discussion".[75] These wide divergences notwithstanding, the collaboration between German, Austrian and Hungarian eugenicists continued until the last month of the war.[76] The Medical Section of the Hungarian Fraternal Military Association (A Magyarországi Bajtársi Szövetség Orvosi Szakosztálya) organized the next meeting of the Fraternal Military Association in Budapest between 21 and 23 September 1918. The first two days of the conference were devoted to the fight against malaria, while the last day, 23 September, was dedicated to racial hygiene and population policy.[77]

In his opening address, Emil Grósz – who had replaced Liebermann as the president of the Hungarian Medical Section – expressed his hope that this meeting "would help to re-establish contacts between men of science" and "contribute to international cooperation to the benefit of intellectual culture, civilization and humanity". There was "no other goal to be pursued but the happiness of humanity".[78] Noble as this vocation certainly was, it was too little, too late. In any event, the conference was, by all accounts, an impressive meeting, with over 100 participants. A fragment from a contemporary newsreel that survived depicts the political importance of this conference: the arrival of Archduke Joseph August, with his wife Augusta and the official delegation at the opening ceremony. In view of the ruinous outcome of the war, such involvement was intended to offer the participant an image of national unity and political strength.

Most Hungarian eugenicists – from both the dissolved Eugenic Committee and the active Hungarian Society for Racial Hygiene and Population Policy – attended the meeting (Apáthy, however, was again absent!). Some of them, like László Benedek and Lajos Naményi, would only become eugenicists in their own right after the war. Interestingly, Iuliu Moldovan, the future founder of the eugenic movement in Romania after 1918, was also in attendance. One can only speculate how significant this conference was on the gestation of his theories on national eugenics and biopolitics.

As announced on the invitation, on the third day of the conference, the first meeting of the Hungarian, Austrian and German eugenicists took place at the Hungarian Academy of Sciences in Budapest. The Military Welfare Office and the Hungarian Society for Racial Hygiene and Population Policy organized the event, assisted by the Austrian Institute for the Statistics of National Minorities (Institut für Statistik der Minderheitsvölker) and the Austrian Society for Population Policy (Österreichische Gesellschaft für Bevölkerungspolitik).[79]

Pál Teleki opened the meeting with an outline of the Military Welfare Office's activities, highlighting the need for collaboration between population policy advisers, eugenicists and health experts.[80] The meeting consisted of five lectures. The first speaker, Max von Gruber, discussed "Rassenhygiene und Bevölkerungspolitik" ("Racial Hygiene and Population Policy"). He was followed by Wilhelm Weinberg, who considered racial hygiene from the perspective of the science of heredity ("Vererbungslehre und Rassenhygiene"). It was then the turn of Wilhelm Hecke to speak on "Bevölkerungspolitik in Österreich" ("Population Policy in Austria"). Finally, the last two speakers were Teleki, who examined the relationship between "Rassenhygiene und Bevölkerungspolitik in Ungarn" ("Racial Hygiene and Population Policy in Hungary"),[81] and Hoffmann, who discussed various conceptual differences between eugenics and racial hygiene.[82]

The Society for Racial Hygiene and Population Policy had thus established itself as an important international eugenic society. That this was the case was confirmed only a few weeks later when, under the patronage of the Austrian Ministry for Social Welfare, an Austro-Hungarian Conference on Child Rearing (Österreichisch-Ungarische Tagung über die Fragen der Kinderaufzucht) was convened in Vienna between 13 and 14 October 1918. Two Hungarian organizations were represented on this occasion: the Society for Racial Hygiene and Population Policy and the Stefánia Association. Austrian participants included Julius Tandler, Ignaz Kaup and Clemens Pirquet. The Hungarian delegation was composed of Sándor Szana, József Madzsar, Vilmos Tauffer, Miklós Berend, Zsigmond Engel and Géza Pap. Topics discussed at the conference included child mortality, the protection of mothers, population policy, midwifery, guardianship and social insurance.[83]

These international conferences on eugenics, population policy and race-protectionism, which occurred in 1918, are revealing for a number of reasons. On the one hand, they produced a diversity of interpretations about eugenics, as well as its immediate social and political purposes, which illustrates yet again the importance this topic had acquired in national politics in Hungary towards the end of the war. On the other hand, due to the specific circumstances of the war, which emphasized the growing importance of the issue of national survival and race-protectionism, these conferences demonstrate that the internationalization of eugenics had continued unabated until the last moment of the conflict, not least due to scientific allegiances and personal relations.

The apprehension about the negative outcomes of a lost war had by then, however, penetrated every fibre of political life in Hungary. Eventually, on 18 October 1918 the Hungarian Prime Minister István Tisza admitted that "We lost the war!" Unknown to Tisza, who was assassinated on 31 October, Hungary lost more than just the war. It would soon lose the peace. Adding to the human loss and a profound disruption of family life, there was a chronic economic crisis, and more threateningly for the state – as it turned out – the disruptive nationalist activism of the ethnic minorities. This shifting configuration of politics had profound implications for the eugenic movement. With Hungarian statehood under threat from internal and external forces, the much-needed national solidarity that eugenicists had so vehemently promoted since the beginning of the twentieth century remained an elusive one.

Eugenic Morality

The disintegration of the Austro-Hungarian Monarchy, following the military defeat of the Central Powers, prompted the leadership of the Society for Racial Hygiene and Population Policy to reassess the social and political role of eugenics. Engaging with this issue, in an article written for the new voice of the Catholic revival, *Magyar Kultúra* (*Hungarian Culture*) in early October 1918, Hoffmann praised the Society for Racial Hygiene and Population Policy, not for its practical work in the field of race improvement, but for its attempts to impress upon Hungarian society the need for a new eugenic morality.

Protecting the future of the national community was a collective obligation, Hoffmann argued. Eugenics' cultural role was to provide the Hungarian nation with the motivation for a social and political renewal, and thus enable the race to withstand moments of defeat and humiliation such as those experienced by the country at the end of the war. Hoffmann noted the generalized disintegration, not only of the political system but also of the social structure that – for centuries – had assured the functioning of Hungarian society. In a typically conservative fashion, he criticized "excessive intellectualism and rationalism", which placed "the happiness of the individual" above that of the collective and which now was associated with progressive and radical intellectuals. Hoffmann's disillusionment was social as well as political.

To substantiate his arguments about national unity, Hoffmann invoked the responsibilities towards the racial community. If a sense of collective destiny was not upheld and regularly promoted, the Hungarian race would "eventually decline, despite its great cultural achievements".[84] Hoffmann continued to advocate the eugenic vision of a numerically strong and biologically superior Hungarian race, but his moralizing arguments expressed wider eugenic anxieties about the future of the country. While aware of Hungary's social and economic vicissitudes, Hoffmann's vision of racial hygiene remained nevertheless strongly hereditarian and interventionist. Hoffmann, like many other eugenicists, had long desired the state to supervise the eugenic transformation of society. Sándor Korányi, Hungary's foremost specialist in internal medicine and a member of the Society for Racial Hygiene and Population Policy, made this obvious when he said, "Mere charity in medical care will no longer solve our national problems. The solution can only be a reversal of the state and society: leadership has to be assumed by the state."[85] For eugenicists like Hoffmann and Korányi, the connection between eugenics and the state was essential to social change and national progress.

The role of the state remained, however, contested. One such criticism came from the literary critic and future anthropologist, Erich Frigyes Podach.[86] In a short article entitled "Állami eugenika" ("State

Eugenics") published in November 1918, Podach questioned the eugenicists' endorsement of state interventionism. He acknowledged the efforts of those "physicians and biologists, writers and thinkers", who – gathered "under the flag of eugenics" – were continually striving "to improve the racial quality of the society and its deplorable condition".[87] However, he disagreed with their "requirement of the intervention of the state" in matters of reproduction and demographic growth. Furthermore, Podach described as "Romantic phantasmagorias" eugenic ideas such as "the breeding of geniuses, the introduction of compulsory medical examinations and subsidies for parents of large families".[88] The immediate medical needs of the Hungarian society after the war should be addressed first, prior to the state and the eugenicists initiating a nation-wide programme of social and biological engineering.

According to Podach, to promote selective eugenic breeding and the control of reproduction in a country characterized by dreadful economic and social conditions following a devastating war was simply "naïve". There was a discrepancy, Podach insisted, between the actual material and medical needs of the population and "ideas and thoughts about human engineering and eugenic selection" proposed by "thinkers, professors, and scientists". Few of them understood, in fact, "the murderous atmosphere of today's life" and that "human suffering" was difficult to convey on "a white sheet of paper, in a shiny clinic, or indeed with scientific exactnesses". It was all very well to marvel at the beautiful racial future as described by the eugenicists, but "one could only fight against tuberculosis, venereal diseases and alcoholism by striving for better living conditions".[89]

Podach was not completely opposed to "the noble ideas of eugenics" – he merely advocated a more pragmatic approach to it, one concerned less with racial hygiene and more with the improvement of society. He thus spoke for many supporters of social and biological improvement in Hungary when he declared,

> The illnesses attacking our society at its roots may only end after a great social reform. As long as the war goes on [...] and the state speaks of a demographic increase but does not want to distribute the land, or assure a just salary for the offspring; as long as the young are intoxicated with the drug of the barbarian past instead of being nurtured with ideals of social morality and honest work; as long as women are pushed into prostitution instead of work and motherhood; as long as capitalism makes it possible for idlers to spread the opium of luxury, games, debauchery and lechery, and humankind

seeks money instead of work and honour, we will not be able to eradicate the poisons plaguing the body of our society.[90]

Podach linked the nation's biological body to its social environment, recognizing that the eugenic transformation of society was more than just a biological improvement. Such views were, of course, not new. As discussed in Chapters 1 and 2, eugenics in Hungary was often expressed through the prism of social welfare and public health. In trying to resolve some of Hungary's immediate post-war problems, including the demobilization of soldiers, overpopulation in urban areas – particularly in Budapest – food and housing shortages, widespread diseases as well as disrupted families and communities, eugenicists advocated broadly inclusive social and medical policies. The state was seen as crucial to the implementation of these policies, however. As Podach rightly pointed out, the role of the state in the management of the population's health had been steadily increasing since the beginning of the war. It only grew more influential in the period afterwards.

Eugenics during the Revolutions

Podach's expectation of a "great social reform" was no doubt a reflection of the profound changes experienced by Hungary at the time. This is an often-told story, and only the essential events are worth repeating here: in October 1918, the Social Democrat (Szociáldemokrata Párt), the Civic Radical (Polgári Radikális Párt) and the Independence (Függetlenségi Párt) parties formed the Hungarian National Council (Magyar Nemzeti Tanács) under Mihály Károlyi's leadership. On 29 October, Károlyi became prime minister, promising to offer the much-anticipated social, national and political revitalization of Hungarian society. Shortly after, on 16 November 1918, the Hungarian People's Republic (Magyar Népköztársaság) was proclaimed.[91]

The change of political regime did not diminish the general preoccupation with the health of the population. Again, eugenicists were among those who most actively promoted public health and education reforms. Some eugenic and public health policies discussed in the previous chapters – namely the protection of mothers and infants, venereal and contagious diseases, and alcoholism – figured prominently on the medical and health agenda of the new regime.[92] To some extent, this was to be expected considering that eugenicists affiliated to the Civic Radical Party, like József Madzsar, played an important political role. The eugenic rhetoric of national renewal, together with a social vision of society based on science and rationalism – both widely promoted since the 1900s by the Society of Social Sciences and its journal *Huszadik Század* – were now couched in an official political language, one that the revolutionary reformers hoped would facilitate the transformation of the new Hungarian republic into a modern state.

Once in power, progressive eugenicists and social hygienists attempted to implement many of the eugenic policies they had promoted during the war. Madzsar, in particular, was very active in shaping the public health doctrine of the new regime to reflect his eugenic and medical concerns. On 8 November 1918, the Association of Progressive Doctors (Progresszív Orvosok Szindikátusa) was created, with Madzsar as president.[93] Other active eugenicists included Géza Lobmayer, László Detre, Dezső Hahn, Lajos Dienes, Zoltán Rónai and Tibor Péterfi – all of them had participated in the debate on eugenics organized by the Society of Social Sciences in 1911 (discussed in Chapter 2).[94] Moreover, a newly reformed Ministry of Labour and Social Welfare (Munkaügyi és Népjóléti Minisztérium) was established on 8 December 1918. It was Madzsar, again, who was appointed as the secretary of state in charge of the Ministry's Department of Health.[95]

New socialist programmes of medical care and health reform were devised to solve the recurrent social and biological problems afflicting Hungarian society.[96] Eugenics has remained central to this revolutionary rhetoric.[97] Educational programmes and concrete eugenic proposals included, for example, the fight against venereal and contagious diseases, the protection of mothers and infants as well as the creation of medical agencies to assist the population with marriage and family planning advice.[98] To consolidate these efforts a Scientific Society of Social Hygiene (Szociálegészségügyi Tudományos Társaság) was eventually established in February 1919.[99]

As a result of these developments, the scope of eugenics was substantially enlarged. It continued to focus on the improvement of hereditary health but did so through the wider application of health-care values and preventive medicine. Thus defined, the eugenic emphasis increasingly shifted from biological to social improvement. While social hygiene took prominence in official health discourses, Madzsar viewed it together with eugenics, and deemed both as viable strategies for the nation's post-war reconstruction. In his official capacity, but also in his publications during this period, he strove to transform medical practice and public policy to accommodate the health requirements of Hungarian society.[100] As discussed in the previous chapters, Madzsar seized upon eugenics as a strategy to exert scientific authority over the transformation of society. He viewed social progress through biological determinism; hence his eugenic preoccupation with the social and economic costs associated with caring for the "unfit". He continued, for example, to endorse sterilization as a measure to prevent the reproduction of "genetically inferior" individuals. In promoting such views, Madzsar's eugenics resembled, in fact, the version of racial hygiene proposed by conservative eugenicists like Hoffmann. Both agreed, in the words of the latter, that "Much stress is laid upon the positive side of the question, i.e. the propagation of the fit, and no steps have yet been taken to cut off the propagation of the unfit."[101]

Such overlapping between social and racial hygiene was, however, a common occurrence at the time, in Hungary and elsewhere. As discussed in the previous chapters, these two schools of social and biological improvement had always coexisted. It was during this period (October 1918–March 1919), however, that theoretical differences between them – alongside the ideological divisions within the eugenic movement – gradually became irreconcilable. Other developments in the political life of the country also contributed to further widen the division between social and racial hygiene as well as between socialist and conservative eugenicists. Certainly, religious and political elites associated with the traditional Hungarian social order observed the popular success of the radical left and their intention of social revolution with increased anxiety. This, in turn, fuelled a certain nationalist rhetoric that connected the radical intellectuals, some of them Jewish, with the corrosion of traditional Hungarian values, hence accusing them of undermining the racial vitality of the Hungarian race.[102]

Both social hygienists and racial hygienists claimed, of course, to embody the set of values and principles essential to the survival of the Hungarian nation and state. It is revealing in this respect that, for example, in January 1919, none other than the journal of the radical and progressive intellectuals, *Huszadik Század*, publicized the Society for Racial Hygiene and Population Policy's planned activities for that year.[103] Nation-wide competitions were announced for "the best drawing, photograph, painting or any kind of graphic reproduction *on the subject of racial hygiene*" (deadline 15 February); "the best *poem on the subject of racial hygiene*" (deadline 15 April); "the best *children's tale on the subject of racial hygiene*" (deadline 15 June) and finally "the best *family genealogy*" (deadline 15 December).

Additional information was then provided for the prospective applicants as to what was accepted as a "subject" for a eugenic photograph, poem or tale, namely "the birth of a child; the increase in size of a family, a group of people, or the nation from generation to generation; racial individuality; the responsibility towards the next generation; racial degeneration and so on".[104] Similarly, explanations were offered to those preparing eugenic genealogies of families. "Only information regarding the author's own family" was accepted. Furthermore, the genealogy "must cover at least three generations, and include family trees, together with physical, psychological and biographical data of each individual family member". The rewarded submissions would be published in the journal *Nemzetvédelem*, the children's tale aside, which would be "published in a children's magazine, if possible".[105]

The last issue of *Nemzetvédelem* was published at the beginning of 1919. It is therefore very unlikely that these competitions materialized in any way. These planned activities nevertheless reveal how unrelentingly persistent the Society for Racial Hygiene and Population Policy was amidst these very troubled political times. This commitment is also clearly noticeable in the programmatic article published as an editorial to the 1919 issue.[106] Devoted to "practical racial hygienic measures", the article presented the Society's eugenic and population policy proposals made during the last years of the war, particularly with respect to rural resettlement, the housing problem, emigration and immigration, family subsidies, inheritance laws, child protection, premarital health certificates and so on.

The tone of the article was visibly optimistic. Recent political changes had not deterred the proponents of racial hygiene in Hungary in their quest for social and biological improvement. Hoffmann, in particular, continued to publish in 1919.[107] He was swimming against the tide, however. In March 1919, the Hungarian Republic of Councils (Magyarországi Tanácsköztársaság) was established. Hoffmann provided a succinct description of the new regime's attitude towards the Society for Racial Hygiene and Population Policy:

> During the period of communism the remaining numbers [*sic*] of the eugenic periodical *Nemzetvédelem* were burned as immoral literature, and the eugenic movement was called one of the most dangerous and reactionary things existing. As one of the chief aims of Bolshevism in Hungary was to exterminate the upper-class families and to establish proletariat rule, the anger of the communists against eugenics can be understood.[108]

Undoubtedly, eugenicists like Hoffmann with his racial interpretation of the Hungarian upper classes were not welcomed by Béla Kun's "dictatorship of the proletariat". Was, however, the Hungarian brand of communism incompatible with other forms of eugenics, such as those – for example – advocated by Madzsar and other progressive and left-wing physicians?

Eugenics and Communism

That communism was not – at least not initially – antithetical to eugenics has long been established.[109] The Bolsheviks promised not only the creation of a new society but ultimately that of a new human being.[110] Science was hailed as the foundation of the emerging communist utopia.[111] Correspondingly, an egalitarian vision of social and gender relations was proposed, alongside a glorification of motherhood as the source of social rejuvenation. As the Hungarian feminist writer Zsófia Dénes put it, motherhood "consumes so much energy and physical strength that the Communist order recognizes its fulfilment as much as it does productive labour".[112]

As in the emerging Soviet Union, control of the national body was fundamental to the project of social engineering envisioned by Hungarian communist leaders. The same contagious enthusiasm and proclivity for total social transformation also existed in Hungary. Revolutionaries portrayed the ideal Hungarian society within their reach. A wide range of medical issues – ranging from personal hygiene and mental health care to child welfare and the protection of mothers – were considered essential to the socialist transformation of Hungarian society.[113] Mária Kovács appropriately described these measures as "a curious mixture of elaborate welfare concepts" and "utopian decrees dressed in professionally impeccable minutiae".[114]

Their short time in power notwithstanding (21 March–1 August 1919), the communists' efforts to establish a new Hungarian society consisted of little more than rhetoric. As Frank Eckelt rightly pointed out,

> In no other aspect of its existence did the Hungarian Soviet Republic follow the straight path of Marxist ideology as rigidly as in the orders, proclamations, and directives which were issued by the Revolutionary Governing Council and the various commissariats and which shaped the internal affairs of Hungary.[115]

Thus, following the Soviet model a Commissariat for Labour and Social Welfare (Munkaügyi és Népjóléti Népbiztosság) was created in March 1919, one whose aim was to supervise the social and medical reforms planned by the new regime.[116] Moreover, all private hospitals and sanatoriums were nationalized, allowing state intervention in financing health and medical services.

The Commissariat for Education (Közoktatásügyi Népbiztosság) also organized lectures on general topics related to health and hygiene, and progressive eugenicists like Madzsar, Dezső Hahn and Tibor Péterfi were invited as guest lecturers. To publicize this revolutionary vision of public health and social hygiene, new medical journals were established, like *Az Orvos* (*The Doctor*) and *Egészségügyi Munkás* (*The Health Worker*). Social hygiene was promoted as an antidote to the problems health experts associated with poor social and living conditions of the population. Working towards the same goal, the Social Museum and the Budapest District Workers' Insurance Fund (Budapesti Kerületi Munkásbiztosító Pénztár) organized a Public Health Exhibition (Népegészségügyi Kiállítás) between 20 April and 5 July 1919.[117] The aim was to promote health awareness and disease prevention, as well as to inculcate the values of personal hygiene.

This educational element of health and the associated welfare ideology were certainly consistent with the eugenic goals formulated by the Eugenic Committee and then the Society for Racial Hygiene and Population Policy since the beginning of the war. Not surprisingly, then, when the most important medical authority in the country became the National Council of Health (Országos Egészségügyi Tanács), which was reorganized in April 1919, the eugenicists dominated it. Madzsar became its president, while Leó Liebermann and Dezső Hahn were appointed as vice presidents.[118] Vilmos Tauffer, Lajos Török (former members of the Eugenic Committee), Emil Grósz and Sándor Korányi (former members of both the Eugenic Committee and the Society for Racial Hygiene and Population Policy) were also invited to join (see Table 3). Another eugenicist, Zoltán Rónai, became the People's Commissar of Justice.

Eugenic discourses and practices were adapted and recalibrated to the new political circumstances. With respect to the general health of the population, the Revolutionary Governing Council initiated an ambitious programme which included the reform of the public health administration, the protection of mothers and children, work hygiene, urban health, physical education, health propaganda, sexual education and so on.[119] Concerned hygienists and eugenicists committed to the elimination of social problems by concentrating on improving the moral character of the working classes. Public health campaigns insisted on the intrinsic relationship between sanitation and morality, and revolutionary publications like *Tanácsköztársaság* (*Soviet Republic*) and *Vörös Újság* (*Red News*) regularly reported on the dangers posed to the health of society by alcoholism and prostitution. To this end, health

Table 3 Members of the National Council
of Health

President	József Madzsar
Vice presidents	Leó Liebermann
	Dezső Hahn
Members	Tibor Péterfi
	Aladár Aujeszky
	János Bókay
	István Bugovszky
	Endre Deér
	László Epstein
	Lajos Fejes
	Sándor Ferenczy
	Emil Grósz
	Sándor Korányi
	Frigyes Korb
	József Lévai
	József Marek
	Miklós Matolcsi
	Jenő Pólya
	Béla Székács
	Vilmos Tauffer
	Lajos Török
	István Weiser

exhibitions were organized to inform the workers of health, social and medical issues, such as the one in Budapest in April–May 1919, which focused on child protection, contagious and venereal diseases, alcoholism, work protection and emigration.[120]

Child welfare was exceptionally methodical. For instance, all children between the ages of 6 and 14 benefitted from free medical examination. Furthermore, trained experts were appointed in schools and entrusted with wider responsibilities for the protection of children.[121] Similarly, committees were established to ensure that the children's moral education was linked to their intellectual and physical development. In an attempt to synchronize all child welfare programmes, the government also announced the creation of special homes for children, of dispensaries and sanatoriums, as well as of custodial institutions for the "disturbed, feeble-minded, socially maladjusted and psychotic child".[122] The discursive construction of the child as the future of the nation was thus used to strengthen existing preoccupations with maternal and family welfare.

Long-standing eugenic anxieties about the nation's moral and physical degeneration resurfaced, but this time the communist authorities were determined to take decisive action. The eugenic encoding of the health of the population can clearly be seen not only in the creation of institutions where children deemed "anti-social" were to be "re-educated", but more importantly in the segregation of the mentally defective. Order no. 117 issued by the Revolutionary Governing Council thus stipulated the "compulsory institutionalization of the mentally ill and of those suffering from mental disability regardless of whether they were curable or not". Furthermore, mental patients were now declared to be "under the supervision and care of the Republic of Councils".[123] This eugenic project – no less so than previous visions of a healthy Hungarian nation – involved the re-creation of society according to the cult of science it promoted.

As eugenicists always required, the state had finally become the embodiment of the agencies and institutions concerned with the population's health, while the nation was seen and valued as biologically improvable through eugenic technologies of social and biological selection. The communist reconfiguration of the traditional private sphere and of individual, gender, and religious rights was one important consequence of this transformation. Essentially, the boundary between private and public spheres was blurred by the idea of public responsibility, which came to dominate both. The formulation of "communist hygiene" during Kun's short-lived regime thus provides a powerful example of how eugenics was re-mobilized to provide the scientific rationale for Hungary's social transformation. To ensure the nation's health was as much a eugenic aspiration as a political necessity.

Social democrats and socialist eugenicists believed that many of Hungary's social problems such as inequality and poverty were caused primarily by an unjust political system; yet, they did not dismiss hereditarian factors. The notion of race was no longer at the centre of this rational utopia, but the idea of a healthy Hungarian nation did not disappear: it was recast as "the health of the proletariat". No one articulated it better than Hungary's foremost anthropologist Lajos Bartucz:

> The eugenic improvement of humanity was, until recently, considered a utopian idea. Today, however, political power is everywhere changing hands from a few privileged elites to the people and the working people. This makes the principal concerns that politicians have, to be the protectors of mothers and infants, and of the working people. Therefore we hope that the eugenic utopian idea of creating better and more beautiful, physically and intellectually superior, individuals may soon materialize.[124]

To a large extent, the eugenic logic evidenced in the cultural and health policies proposed and implemented by the democratic and communist regimes was one embedded in previous social and biological narratives of national improvement. Yet, communist discourses on the health of the population also added new ideas to an existing arsenal of eugenic knowledge in Hungary, one that had expanded considerably since the beginning of the twentieth century. Rather than neglect this difficult period in the history of Hungarian eugenics, it is important to consider it in its own terms.

As discussed in this chapter, the period between October 1918 and August 1919 illustrates the way in which Hungarian eugenicists across the political spectrum tried to overcome their ideological differences. The specific nature of Hungarian eugenics was in holding these incompatibilities together, albeit ultimately unsuccessfully. Hope, however, was not completely lost. According to Hoffmann,

> The whole country needs "race regeneration", not so much in the sense of eugenics, but sound morals, order and law, healthy family life, and regard for future generations. Everybody's whole time and energy is devoted to the reorganization of the country and to avert the consequences of a so-called peace. Later, when conditions change, the time will come to continue the work of eugenics.[125]

Perhaps there is no better way to illustrate the return of this conservative and highly nationalistic eugenic rhetoric, fuelled by the counterrevolution, than to invoke Dezső Szabó's iconic *Az elsodort falu* (*The Village Swept Away*).[126] Published in 1919, at a time when Hungary's territorial dissolution had not yet coalesced into the tumultuous political settlement of the interwar years, this novel supplied, in a celebrated literary work, a by-now-familiar version of the eugenic programme of national renewal promoted for almost two decades in Hungary: to return to the village, cultivate the land and breed the future generation a superior race. The far-reaching implications of this longing for racial solidarity and purity was, on the one hand, the expression of a generalized feeling of betrayal and 'back-stabbing' that most Hungarians felt after the war.[127] On the other hand, however, Szabó's novel reflected the national imagining of the future Hungarian. He would be someone who – like the novel's hero, János Böjthe – would personify the redeeming eugenic qualities and the much-needed vitality to re-create the once-strong Hungarian race.[128]

Conclusions

In 1919, Hungary was a country in ruins. The *fin de siècle* optimism described at the beginning of this book had all but vanished. The end of the war also delivered a deathblow to the Austro-Hungarian Empire. In a proclamation issued on 11 November 1918, the last Habsburg Emperor and King of Hungary, Charles I (IV), recognized his Austrian subjects' right to ethnic self-determination. Within a week, the Austro-Hungarian Empire was no more; two separate states, the Republic of German-Austria and the Hungarian Democratic Republic, emerged from its demise.

As noted in Chapter 7, what followed was a quick succession of political regimes, from democratic to communist, and then a civil war, culminating in the restoration of the Kingdom of Hungary under the Regent Miklós Horthy in March 1920. During this revolutionary period, there were several unsuccessful attempts to reach a political solution to the so-called nationality question – including Oszkár Jászi's proposal for the transformation of Hungary into a Danubian federation[1] – in order to prevent the partition of the country. The hopes of saving Hungary's territorial integrity were finally dashed at the Trianon Conference in June 1920, notwithstanding the political ingenuity, diplomatic ability and scientific prestige shown by the leaders of the Hungarian delegation such as Apponyi and Teleki. The latter, in particular, earnestly promoted the links between eugenics, history, geography, economy and ethnography.[2] A short pamphlet, used as propaganda material, entitled *The Consequences of the Division of Hungary from the Standpoint of Eugenics*, is particularly illuminating.[3]

It was no mistake that this explicitly eugenic text was added to the propaganda materials intended to convince international public opinion of Hungary's injustice. True to their credo, Hungarian eugenicists never abandoned the hope of seeing their country survive the war and

236

the subsequent political changes. As demonstrated throughout this book, Teleki was among the earliest and most committed advocates of eugenics in Hungary, as the president of both the Eugenics Committee (1914–17) as well as the Society for Racial Hygiene and Population Policy (1917–1920). Another supporter of the eugenic cause was none other than the head of the Hungarian delegation, Albert Apponyi, also the president of the Stefánia Association for the Protection of Mothers and Infants. Both Apponyi and Teleki relied upon distinctly eugenic arguments to support their claims that Hungary was a well-integrated territorial, social, economic and biological unit.[4]

The eugenic arguments put forward in this propaganda text expressed an idealized Hungarian national ontology, one whose legitimacy was now fundamentally placed in question. "Hungary", it was suggested at the beginning, "never was a unit in an ethnological sense, for the land was always inhabited by different nationalities". No explicit reference was made to the existence of a distinct Hungarian race, one nevertheless destined to reign over the Carpathian basin due to its innate cultural and biological qualities. "In the course of time", moreover, "these nationalities did not remain isolated from each other, but always mingled more or less. The Hungarian-speaking population consists to a great extent of the descendants of the different Slav nationalities, which on the other hand assimilated a great many Hungarians." Therefore, *"the division of the country cannot be justified by ethnographic or racial differences"*.[5] Linguistic differences among Hungary's various ethnic communities were recognized as such, even if they were rarely deemed to be a "sufficient reason" to warrant claims for political and territorial recognition.

The process of assimilation into Hungarian culture was even more pressing when considering that "the *middle and upper class* [are] now speaking Hungarian all over the land". This class had, for centuries, *"assimilated the ambitious elements of all [of] the nationalities"*.[6] This particular historical experience was the very edifice upon which the Hungarian nation had formed and expanded in modern times. This eugenic interpretation further argued that "The middle and upper class, which is a very valuable material from a racial point of view, is therefore of similar origin all over the land. It consists of the body and blood of all [of] the people." Tellingly, this view was then situated within a classical eugenic framework: the correlation between cultural achievement and social status.

There were two important corollaries for this line of reasoning. First, it was assumed that the middle and upper classes would be, "willingly or unwillingly", expelled from their territorial, cultural and political positions. Accordingly, "the *territory given over to the nationalities would lose its intelligence*, the cream of the socially selected" – the loss of which could not easily be remedied in the short term. Second, it was assumed that a smaller Hungary "would on the other hand be *overcrowded with intelligence*".[7] To be sure, such a scenario was not necessarily damaging to the future of the Hungarian nation. In fact, "this [overpopulation] may in the long run perhaps not constitute a serious danger, as after a few generations this valuable human material would be able to find its proper place again, but in the meantime it would have to suffer the miseries of a fallen proletariat".[8]

There was, however, an additional problem, once again formulated in distinctly eugenic terms:

> It is also to be feared that this intelligence crowded in a small territory and not being able to make a decent living would lose its love of life, would not raise enough children and would to a great extent soon become extinct. This danger is all the more imminent as the greatest part of these families move to Budapest, where the conditions for raising children are not favourable at all.[9]

There were further consequences, as well; for example, an overcrowded capital, Budapest, "would be too large for a crippled Hungary: it would become the monstrous head of a dwarf and would *decline rapidly*".[10]

Corresponding to these and other immediate dysgenic effects, any territorial loss and population displacement was also perceived to have significant long-term eugenic consequences. "The increase of the population, very much to be desired after this bloody war, cannot be hoped for", particularly "if the territory is narrowed down and economical [*sic*] regeneration made more difficult." Moreover, "the desire to raise a large family will be greatly reduced, if the possibilities of gaining suitable stations of life are diminished by territorial destruction".[11] These were eugenic symptoms of a widespread population crisis, ominously diagnosed as being capable of eroding Hungary's moral and social values.

While resembling earlier eugenic representations of the nation reviewed in preceding chapters, this short text nevertheless adds a distinctive note of alarm, one no doubt prompted by the country's unfolding tragedy at the time. It is precisely this premonition perceiving any "division of Hungary" as a process carrying with it profound eugenic consequences. The prognosis for the future of the Hungarian nation was thus clearly pessimistic: "The birth rate would rapidly fall and voluntary 'race suicide' would terminate the cruel work of [the] war and of short-sighted, misinformed international politics." But there was also a warning to the future: "Will the Hungarian mothers raise enough children to regain what has been taken by force and injustice? Then this means a new *irredenta*, a new Balkan-question for the coming generations. New wars and revolutions instead of justice and peace."[12]

The inclusion of this text in Hungary's propaganda materials further demonstrates the mainstream political status eugenics had achieved in the country by the end of the First World War. As maintained throughout this book, the development and influence of eugenic thinking in early twentieth-century Hungary should not be underestimated. Nor should eugenics be construed as a *vulgar* trivialization of Darwinism, or as merely *pseudo-science* conducted with insufficient intellectual endorsement from Hungarian scientific and political elites. On the contrary, there is now considerable evidence to argue that between 1900 and 1919 eugenics – in Hungary, as elsewhere in Europe and the USA – occupied a central place within a number of theories about the nation's social and biological improvement. The appeal of eugenics was based upon an ambitious programme, predicated upon constructing a modern Hungarian state, one characterized by a broad conception of public welfare, health care and social hygiene, as well as a nationalist vision of biological renewal. The emergence of eugenics in early twentieth-century Hungary therefore reflects this diversity of opinion, revealing striking interdisciplinary connections between various cultural and political discourses regarding the nation's biological future then circulating.

That eugenics proved successful in Hungary during the first two decades of the twentieth century was, moreover, due to a number of factors. First and foremost, there was a gradual recognition of the importance of applying evolutionary theories of social and biological improvement to Hungarian society. Second, professional experts such as the eugenicists were recognized as indispensable agents in the mediation and propagation of that scientific knowledge. The eugenicists discussed in this book belonged to a generation of accomplished intellectuals, scientists and politicians, all desiring a state committed to the nation's general welfare. Eugenics was fundamentally associated not only with an individual's personal history but – no less – with that of the family, the community, the nation and, ultimately, the race. It was this multiple, interwoven nature of identity that Hungarian eugenicists aspired to delineate, and over which they aimed to exercise their scientific authority between 1900 and 1919. Energetic work in the fields of public health, preventive medicine and social hygiene had been carried out with one overriding objective in mind: the making of a modern

Hungarian nation and state. Yet this was not simply a biological project of breeding stronger and healthier individuals. The Hungarian nation – eugenicists argued – was in need of both physical and moral rejuvenation, as one without the other would ultimately render any long-term eugenic programme of social and biological improvement ineffectual.

The growing interest in eugenics in Hungary after 1900, importantly, arose out of a complex attempt to understand the predicaments of the present, and to ensure a better future for the national community. As an accepted scientific discourse, eugenic thinking deliberately engaged with trends from culture, society and politics; publicized its legitimacy through accepting modernity as inevitable, while also seeking to control its more undesirable and less tolerable elements within Hungarian society. The close relations between medical, natural and social sciences – particularly between medicine, anthropology and sociology – served to underscore the extent to which eugenics attracted a wide range of Hungarian intellectuals in the first two decades of the twentieth century. As the programme of cultural and biological awakening became linked to notions of social change and racial improvement, eugenics increasingly seemed to provide a vital instrument with which to measure the allegedly deteriorating health of the Hungarian nation.

Any privileging of the idea of the nation should not, however, underplay other concerns that shaped, and in some cases even constituted, eugenic preoccupations with social and biological improvement. At the beginning of the twentieth century, eugenics was perhaps the leading social and biological narrative of national improvement for the progressive Hungarian middle class. As recounted in Chapters 1 and 2, eugenics entered scientific and popular discourses at the very moment when various intellectuals and scientists in Hungary were experimenting with Darwinism, positivism and a number of emerging sociologies of progress. Biologists may have been more familiar with theories of evolution and heredity, and physicians with medical pathologies than social commentators, but it was in the realm of the social sciences that eugenicists were first able to command a cultural authority, deliberately carving out a space for the public acceptance of eugenics. As shown in Chapters 2 and 3, the Society of Social Sciences and the Association for Social Sciences played a central role in facilitating this process of cultural adaption and dissemination. For members of these cultural societies, eugenics was first and foremost an eclectic theory of human evolution, one centred on preventive medicine, social welfare and population management.

As part of a wider reform agenda, eugenicists also envisioned a modern Hungarian state engineered and regulated so as to ensure a growing, healthy population that would, in turn, increase national cohesion. As established in Chapter 4, this social and medical appropriation of eugenics – advanced by a vibrant print culture of journals and a growing reading public, especially in urban areas – was gradually replaced by a more deterministic understanding of eugenics, which favoured heredity over the environment. More than any other, the Association for Social Sciences assiduously promoted this interpretation of eugenics. Eugenicists of this sociopolitical orientation argued for racial instead of social welfare, championing a biologically determined explanation of Hungary's historical evolution. Improving the health of the population was often couched in a racial language that gained growing acceptance before and especially during the First World War. Eugenic narratives of social and biological improvement became closely connected.

Just as elsewhere at the time, the debate between the social (nurture) and the biological (nature), hence also defined the development of Hungarian eugenics. To be sure, the theoretical focus of specific eugenic arguments involved in this debate varied according to the professional and intellectual background of the individual author; yet, nevertheless a unified corpus of eugenic ideas gradually materialized, accounting for the various indicators of social and biological degeneration in Hungary – most notably alcoholism, venereal and contagious diseases. The eugenic descriptions of these simultaneous social and medical problems then formed the basis for systematic attempts to fashion Hungarian society according to health norms, demographic predictions and idealized representations of the family and gender roles. With respect to the latter, increasingly this eugenic vision of a modern state assigned a new importance to Hungarian women and to their reproductive role in society. As observed in Chapter 4, a critical reappraisal of the role of motherhood was central to the eugenic project. Hungarian eugenicists viewed women as mothers of the nation, often on the basis of traditional family values serving to justify their social, economic and political subordination. Such views were naturally rejected by Hungarian feminists, who passionately campaigned for women's emancipation – while still revealing an interest in the eugenic promise regarding rational reproduction and interventionist forms of motherhood.

The process of articulating an institutionalized eugenic movement in Hungary was difficult to achieve but ultimately successful. Hungarian eugenicists benefitted greatly from existing scientific, cultural and personal networks and professional exchange, often expressed in the form of a learned society or an association, such as the above-mentioned Society of Social Sciences and the Associations for Social Sciences. These societies transcended academic disciplines and favoured an open dialogue regarding Hungary's national future, which ultimately allowed for a distinct eugenic community to emerge at the beginning of the twentieth century. As highlighted in Chapters 2 and 3, the institutionalization of eugenics in Hungary was preceded by two significant public debates: that on eugenics organized by the Society of Social Sciences in 1911, and another on public health convened by the Association for Social Sciences in the next year. These key events brought together eugenicists from a variety of cultural and political orientations, reinforcing both the conceptual strength and the wide acceptance of eugenic ideas in Hungary. Importantly, the Association for Social Sciences now played a leading role in shaping the development of eugenics as an organized movement – culminating in the creation of the Eugenic Committee of Hungarian Societies in January 1914.

As detailed in Chapter 5, the outbreak of the First World War on 28 July 1914 had a significant impact on the evolution of eugenic thinking in Hungary, highlighting its immediate political relevance. Eugenicists pointed out that virtually all of the demographic, social and biological transformations that they had been observing since the beginning of the twentieth century had escalated during the war. They appealed to political elites and the government to use eugenics as a means of promoting social and biological protection, highlighting the need for a rigorous health policy, which was able to integrate explicit eugenic principles. During the war, in fact for the first time, eugenics extended beyond scholarly publications to infiltrate wider Hungarian political discourses. An intense debate ensued not only about the nation's health and protection but, by the latter stages of the First World War, about Hungary's biological survival as a country.

The war not only reconfigured national politics in Hungary, but also conscripted eugenics in the battle for national regeneration. A sense of racial endangerment became widespread and the state itself was construed as the guardian of the nation. It was increasingly argued that the health of the population should be protected through state intervention and regulation. Equally important, such developments also assisted the transformation of eugenics into a specialized discourse about social and biological improvement. Not only did Hungarian eugenicists establish two eugenic societies – in 1914 and 1917, respectively – but many leading figures like Géza Hoffmann and Pál Teleki enjoyed an international reputation, contributing extensively to the development of eugenics in both Hungary and abroad.[13] Moreover, there were eugenicists who during the First World War succeeded in influencing wide-ranging political decisions in Hungary; then yet again, this was partly because some of them also held political offices and were members of various governments.

Offering crucial intellectual underpinnings for these ideas was the assimilation of eugenic models emanating from other European countries, particularly Britain and Germany. The Hungarian case convincingly illustrates the international reach of eugenic ideas between the end of the nineteenth century and the First World War. Yet this is not simply a story of imitation and appropriation. Eugenic developments in Hungary were comparable with those in other European countries at the time – particularly Germany and Britain – but they invariably exhibited specific characteristics. Most fundamental amongst these is the coexistence within the same movement of eugenicists situated across the political spectrum – republican socialists, liberals and feminists, together with religious conservatives and royalists. Both before and after the war, it merits restating, there were many active eugenicists of Jewish origin in Hungary. This was because eugenics was a modern, rational and scientific interpretation of social and biological improvement that brought any number of different social, religious and political groups together. Engaged in the same debates over the future of the country, they were brought into direct contact in exploring common ways to bring about positive eugenic change. Radical socialists, assimilated Jews, conservative aristocrats and racial nationalists may have viewed social and national welfare quite differently, but there was a shared recognition nonetheless holding that the modern transformation of Hungarian society was both desirable and imminent.

As a cluster of social, biological, and cultural ideas of national improvement, eugenics also promoted a regenerative racial programme. Hungarian eugenicists located a dominant social or ethnic group as the repository of the nation's racial qualities. Correspondingly, they then deployed biological, social, cultural and political means in order to assess and eliminate those factors seen as contributing to the nation's perceived degeneration. Ideas of racial protectionism were clearly compelling at the time, and most eugenicists believed that national success and the cultivation of superior racial qualities were closely intertwined. As repeatedly argued throughout this book, the attempted biological renewal of the Hungarian national community through eugenic means further reflected the importance nationalism was afforded by cultural and political elites between 1900 and 1919. Eugenicists on the left no less than those on the right may have fundamentally disagreed about the importance of race in shaping human history, but they agreed nevertheless on the Hungarian nation's hegemonic future in the Carpathian basin.

Like much of Europe, the First World War proved to be the catalyst that effectuated the shift of focus from society to the nation. This was so much so that, by 1917, the idea of a wounded national community became commonplace in eugenic rhetoric and, increasingly, practice. As the war progressed, so too did its ravaging effects on Hungarian society. The symbol of a nation-under-siege became the centre of political debates about eugenic welfare. The armoury of state interventionism during the war – the protection against the spread of venereal diseases, the social reintegration of invalid and injured soldiers, the protection of mothers and orphans and so on – was systematically deployed in promoting an increasingly nationalistic eugenic cause. State interventionism was pivotal in establishing eugenic programmes of social and biological protectionism, as well as in underwriting the eugenic vision of a healthy Hungarian nation. The creation of a number of eugenically oriented societies and associations such as the Stefánia Association for the Protection of Mothers and Infants, the Association of National Protection against Venereal Diseases, the National Military Welfare Office and, especially, the Society for Eugenics and Population Policy were to have momentous implications. In both practical and symbolic terms these institutions pointed towards a definitive integration of eugenics within official Hungarian politics.

Following the end of the First World War, Hungary experienced unprecedented social and national changes. The country's new political orientation led to unavoidable tensions within the eugenic movement. During the revolutionary years, however, eugenicists attempted to bridge these divergent cultural and political perspectives on Hungary's transformation into a modern state by focusing on social hygiene as a mediator between socialist eugenics and conservative racial hygiene. Even under Kun's communist regime, eugenics remained deeply embedded within the fabric of debates regarding social, medical and public health reform – especially in proposed "treatments" for the nation's body. Insofar as general support for eugenics came in response to the social and biological devastation caused by the war, a principal challenge for the Hungarian state was less in initiating eugenic policies than in sustaining them. This was to be a recurrent problem under the democratic and communist regimes of 1918 and 1919, when eugenics became part of a new technocratic credo – heralded as a rational means of modernizing the Hungarian society and state. As emphasized in Chapters 6 and 7, alongside their struggle for political and social legitimacy, eugenicists attempted health reforms amid particularly dire circumstances: a country ravaged by war and torn between conflicting social and ethno-national movements.

By 1919, the fear that the Hungarian nation was becoming the site of racial dissolution took on a new sense of urgency. With the country's future seemingly at stake, eugenic ideas of social and biological improvement were again recruited in the reconstitution of a collective Hungarian identity. The aforementioned eugenic pamphlet, distributed to the international community before the Peace Treaties, was not only to be the last episode in the history of eugenics charted in this book, it was simultaneously the utopian expression of a failed state-building project. The end of the war in 1918, the two revolutions that followed – democratic and then communist – as well as the counterrevolution and the consequences of Trianon, were to herald fundamental changes for the future development of eugenics in Hungary. The eugenic vision of a modern Hungarian state, nevertheless, one reflecting the historical right and the racial potency of the Hungarian nation, was now fatally impaired. Yet it did not disappear. Faithful to its promise of national renewal and biological regeneration, eugenics re-emerged – stronger than ever – during the interwar period and the Second World War.

Epilogue

The Swedish scientist Pontus Fahlbeck was one of the acquaintances Pál Teleki made at the meeting of the International Society for Racial Hygiene, in Dresden in August 1911. They had stayed in contact and, at the end of the war, Fahlbeck arranged for a letter from Teleki to be delivered to the president of the Eugenics Education Society, Leonard Darwin. *The Eugenics Review* published it in January 1919.

There was something both heroic and hopeless about this letter. In the name of the "Hungarian Society for Eugenics (Race-Hygiene)", Teleki first protested "against the intended partition of Hungary, not in the interest of the Magyar or any other race, but for the progress of the entire human race". Next, he courteously invoked "the great heritage which the genius of Francis Galton left to mankind", while noting that the founder of eugenics "always maintained that the upper social strata comprise the best of the race". Across Hungary the "upper social classes" were Hungarian. To partition the country would only contribute to the destruction of this social class, Teleki believed. While "the people belonging to other nationalities" could "well exist under the Hungarian government", it was "evident that the [Hungarian] educated and higher social classes" could not survive "in an alien environment".

This eugenic interpretation of Hungary's social and national order flowed naturally from Teleki's unfailing belief in Hungarian cultural and racial superiority. To assume that Hungary could be partitioned in any way was just unimaginable. As both a representative of the Hungarian "upper social strata" and the president of the Hungarian Society for Racial Hygiene and Population Policy, Teleki therefore asked

the President of the Eugenics Education Society to take action as far as he is able and willing to do so, in order that competent individuals and the public in England consider the special situation of Hungary and do not demand a solution to the nationality question which destroys the best families of the nation.

In the name of shared eugenic values, Teleki hoped to generate a change in the prevalent political attitude in Britain against Hungary. His gesture was a desperate, noble, but ultimately, futile one.

The history of Hungarian eugenics portrayed in this book concluded as it began: with a letter to a British eugenicist. In 1907, a confident Oszkár Jászi notified Francis Galton of his admiring Hungarian supporters. Only 12 years later a distressed Teleki invoked Galton to his successors, hoping that they – more than others – would understand the eugenic rhetoric of class, race and empire. They did, of course. Yet, no British eugenicist was willing to act on it. As the editors of the journal cautioned the reader, "the question discussed here is largely political, and on it we express no opinion".[1] No eugenic society, in Britain or elsewhere, could avert the national tragedy that was to engulf the Hungarian eugenicists at this time.

Biographical Information

With few exceptions, posterity has not been kind to Hungarian eugenicists who are discussed in this book. Géza Hoffmann died prematurely in 1921 from an infectious disease. István Apáthy followed in 1922, having suffered the humiliation of being imprisoned for one-and-a-half years in Transylvania by the Romanian authorities, after initially sentencing him to death.

Moreover, the purges of Hungarian universities of communists and Jews, following the collapse of Kun's communist regime, significantly affected the eugenicists on the left of the political spectrum. József Madzsar, Tibor Péterfi and Lajos Dienes, for example, were deposed of their individual professorship at the Faculty of Medicine in Budapest and eventually left the country. In 1919, Jenő Vámos emigrated from Hungary as well. Furthermore, Leó Liebermann, Rezső Bálint, Sándor Korányi and Emil Grósz were similarly harassed during the early 1920s due to their Jewish origin. Others, like René Berkovits and Ernő Deutsch, survived the interwar period only to fall victims to the Nazi Holocaust. Madzsar died in the Gulag.

Due to space restrictions in this text only Apáthy and Hoffmann benefit from having an adequate biographical analysis. However, it is important to note that at least some selective biographical information is provided for the other eugenicists who are discussed in this book, as well.

Angyal, Pál (Pécs, 1873 – Budapest, 1949): lawyer and university professor.
Apáthy, István (Pest, 1863 – Szeged, 1922): zoologist and neurologist.
Count Apponyi, Albert (Vienna, 1846 – Geneva, 1933): social scientist and politician.
Balás, Károly (Balassagyarmat, 1877 – Körmend, 1961): economist and university professor.
Bálint, Rezső (Budapest, 1874 – Budapest, 1929): Jewish neurologist, psychiatrist and university professor.
Bársony, János (Nagykároly, 1860 – Budapest, 1926): obstetrician, gynaecologist and university professor.
Berend, Miklós (Nagykálló, 1870 – Budapest, 1919): Jewish paediatrician.
Berkovits, René (Nagyvárad, 1882 – ? 1944): Jewish physician.
Buday, László (Pécs, 1873 – Budapest, 1925): statistician and university professor.
Dalmady, Zoltán (Budapest, 1880 – Budapest, 1934): balneologist and pioneer of sports medicine.
Detre, László (Nagysurány, 1874 – Washington, 1939): Jewish bacteriologist, histopathologist and microbiologist.

Deutsch, Ernő (Budapest, 1872 – Budapest, 1944): Jewish paediatrician.
Dienes, Lajos (Tokaj, 1885 – Boston, 1974): biologist and bacteriologist.
Doktor, Sándor (Beregrákos, 1864 – Keszthely, 1945): physician.
Donáth, Gyula (Baja, 1849 – Budapest, 1944): Jewish neurologist and anti-alcohol campaigner.
Fenyvessy, Béla (Budapest, 1873 – Pécs, 1954): Jewish social hygienist and university professor.
Gaál, Jenő (Pusztagerendás, 1846 – Budapest, 1934): economist and social scientist.
Geőcze, Sarolta (Bacskó, 1862 – Budapest, 1928): sociologist, teacher and feminist.
Glücklich,Vilma (Vágújhely, 1872 – Vienna, 1927): feminist and teacher.
Grósz, Emil (Nagyvárad, 1865 – Budapest, 1941): Jewish ophthalmologist and university professor.
Count Dessewffy, Emil (Szent-Mihály, Turóc megye, 1873 – Királytelekpuszta, Zemplén megye, 1935): politician and diplomat.
Fülöp Zsigmond (Budapest, 1882 – Budapest, 1948): science writer and translator.
Hahn, Dezső (? 1876 – ? 1921): Jewish social hygienist and venereologist.
Hajós, Lajos (? – ? 1930): neurologist and university professor.
Hoffmann, Géza (Nagyvárad, 1885 – Budapest, 1921): diplomat.
Hutyra, Ferenc (Zsibra, 1860 – Budapest, 1934): veterinarian, pathologist and university professor.
Jászi, Oszkár (Nagykároly, 1875 – Oberlin, USA, 1957): Jewish politician and social philosopher.
Jendrassik, Ernő (Kolozsvár, 1858 – Budapest, 1921): neurologist and psychiatrist.
Baron Korányi, Sándor (Pest, 1866 – Budapest, 1944): physiologist, internist and pathologist.
Kozáry, Gyula (Miháld, 1864 – Kőszeg, 1925): Roman Catholic theologian and teacher.
Kovács, Alajos (Gyöngyöspüspöki, 1877 – Budapest, 1963): statistician.
Lenhossék, Mihály (Pest, 1863 – Budapest, 1937): medical anthropologist, anatomist and neurologist.
Liebermann, Leó Sr (Debrecen, 1852 – Budapest, 1926): Jewish social hygienist, immunologist and chemist.
Lobmayer, Géza (Budapest, 1880 – Budapest, 1940): urologist and surgeon.
Madzsar, József (Nagykároly, 1876 – Arkhangelsk, Soviet Union, 1940): physician and politician.
Nékám, Lajos (Pest, 1868 – Budapest, 1957): dermatologist, bacteriologist and university professor.
Count Edelsheim-Gyulai, Lipót (Salzburg, 1863 – Budapest, 1926): lawyer and child-protection activist.
Paikert Alajos, (Nagyszombat, 1866 – Budapest, 1948): agronomist and lawyer.
Péterfi, Tibor (Dés, 1883 – Budapest, 1953): Jewish biologist and physiologist.
Rónai, Zoltán (Budapest, 1880 – Brussels, 1940): lawyer and politician.
Schuschny, Henrik (Prague, 1857 – Budapest, 1929): hygienist and school physician.
Schwimmer, Rosika (Budapest, 1877 – New York, 1948): Jewish feminist campaigner.

Baron Szterényi József (Lengyeltóti, 1861 – Budapest, 1941): Jewish politician.

Szana, Sándor (Temesvár, 1868 – Budapest, 1926): physician and paediatrician.

Szentpéteri Kun, Béla (Monostorpályi, 1874 – Debrecen, 1950): lawyer and university professor.

Baron Szentkereszty Béla (Nagyszeben, 1851 – Budapest, 1925): politician.

Tauffer, Vilmos (Kolozsvár, 1851 – Budapest, 1934): obstetrician and gynaecologist.

Count Teleki, Pál (Budapest, 1879 – Budapest, 1941): geographer and politician.

Tomor, Ernő (Kolozsvár, 1884 – sometime after 1942): internist and tuberculosis specialist.

Török, Lajos (Budapest, 1863 – Budapest, 1945): Jewish dermatologist.

Vámos, Jenő (Miskolc, 1882 – Budapest, 1950): Jewish physician and veterinarian.

Notes

Prologue

1. As Milly (Mrs J. C. Baron Lethbridge) reminisced, Galton would write on Monday or Tuesday and she would reply on Friday. See Karl Pearson, ed., *The Life, Letters and Labours of Francis Galton*, vol. IIIB (Cambridge University Press, 1930), 447.
2. Ibid., 47. The journal in question was *Huszadik Század* (*Twentieth Century*).
3. O. Jászi's letter to F. Galton, dated 26 November 1907. University College London, Galton Papers, Fond 267/3.

Introduction

1. Ödön Frank, ed., *A millenniumi közegészségi és orvosügyi kongresszus tárgyalásai* (Budapest: Franklin, 1897).
2. József Jekelfalussy, *The Millennium of Hungary and Its People* (Budapest: Pesti könyvnyomda, 1897), i.
3. Lee Congdon, "Endre Ady's Summons for National Regeneration in Hungary, 1900–1919", *Slavic Review* 33, 2 (1974): 302.
4. See Susan Zimmermann, *Prächtige Armut: Fürsorge, Kinderschutz und Sozialreform in Budapest: Das "sozialpolitische Laboratorium" der Doppelmonarchie im Vergleich zu Wien 1873–1914* (Sigmaringen: J. Thorbecke, 1997).
5. See István Laczkó, *A magyar munkás- és társadalombiztosítás története* (Budapest: Táncsics Kiadó, 1968), and András Sipos, *Várospolitika és városigazgatás Budapesten 1890–1914* (Budapest: Fővárosi Levéltár, 1996). For a recent discussion, see Susan Zimmermann, *Divide, Provide and Rule: An Integrative History of Poverty Policy, Social Reform, and Social Policy in Hungary under the Habsburg Monarchy* (Budapest: CEU Press, 2012).
6. See, especially, Péter Hanák, *The Garden and the Workshop: Essays on the Cultural History of Vienna and Budapest* (Princeton University Press, 1998).
7. "Felolvasó ülés: Kende Mór, A degenerációról", *Egészség* 15, 1 (1901): 14–15. This short lecture introduced some of the arguments discussed at length in Kende's book on *Die Entartung des Menschengeschlechts, ihre Ursachen und die Mittel zu ihrer Bekämpfung* (Halle: Carl Marhold, 1901).
8. See Zsigmond Gerlóczy, ed., *Jelentés az 1894. szeptember hó 1-től 9-ig Budapesten tartott VIII-ik nemzetközi közegészségi és demografiai congresusról és annak tudományos munkálatairól*, vol. 7 (Budapest: Pesti könyvnyomda, 1896), 597–737. Speakers included the psychiatrist Richard von Krafft-Ebing; the military physician Philipp Peck; the neuropathologist Gyula Donáth; Francis Warner, a consulting physician to the London hospital; and the psychiatrist Gusztáv Oláh.
9. István Scherer, ed., *Nemzetközi Gyermekvédő Kongresszus Naplója* (Budapest: Pesti könyvnyomda, 1900).

10. See, for example, Ernő Jendrassik, "Az átöröklődő idegbajok (Elsődleges degeneratiók)", in Árpád Bókay, Károly Kétli and Frigyes Korányi, eds, *A belgyógyászat kézikönyve*, vol. 6 (Budapest: Dobrowsky, 1900), 994–1018, and idem, "Mi az oka annak, hogy több fiú születik, mint leány? És más öröklési problémákról", *Magyar Orvosi Archivum* 12 (1911): 331–43. See also Andrew Czeizel, "A Historical Evaluation of the Doctrine of Heredodegeneration", *Orvostörténeti Közlemények* 24, 4 (1978): 157–79, esp. 157–63.

11. See Marius Turda, *Modernism and Eugenics* (Basingstoke: Palgrave Macmillan, 2010), 13–39.

12. Michel Foucault, *The History of Sexuality*, vol. 1 (New York: Vintage Books, 1990), 118.

13. My interpretative strategy evokes similar attempts to place the history of science within a particular national context by, for example, Michael D. Gordin and Karl Hall, "Introduction: Intelligentsia Science Inside and Outside Russia", *Osiris* 23 (2008): 1–19, and Mitchell G. Ash and Jan Surman, eds, *The Nationalization of Scientific Knowledge in the Habsburg Empire, 1848–1918* (Basingstoke: Palgrave Macmillan, 2012).

14. In viewing eugenicists as important contributors to Hungarian national culture I am indebted to the interpretation provided by Stefan Collini, *Public Moralists: Political Thought and Intellectual Life in Britain, 1850–1930* (Oxford: Clarendon Press, 1991).

15. Francis Galton, "Eugenics: Its Definition, Scope and Aims", in idem, *Essays in Eugenics* (London: Eugenics Education Society, 1909), 42.

16. Francis Galton, *Memories of My Life*, 2nd edn (London: Methuen, 1908), 321.

17. C. W. Saleeby, *The Methods of Race-Regeneration* (London: Cassell, 1911), 45.

18. See Alfred Ploetz, *Grundlinien einer Rassen-Hygiene. Die Tüchtigkeit unserer Rasse und der Schutz der Schwachen*, vol. 1 (Berlin: S. Fischer, 1895). For a discussion of Ploetz's definition of racial hygiene, see Sheila Faith Weiss, "The Race Hygiene Movement in Germany", *Osiris* 3 (1987): 199–203, and Paul J. Weindling, *Health, Race and German Politics between National Unification and Nazism, 1870–1945* (Cambridge University Press, 1989), 64–80.

19. Veronika Lipphardt, "'Jüdische Eugenik'? Deutsche Biowissenschaftler mit jüdischem Hintergrund und ihre Vorstellungen von Eugenik (1900–1935)", in Regina Wecker et al., eds, *Wie nationalsozialistisch ist die Eugenik? – What Is National Socialist about Eugenics?* (Vienna: Böhlau Verlag, 2009), 151–64, and Kamila Uzarczyk, "'Moses als Eugeniker?' The Reception of Eugenic Ideas in Jewish Medical Circles in Interwar Poland", in Marius Turda and Paul J. Weindling, eds, *Blood and Homeland: Eugenics and Racial Nationalism in Central and Southeast Europe, 1900–1940* (Budapest: CEU Press, 2007), 283–97.

20. See the chapters in Alison Bashford and Philippa Levine, eds, *The Oxford Handbook of the History of Eugenics* (New York: Oxford University Press, 2010).

21. See Marius Turda, "Nationalizing Eugenics: The Hungarian Public Debate of 1910–1911", in Ash and Surman, eds, *The Nationalization of Scientific Knowledge*, 183–208; Magdalena Gawin, *Rasa i nowoczesność: historia polskiego ruchu eugenicznego, 1880–1952* (Warsaw: Wydawnicwo Neriton, 2003); Olga Blach, "Eugenics: A Side Effect of Progressivism? Analysis of the Role of Scientific and Medical Elites in the Rise and Fall of Eugenics in Pre-War Poland", *Vesalius* 16, 1 (2010): 10–15; Teresa Ziółkowska, "The Origins of the

Poznań Eugenic Society and Its Significance for the Development of Physical Culture in Poland", *Studies in Physical Culture and Tourism* 9 (2002): 65–79; Gerhard Baader, Veronika Hofer and Thomas Mayer, eds, *Eugenik in Österreich. Biopolitischen Strukturen von 1900 bis 1945* (Vienna: Czernin Verlag, 2007); and the contributions on Hungarian, Polish and Czech eugenics in Turda and Weindling, eds, *Blood and Homeland*.

22. See Monika Löscher, *"...der gesunden Vernuft nicht zuwider..."? Katholische Eugenik in Österreich vor 1938* (Innsbruck: Studienverlag, 2009); Paul J. Weindling, "A City Regenerated: Eugenics, Race, and Welfare in Interwar Vienna", in Deborah Holmes and Lisa Silverman, eds, *Interwar Vienna: Culture between Tradition and Modernity* (Rochester, NY: Camden House, 2009), 81–113; Britta I. McEwen, "Welfare and Eugenics: Julius Tandler's *Rassenhygienische* Vision for Interwar Vienna", *Austrian History Yearbook* 41 (2010): 170–90; and Marius Turda, "In Pursuit of Greater Hungary: Eugenic Ideas of Social and Biological Improvement, 1940–1941", *The Journal of Modern History* 85, 3 (2013): 558–91.

23. See Endre Kárpáti, ed., *Madzsar József válogatott írásai* (Budapest: Akadémiai Kiadó, 1967); Endre Réti, ed., *A magyar orvosi iskola mesterei* (Budapest: Medicina, 1969); idem, "Darwin's Influence on Hungarian Medical Thought", in József Antall, ed., *Medical History in Hungary* (Budapest: Medicina, 1972), 157–67; Endre Czeizel, "A biométerek és a mendelisták ellentéte", *Orvosi Hetilap* 113, 4 (1972): 213–17; idem, "Az eugenika létrejötte, kompromittálása és jövője", *Orvosi Hetilap* 113, 6 (1972): 331–4; and Győző Birtalan, "Új problémakörök: szociális hygiéne, egészségügyi biztosítás, eugenika", in idem, *Európai orvoslás az újkorban, 1640–1920* (Budapest: Hungarica, 1988), 128–30. For a different interpretation, see Sándor Lénárd, "Az eugenikáról", *Kortárs* 29, 2 (1985): 112–17.

24. This is aptly illustrated by, for example, György Litván and László Szűcs's editorial inclusion of only one article on eugenics – József Madzsar, "Gyakorlati eugenika" – in their seminal anthology of texts published in the journal *Huszadik Század*. See György Litván and László Szűcs, eds, *A szociológia első magyar műhelye: A Huszadik Század köre*, vol. 2 (Budapest: Gondolat, 1973), 521–4.

25. See Mária M. Kovács, *Liberal Professions and Illiberal Politics. Hungary from the Habsburgs to the Holocaust* (Washington: Woodrow Wilson Center Press, 1994); and idem, *Liberalizmus, radikalizmus, antiszemitizmus* (Budapest: Helikon, 2001).

26. Most notably by János Gyurgyák, *A zsidókérdés Magyarországon. Politikai eszmetörténet* (Budapest: Osiris, 2001); idem, *Ezzé lett magyar hazátok. A magyar nemzeteszme és nacionalizmus története* (Budapest: Osiris, 2007); and idem, *Magyar fajvédők* (Budapest: Osiris, 2012).

27. See Miklós Szabó, *Politikai kultúra Magyarországon, 1896–1986* (Budapest: Medvetánc, 1989), and idem, *Az újkonzervativizmus és a jobboldali radikalizmus története, 1867–1918* (Budapest: Új Mandátum, 2003).

28. See Balázs Ablonczy, "Az eugenika vonzásában: A társadalom biológiai tervezése", *Rubicon* 15, 2 (2004): 15–18; idem, *Pál Teleki* (Budapest: Osiris, 2005); and idem, "Bethlen István és Teleki Pál konzervativizmusa", in Ignác Romsics, ed., *A magyar jobboldali hagyomány, 1900–1948* (Budapest: Osiris Kiadó, 2009), 167–85.

29. Attempts to correct this neglect have nevertheless been made by László Perecz, "'Fajegészségtan', balról jobbra. Az eugenika század eleji recepciójához: Madzsar és Pekár", *A totalitarizmus és a magyar filozófia* (Debrecen: Vulgo, 2005), 200–12; Imre Szebik, "Eugenika a 19. és a 20. század fordulóján", *'2000': Irodalmi és Társadalmi Havi Lap* 14, 4 (2002): 68–74; Ildikó Farkas, "Eugenika", *História* 25, 4 (2003): 28–9; Marius Turda, "The Biology of War: Eugenics in Hungary, 1914–1918", *Austrian History Yearbook* 40 (2009), 238–64; idem, "'A New Religion': Eugenics and Racial Scientism in Pre-First World War Hungary", *Totalitarian Movements and Political Religions* 7, 3 (2006): 303–25; and idem, "Heredity and Eugenic Thought in Early Twentieth-Century Hungary", *Orvostörténeti Közlemények* 52, 1–2 (2006): 101–18.

1 A New Dawn

1. "Spencer Herbert levele a Huszadik Századhoz", *Huszadik Század* 1, 1 (1900): 1.
2. Quoted in Mary Gluck, *Georg Lukács and His Generation 1900–1918* (Cambridge, MA: Harvard University Press, 1985), 99.
3. In both cultural and political terms, the *Társadalomtudományi Társaság* belonged to a wider European phenomenon, one termed the "first Darwinian left" by David Stack in his *The Darwinian Left: Socialism and Darwinism, 1859–1914* (London: New Clarion, 2003).
4. See, for example, Attila Pók, ed., *A Huszadik Század körének történetfelfogása* (Budapest: Gondolat, 1982); idem, "The Social Function of Sociology in Fin-de-Siècle Budapest", in György Ránki, ed., *Hungary and European Civilization* (Budapest: Akadémiai Kiadó, 1989), 265–83; as well as his *A magyarországi radikális demokrata ideológia kialakulása. A "Huszadik Század" társadalomszemlélete (1900–1907)* (Budapest: Akadémiai Kiadó, 1990).
5. Oszkár Jászi to Bódog Somló (dated 8 October 1899). Quoted in Mary Gluck, "Politics versus Culture: Radicalism and the Lukács Circle in Turn of the Century Hungary", *East European Quarterly* 14, 2 (1980): 129.
6. The intellectual context of this society has been the subject of detailed historical research. See, for example, József Saád, "The Centenary of Hungarian Sociology", *Review of Sociology* 8, 1 (2001): 99–111; Zoltán Horváth, *Magyar századforduló. A második reformnemzedék története, 1896–1914*, 2nd edn (Budapest: Gondolat, 1974); Janos Hauszmann, *Bürgerlicher Radikalismus und demokratisches Denken im Ungarn des 20. Jahrhunderts: Der Jászi-Kreis um 'Huszadik Század* (Frankfurt: Peter Lang, 1988); and György Litván, *A Twentieth-Century Prophet: Oscar Jászi, 1875–1957* (Budapest: Central European University Press, 2006).
7. Philip Abrams, *The Origins of British Sociology, 1834–1914* (University of Chicago Press, 1968) and Christopher Adair-Toteff, *Sociological Beginnings: The First Conference of the German Society for Sociology* (Liverpool University Press, 2005).
8. For an interesting view, see Francisco Louçã, "Emancipation through Interaction: How Eugenics and Statistics Converged and Diverged", *Journal of the History of Biology* 42, 4 (2009): 649–84.
9. R. J. Halliday, "The Sociological Movement, the Sociological Society and the Genesis of Academic Sociology in Britain", *The Sociological Review* 16, 3

(1968): 377–98. For a recent discussion, see Chris Renwick, "From Political Economy to Sociology: Francis Galton and the Social-Scientific Origins of Eugenics", *The British Journal for the History of Science* 44, 3 (2011): 343–69, esp. 363–7.

10. Terry N. Clark, *Prophets and Patrons: The French University and the Emergence of Social Sciences* (Cambridge, MA: Harvard University Press, 1973), 2. See also Daniela S. Barberis, "In Search of an Object: Organicist Sociology and the Reality of Society in *fin-de-siècle* France", *History of the Human Sciences* 16, 3 (2003): 51–72.

11. Weindling, *Health, Race and German Politics*, 139.

12. Karl Pearson, *The Groundwork of Eugenics* (London: Dulau, 1909).

13. Alfred Ploetz, "Die Begriffe Rasse und Gesellschaft und die davon abgeleiteten Disziplinen", *Archiv für Rassen- und Gesellschaftsbiologie* 1, 1 (1904): 1–27. A plurality of eugenic definitions existed simultaneously: the other prominent German eugenicist, Wilhelm Schallmayer, for instance, favoured the term *Rassehygiene* (race hygiene), while the social hygienist, Alfred Grotjahn, proposed the term *Fortpflanzungshygiene* (reproductive hygiene).

14. The Transylvanian Saxon, and one of the first members of the Gesellschaft für Rassenhygiene, Heinrich Siegmund, provides one such example of this early internationalization of German eugenics. Siegmund's eugenic treatise, *Zur sächsischen Rassenhygiene*, was published in 1901, thus making it the first eugenic text to be published in Hungary.

15. This claim, in fact, demonstrates the malleable nature of eugenics as a set of biological and social practices that were construed culturally and ideologically to sanction different (and often incompatible) visions of society. For other contexts, see Kathy J. Cooke, "The Limits of Heredity: Nature and Nurture in American Eugenics before 1915", *Journal of the History of Biology* 31, 2 (1998): 263–78, and C. L. Bacchi, "The Nature–Nurture Debate in Australia, 1900–1914", *Historical Studies* 19, 75 (1980): 199–212.

16. Oszkár Jászi, "Tíz év", *Huszadik Század* 11, 1–2 (1910): 8.

17. See Loránt Tilkovszky, *Pál Teleki (1879–1941): A Biographical Sketch* (Budapest: Akadémiai Kiadó, 1974).

18. Pál Teleki, "Társadalomtudomány biológiai alapon", *Huszadik Század* 5, 4 (1904): 318–23.

19. Ibid., 318.

20. Ibid., 319.

21. Ibid., 320.

22. Ibid., 320–1.

23. Ibid., 321.

24. A similar agenda animated another influential cultural society in early twentieth-century Hungary: the Galilei Kör (Galileo Circle), established in 1908 by the mathematician George Pólya and the economist and philosopher Károly Polányi. See Zsigmond Kende, *A Galilei kör megalakulása* (Budapest: Akadémiai Kiadó, 1974), and György Litván, "*Magyar gondolat-szabad gondolat". Nacionalizmus és progresszió a század eleji Magyarországon* (Budapest: Magvető, 1978).

25. One example is the debate on social development and various political ideologies, including liberalism, anarchism, conservatism and Christian socialism, and published as *A társadalmi fejlődés iránya* (Budapest: Politzer Zsigmond, 1904).

26. Stephen Jay Gould, *The Mismeasure of Man* (New York: W. W. Norton, 1981), 20.
27. See R. S. Cowan, "Nature and Nurture: The Interplay of Biology and Politics in the Work of Francis Galton", *Studies in the History of Biology* 1 (1977): 133–208, and Marius Turda, "Race, Science and Eugenics in the Twentieth Century", in Bashford and Levine, eds, *The Oxford Handbook of the History of Eugenics*, 98–127.
28. Francis Galton, "Hereditary Improvement", *Fraser's Magazine* 7, 37 (1873): 116.
29. Ibid., 116.
30. Francis Galton, *Inquiries into Human Faculty and Its Development* (London: Macmillan, 1883), 17.
31. Ploetz's letter to Galton (17 August 1905) is taken from Karl Pearson, ed., *The Life, Letters and Labours of Francis Galton*, vol. IIIB (Cambridge University Press, 1930), 546.
32. See Christopher Adair-Toteff, *Sociological Beginnings. The First Conference of the German Society for Sociology* (Liverpool University Press, 2005).
33. Alfred Ploetz, "Die Begriffe Rasse und Gesellschaft und die davon abge-leiteten Disziplinen", in *Verhandlungen des Ersten Deutschen Soziologentages vom 9.–22. Oktober 1910 in Frankfurt a. M* (Tübingen: J. C. B. Mohr, 1911), 111–47.
34. See Max Weber, "On Race and Society", *Social Research* 38, 1 (1978): 30–41, and *Verhandlungen des Ersten Deutschen Soziologentages*, 151–164. See also Sheila F. Weiss, *Race Hygiene and National Efficiency: The Eugenics of Wilhelm Schallmayer* (Berkeley: University of California Press, 1987), 106–10; Hartmut Schleiff, "Der Streit um den Begiff der Rasse in der frühen Deutschen Gesellschaft für Soziologie als ein Kristallisationspunkt ihrer method-ologischen Konstitution", *Leviathan* 37, 3 (2009): 367–88; and Michal Y. Bodemann, "Ethnos, Race and Nation: Werner Sombart, the Jews and Classical German Sociology", *Patterns of Prejudice* 44, 2 (2010): 117–36.
35. Weber, "On Race and Society", 37.
36. Ibid.
37. Pearson, *The Groundwork of Eugenics*, 19–20.
38. "Max Weber, Dr. Alfred Ploetz, and W. E. B. Du Bois (Max Weber on Race and Society II)", *Sociological Analysis* 34, 4 (1973): 310.
39. Karl Pearson, *The Scope and Importance to the State of the Science of National Eugenics* (London: Dulau, 1909), 12. Pearson further attempted to clarify the meaning of race in his 1911 *The Academic Aspect of the Science of National Eugenics*.
40. See Turda, *Modernism and Eugenics*, 6–8.
41. László Epstein, ed., *Első Országos Elmeorvosi Értekezlet Munkálatai* (Budapest: Pallas, 1901). It is important to note that most participants were affiliated, directly or indirectly, with the Hungarian Royal State Institute of Psychiatry and Neurology in Budapest-Lipótmező (Budapest-Lipótmezei Magyar Királyi Állami Elme- és Ideggyógyintézet), the most important institution in Hungary devoted to the mentally ill.
42. For an overview, see Kálmán Pándy, *Gondoskodás az elmebetegekről más államokban és nálunk* (Gyula: Vértesi Arnold, 1905). See also Ferenc Pisztora, "Oláh Gusztáv élete és életművének jelentősége a magyar pszichiátria számára", *Ideggyógyászati Szemle* 36, 1 (1983): 1–11, and Emese Lafferton,

"A magántébolydától az egyetemi klinikáig – a magyar pszichiátria történetének vázlata európai kontextusban, 1850–1908", in Vera Békés, ed., *A kreativitás mintázatai: magyar tudósok, magyar intézmények a modernitás kihívásában* (Budapest: Áron Kiadó, 2004), 34–73.

43. Béla Révész, "Der Einfluss des Alters der Mutter auf die Körperhöhe", *Archiv für Anthropologie* 4, 3 (1906): 160–7, and idem, "Rassen und Geisteskrankheiten. Ein Beitrag zur Rassenpathologie", *Archiv für Anthropologie* 6, 3 (1907): 180–7.
44. See István Hollós, "Az öröklésnek az elmebetegségek fellépésére való jelentősége", *Budapesti Orvosi Újság*, 2, 20 (1904): 420–3.
45. Bernát Alexander and Mihály Lenhossék, eds, *Az ember testi és lelki élete, egyéni és faji sajátságai*, 2 vols (Budapest: Athenaeum, 1905, 1907).
46. Alexander and Lenhossék, eds, *Az ember*, vol. 1, 1.
47. Zoltán Dalmady, "Betegség és egészség", in Alexander and Lenhossék, eds, *Az ember*, vol. 2, 461–536.
48. The chapter on "Betegség és egészség" was followed by Dalmady's two additional commentaries: one on the protection against diseases; the other on how to live in accordance with a modern philosophy of health.
49. Károly Balás, *A népesedés* (Budapest: Politzer, 1905), 63–78.
50. Ibid., 81–90.
51. Sándor Szana, *Az állami gyermekvédelem fejlesztéséről* (Temesvár: Uhrmann Henrik, 1903); idem, "Die Pflege kranker Säuglinge in Anstalten", *Wiener Klinische Wochenschrift* 18, 2 (1904): 46–52; and idem, "Fürsorge für in öffentliche Versorgung gelangende Säuglinge", in *Bericht über den XIV. Internationalen Kongress für Hygiene und Demographie. Berlin. 23–29 September 1907* (Berlin: Verlag von August Hirschwald, 1908), 439–52.
52. See Sándor Szana, *A züllött. Gyermek socialhygieniájának magyar rendszere* (Budapest: Pallas nyomda, 1910).
53. Lajos Hajós "Az egészség társadalmi védelme", *Huszadik Század* 2, 8 (1901): 81–90; idem, "Az egészség társadalmi védelme", *Huszadik Század* 2, 9 (1901): 185–92; idem, "Az egészség társadalmi védelme", *Huszadik Század* 2, 10 (1901): 286–92; and idem, "Az egészség társadalmi védelme", *Huszadik Század* 2, 11 (1901): 370–5.
54. Michel Foucault, *"Society Must Be Defended": Lectures at the Collège de France, 1975–1976* (New York: Picador, 2003), 240.
55. Liebermann's son, also a physician, bore his father's name. Leó Liebermann Jr committed suicide in 1938. In this study, I only refer to Leó Liebermann Sr.
56. Taav Laitinen, "Der Einfluss des Alkohols auf die Widerstandsfähigkeit des menschlichen und tierischen Organismus mit besonderer Berücksichtigung der Vererbung", in *Leitgedanken der Referate. Nemzetközi Alkoholizmus Elleni X. Kongresszus. Budapest, 1905. Szept. 11–16* (Budapest: Hausdruckerei der Haupt- u. Residenzstadt, 1905), 2.
57. The lecture was published in three consecutive issues of *Wiener Medizinischen Wochenschrift* in 1907 and then republished as a pamphlet. See Gyula Donáth, *Der Arzt und die Alkoholfrage* (Vienna: Moritz Perles, 1907). He had raised similar arguments about the role of the state in combating degeneration in his paper given to the 8th International Congress of Hygiene and Demography. See also Gyula Donáth, "Der physische Rückgang der Bevölkerung in den modernen Culturstaaten, mit besonderer Rücksicht auf Oesterreich-Ungarn", in Gerlóczy, ed., *Jelentés*, vol. 7, 605–17.

58. Donáth, *Der Arzt und die Alkoholfrage*, 2.
59. This agenda was most effectively pursued by the National Anti-Alcohol Association (Országos Magyar Alkoholellenes Egyesület). A short description of the Society's activities, dated from 1907, can be found at the Archive of the Semmelweis Orvostörténeti Múzeum, Könyvtár és Levéltár (hereafter SOM), papers of the Budapesti Királyi Orvosegyesület. The Society also published a weekly paper, *Az alkoholizmus ellen: Az 'Alkoholellenes Szövetség' havi folyóirata*. Another journal devoted to alcoholism and its social and biological consequences (*Az Alkoholismus*) was established in 1905 and edited by József Hollós and József Madzsar. See József Honti, "Hollós József dr. és Madzsar József dr. kapcsolata", *Orvosi Hetilap* 117, 40 (1976): 2436–8.
60. Donáth, *Der Arzt und die Alkoholfrage*, 16–17. A similar argument was put forward by Fülöp Stein, *Az alkoholkérdés mai állásáról* (Budapest: Posner Károly, 1910).
61. Pál Angyal "A castratio és sterilitatis procuratio", *Jogállam* 1, 5 (1902): 403–9.
62. Rusztem Vámbéry, *A házasság védelme a büntetőjogban* (Budapest: Politzer, 1901).
63. Dezső Buday, *A házasság társadalmi védelme* (Budapest: Politzer Zsigmond, 1902).
64. József Illés, "Magyar házassági vagyonjog és önálló nemzeti fejlődés", *Huszadik Század* 1, 11 (1900): 381–6.
65. Extending his ideas of social protectionism to population policies, Buday connected these with eugenic utilitarianism in his *Az egyke* (Budapest: Deutsch Zsigmond, 1909). See also Ede Harkányi, *Babonák ellen* (Budapest: Grill Károly, 1907) and Emil Lantos, "Az anyaság és a csecsemő védelme", *Budapesti Orvosi Újság* 21, 2 (1908): 19–21.
66. Manó Szántó, *A fakultatív sterilitás kérdéséről* (Budapest: Nagel Ottó, 1905).
67. Géza Illyefalvi-Vitéz, *Születési és termékenységi statisztika* (Budapest: Grill, 1906).
68. For a general discussion, see Randall Hansen, Desmond King, *Sterilized by the State: Eugenics, Race, and the Population Scare in Twentieth-Century North America* (New York: Cambridge University Press, 2013).
69. Robert R. Rentoul, "Proposed Sterilization of Certain Mental Degenerates", *The American Journal of Sociology* 12, 3 (1906): 319–27.
70. The lectures were published as Francis Galton, "Eugenics: Its Definition, Scope and Aims", *Sociological Papers* 1 (1904): 45–50; idem, "Restrictions in Marriage", *Sociological Papers* 2 (1905): 3–13; idem, "Studies in National Eugenics", *Sociological Papers* 2 (1905): 14–17; "Mr Galton's Reply", *Sociological Papers* 2 (1905): 49–51; and idem, "Eugenics as a Factor in Religion", *Sociological Papers* 2 (1905): 52–3.
71. See Francis Galton, "Entwürfe zu einer Fortpflanzungs-Hygiene", *Archiv für Rassen- und Gesellschaftsbiologie* 2, 5–6 (1905): 812–29.
72. Alfred Ploetz's editorial comment to Galton, "Entwürfe zu einer Fortpflanzungs-Hygiene", 812.
73. József Madzsar, "A szaporodás higiénéje", *Huszadik Század* 7, 4 (1906), 366–7.
74. As early as the 1980s, scholars like Michael Freeden, Greta Jones and Diane Paul have urged (albeit quite differently) against interpreting eugenics exclusively as a distorted version of conservative racism. See Michael Freeden, "Eugenics and Progressive Thought", *The Historical Journal* 22, 3 (1979): 645–71; Greta Jones, "Eugenics and Social Policy between the Wars", *The Historical*

Journal 25, 3 (1982): 717–28; Diane Paul, "Eugenics and the Left", *Journal of the History of Ideas* 45, 4 (1984): 567–90; and Michael Schwartz, *Sozialistische Eugenik: Eugenische Sozialtechnologien in Debatten und Politik der deutschen Sozialdemokratie, 1890–1933* (Bonn: J.H.W. Dietz, 1995).

75. Gyula Kozáry, *Az átöröklés problémája* (Budapest: Athenaeum, 1894), and idem, *Átöröklés és nemzeti nevelés. Természetbölcseleti és neveléstani tanulmány* (Budapest: Athenaeum, 1905).

76. Gyula Kozáry, "Az Eugenics kérdése (Lapok az Eugenics kérdésének történetéből)", part 1, *Athenaeum* 15, 2 (1906): 242–8; "Az Eugenics kérdése (Lapok az Eugenics kérdésének történetéből)", part 2, *Athenaeum* 15, 3 (1906): 351–7; "Az Eugenics kérdése (Lapok az Eugenics kérdésének történetéből)", part 3, *Athenaeum* 15, 4 (1906): 458–64; and "Az Eugenics kérdése (Lapok az Eugenics kérdésének történetéből)", part 4, *Athenaeum* 16, 1 (1907): 59–69.

77. Kozáry, "Az Eugenics kérdése", part 1, 242–8.

78. Kozáry, "Az Eugenics kérdése", part 2, 351–7.

79. Kozáry, "Az Eugenics kérdése", part 3, 458–64.

80. Georges Vacher de Lapouge, "The Fundamental Laws of Anthropo-sociology", *The Journal of Political Economy* 6, 1 (1897): 54. See also Géza Czirbusz, *Nemzetek alakulása. Anthropo-geographiai szempontból* (Nagybecskerek: Pleitz Pál, 1910).

81. Kozáry, "Az Eugenics kérdése", part 4, 59–69.

82. Ibid., 67.

83. Ibid., 67–8.

84. Ibid., 68.

85. Ibid., 68–9.

86. Such attention, however, did not generate any cooperation between Hungarian and Saxon eugenicists in Transylvania. Saxon eugenics – as Tudor Georgescu has pointed out – had developed largely in isolation from the Hungarian and Romanian eugenic movements. See his "Ethnic Minorities and the Eugenic Promise: The Transylvanian Saxon Experiment with National Renewal in Interwar Romania", *European Review of History* 17, 6 (2010): 861–80.

87. See *Jelentés a Társadalmi Muzeum berendezéséről és annak első évi munkásságáról* (Budapest: A Társadalmi Muzeum kiadása, 1903).

88. "Sanitary Reform Bureau in the Hungarian Ministry of Public Welfare", 1927, p. 1, folder 43, box 5, series 1.1, Record Group (RG) 1.1, Rockefeller Archive Center, Sleepy Hollow, New York.

89. Janet R. Horne, *A Social Laboratory for Modern France: The Musée Social and the Rise of the Welfare State* (Durham, NC: Duke University Press, 2002), 79.

90. *Jelentés a Társadalmi Muzeum*, 49–50 and 71–72.

91. Both the National League for the Protection of Children – under the leadership of Lipót Edelsheim-Gyulai and Sándor Karsai – and the Hungarian Society for Child Study, led by László Nagy, were established in 1906. For an overview, see Zoltán Bosnyák and L. Edelsheim-Gyulai, *Le droit de l'enfant abandonné et le système hongrois de protection de l'enfance* (Budapest: Athenaeum, 1909); Béla Chyzer, *A gyermekmunka Magyarországon* (Budapest: Az Országos Gyermekvédő Liga kiadása, 1909); Zsigmond Engel, *Grundfragen des Kinderschutzes* (Dresden: Bohmert, 1911); and R. B., "Protection de l'enfance en Hongrie", *L'Enfant* 21, 204 (1912): 185–6. See also Zimmermann, *Divide, Provide and Rule*, 48–56.

92. See Menyhért Szántó, "The Labour Insurance Law in Hungary", *The Economic Journal* 18, 72 (1908): 631–6.
93. Henrik Pach, *Magyar munkásegészségügy* (Budapest: Benkő Gyula, 1907).
94. "Az olvasóhoz!" *Fajegészségügy* 1, 1 (1906): 1.
95. See, for instance, Béla Schmidt, "A tüdővész elleni védekezés rationalis útja", *Fajegészségügy* 2, 1 (1907): 1–3; "Antialkoholos mozgalom", *Fajegészségügy* 2, 2 (1907): 1–2; "Munkásvédelem", *Fajegészségügy* 2, 2 (1907): 3–4; and "Baleset elleni védelem", *Fajegészségügy* 2, 3 (1907): 1–2.
96. Péter Buro, "Társadalmi és faji egészségtan", *Egészség* 21, 7 (1907): 195–9.
97. Ibid., 195.
98. Ibid., 197.
99. Ibid., 198.
100. Ferenc Torday, "A jövő nemzedék egészségének biztosításáról", *Egészség* 22, 6 (1908): 166–81. See also Ferenc Torday, *Das staatliche Kinderschutzwesen in Ungarn* (Langensalza: H. Beyer, 1908).
101. See the comprehensive *L'Hygiène Publique en Hongrie* (Budapest: Wodianer, 1909).
102. The congress was covered extensively by specialized medical journals in Hungary and abroad. See, for example, the successive coverage in "The Sixteenth International Congress of Medicine", *The British Medical Journal* 2, 2541–7 (1909): 706–9; 297–801; 887–90; 990–2; 1076–7; 1163–6; and 1233–4.
103. Max von Gruber, "Vererbung, Auslese und Hygiene", in Emil Grósz, ed., *XVIe Congrès International de Medicine. Compte-Rendu. Volume général* (Budapest: Franklin, 1910), 228–52. The lecture was also published in *Deutsche Medizinischer Wochenschrift* 46 (1909): 2049–53. Many of the arguments presented here were first outlined in Gruber's 1903 address to the German Society for Public Health (Deutsche Gesellschaft für Volkshygiene) and then published as "Führt die Hygiene zur Entartung der Rasse?", *Münchener Medizinischer Wochenschrift* 50 (1903): 1713–18, 1781–5.
104. Gruber, "Vererbung, Auslese und Hygiene", 247.
105. Ibid., 250.
106. Ibid., 249.
107. Ibid., 251.
108. Ibid., 252.
109. Observers at the time were explicit on the assumed connection between heredity and national progress. On this point, and with specific reference to gender relations, see Ede Harkányi, *A holnap asszonyai* (Budapest: Politzer Zsigmond, 1905). With respect to Lombroso's theories, see Zoltán Rónai, "Cesare Lombroso antroposzociológiája", *Huszadik Század* 10, 12 (1909): 453–63.
110. One such example is the blend of cultural physiology and racial sociology developed by Károly Méray-Horváth in his *Die Physiologie unserer Weltgeschichte und der kommende Tag: Die Grundlagen der Sociologie* (Budapest: S. Politzer, 1902). See also Bódog Somló, *Politika és szociológia. Méray rendszere és prognózisai* (Budapest: Deutsch Zsigmond, 1906).
111. István Apáthy outlined the cultural programme of the new society in his *A Magyar Társadalomtudományi Egyesület legelső teendői* (Pécs: Taizs József könyvnyomdája, 1908).

112. István Apáthy, "A nemzetalkotó különbözésről általános fejlődéstani szempontból", *Magyar Társadalomtudományi Szemle* 1, 7 (1908): 597–615, and idem, "A darwinismus bírálata és a társadalomtan", *Magyar Társadalomtudományi Szemle* 2, 4 (1909): 309–39.

113. Károly Balás, "A neomalthusianismusról", *Magyar Társadalomtudományi Szemle* 1, 6 (1908): 500–32; idem, "A népesedés és a szociális kérdés", *Magyar Társadalomtudományi Szemle* 2, 2 (1909): 121–49; and idem, *A család és a magyarság* (Budapest: Grill Károly, 1908).

114. The journal *A Cél* has recently attracted the attention of a young generation of historians in Hungary. See Ibolya Godinek's excellent study, "Fajvédő eszme a Cél című folyóiratban" (BA dissertation, ELTE, Történeti Intézet, 2011).

115. See Gyula Sebestyén, "Turáni Társaság", *Ethnographia* 21, 6 (1910): 324–6. The best work on the Turanic Society remains that of Joseph A. Kessler, "Turanism and Pan-Turanism in Hungary, 1890–1945" (PhD dissertation, University of California, Berkeley, 1967).

116. Karl Pearson, *National Life from the Standpoint of Science* (London: Adam and Charles Black, 1901), 13–14.

117. The translation was published as Francis Galton, "A valószínűség, mint az eugenetika alapja", *Huszadik Század* 8, 12 (1907): 1013–29.

118. Jászi was not the first Hungarian scientist to write letters to Galton. The renowned statistician József Kőrösy met Galton at the 7th International Congress of Hygiene and Demography in London in 1891. Impressed by Kőrösy's research, Galton used it in his 1894 lecture to the Royal Society and a three-year correspondence ensued between the two scientists. See University College London, Galton Papers, Fond 270. See also Francis Galton, "Results Derived from the Natality Table of Kőrösi by Employing the Method of Contours or Isogens", *Nature* 49 (1894): 570–1, and Alan S. Parkes, "The Galton–Kőrösi Correspondence", *Journal of Biosocial Science* 3 (1971): 461–72.

119. Francis Galton, "Probability, the Foundation of Eugenics", in Francis Galton, *Essays in Eugenics* (London: Eugenics Education Society, 1909), 98–9.

120. Francis Galton, "Eugenics: Its Definitions, Scope and Aims", *The American Journal of Sociology* 10, 1 (1904): 5.

2 Debating Eugenics

1. See Lajos Horváth, "Az átöröklés törvényei", *Magyar Társadalomtudományi Szemle* 4, 1 (1911): 57–67 and Zsigmond Fülöp, "Átöröklés és kiválasztás a népek életében", *Huszadik Század* 12, 2 (1911): 239–43.

2. See, for example, Donald Mackenzie, "Sociobiologies in Competition: The Biometrician–Mendelian Debate", in Charles Webster, ed., *Biology, Medicine and Society 1840–1940* (Cambridge University Press, 1981), 243–88, and Peter J. Bowler, *The Mendelian Revolution: The Emergence of Hereditarian Concepts in Modern Science and Society* (London: Athlone Press, 1989).

3. See, for example, Oszkár Jászi, "Lamarckisták és Darwinisták", *Huszadik Század* 1, 8 (1900): 153–5; Bódog Somló, "Kidd Benjámin áldarwinizmusa", *Huszadik Század* 5, 1 (1904): 29–37; Zsigmond Fülöp, "Bizonyítékok a

szerzett sajátságok örökölhetőségére", *Huszadik Század* 8, 5 (1907): 484–6; József Madzsar, *Darwinizmus és Lamarckizmus* (Budapest: Deutsch Zsigmond, 1909); Lajos Méhely, *Az éllettdudomány bibliája* (Budapest: A Pesti Lloyd Társulat, 1909); and, István Apáthy, *A Darwinismus birálata és a társadalomtan* (Budapest: Pesti könyvnyomda, 1910). Recent Hungarian scholarship includes Gábor Palló, "Darwin utazása Magyarországon", *Magyar Tudomány* 54, 6 (2009): 714–26, and Katalin Mund, "The Reception of Darwin in Nineteenth-Century Hungarian Society", in Eve-Marie Engels and Thomas F. Glick, eds, *The Reception of Charles Darwin in Europe*, vol. 2 (London: Continuum, 2008), 441–62.

4. See also Donald A. MacKenzie, *Statistics in Britain, 1865–1930: The Social Construction of Reality* (Edinburgh University Press, 1981), and Pauline M. H. Mazumdar, *Eugenics, Human Genetics and Human Failings: The Eugenics Society, Its Sources and Its Critics* (London: Routledge, 2002).

5. Lajos Dienes, "Biometrika", *Huszadik Század* 11, 1 (1910): 50.

6. Ibid., 51.

7. József Madzsar, "Gyakorlati eugenika", *Huszadik Század* 11, 2 (1910): 115.

8. Ibid., 115.

9. Ibid., 116.

10. Ibid., 117.

11. Zsigmond Fülöp, "Eugenika", *Huszadik Század* 11, 9 (1910): 162.

12. Ibid., 163 (italics in original).

13. Ibid., 168.

14. Ibid., 170–1.

15. Ibid., 173.

16. Ibid., 175.

17. Zsigmond Fülöp, "Francis Galton", *Huszadik Század* 12, 3 (1911): 335–6.

18. A wealth of literature now exists on the relationship between specialized knowledge and the ideology of professionalization. For an approach similar to the one attempted here, see Ian R. Dowbiggin, *Madness: Professionalization and Psychiatric Knowledge in Nineteenth Century France* (Berkeley: University of California Press, 1991). For a discussion of the relationship between medical experts and political power in comparative perspective, see Dorothy Porter, ed., *The History of Public Health and the Modern State* (Amsterdam: Rodopi, 1994).

19. The first invitation (dated 27 January 1911) can be found in the Országos Széchényi Könyvtár (hereafter OSZK), Manuscript Collection, Fond *Magyar Társadalomtudományi Egyesület iratai*, no. 2454, file 145. The second invitation (dated 31 January 1911) can be consulted at SOM, papers of the Budapesti Királyi Orvosegyesület. As indicated in his reply to Jászi, Apáthy was unable to attend the first two lectures on eugenics.

20. József Madzsar, "Fajromlás és fajnemesítés", *Huszadik Század* 12, 2 (1911): 145.

21. Ibid.

22. Ibid., 155.

23. Ibid., 158–9.

24. Ibid., 159.

25. Ibid., 160.

26. Lajos Dienes, "A fajnemesítés biometrikai alapjai", *Huszadik Század* 12, 3 (1911): 291–307.

27. Ibid., 307.
28. Zsigmond Fülöp, "Az eugenetika követelései és korunk társadalmi viszonyai", *Huszadik Század* 12, 3 (1911): 308–19.
29. Ibid., 308.
30. Ibid., 309.
31. Ibid.
32. Ibid.
33. Ibid., 310.
34. Ibid., 318.
35. Interventionist plans reflecting Fülöp's eugenic welfare were, in fact, discussed and introduced in Britain at the beginning of the twentieth century. One example is the Mental Deficiency Act of 1913. See Mathew Thompson, *The Problem of Mental Deficiency: Eugenics, Demography, and Social Policy in Britain, c. 1870–1959* (Oxford University Press, 1998). I am grateful to Elizabeth Hurren for underlining this point.
36. "A fajnemesítés (eugénika) problémái", *Huszadik Század* 12, 6 (1911): 694–709; "A fajnemesítés (eugénika) problémái", *Huszadik Század* 12, 7 (1911): 29–44; "A fajnemesítés (eugénika) problémái", *Huszadik Század* 12, 8–9 (1911): 157–70; and "A fajnemesítés (eugénika) problémái", *Huszadik Század* 12, 10 (1911): 322–36.
37. "I. Kérdések a szociálbiológia köréből; II. Kérdések a gyakorlati eugenika köréből". OSZK, Manuscript Collection, Fond *István Apáthy iratai*, no. 2453, file 144 (dated 23 March 1911).
38. "A fajnemesítés (eugénika) problémái", 39–40.
39. Ibid., 44.
40. Ibid., 158–9.
41. For a full discussion of this term see Chapter 4. Suffice to say here that although similar to the German term *Rassenhygiene*, the Hungarian word *fajegészségtan* had a broader scope than *eugenics*.
42. "A fajnemesítés (eugénika) problémái", 700.
43. Ibid., 701.
44. Ibid., 708–9.
45. Ibid., 709.
46. The term "sociobiological strategy" is used here in the sense formulated by MacKenzie in *Statistics in Britain*, 150.
47. See, for example, Julius Pikler, "Über die biologischen Funktion des Bewußtseins", *Archiv für Rassen- und Gesellschaftsbiologie* 8, 2 (1911): 227–30.
48. Ibid. 165.
49. Ibid., 168.
50. Ibid.
51. Ibid., 170.
52. Or indeed elsewhere in Europe at the time. For instance, Karl Pearson – one of the authors often invoked during the Hungarian debate on eugenics – considered socialism the perfect vehicle for the achievement of eugenic ideals. See Donald MacKenzie, "Karl Pearson and the Professional Middle-Class", *Annals of Science* 36 (1979): 125–43.
53. See Thomas Linehan, *Modernism and British Socialism* (Basingstoke: Palgrave, 2012), esp. 116–31.
54. "A fajnemesítés (eugénika) problémái", 29.

55. Ibid., 157.
56. Ibid., 158.
57. "A fajnemesítés (eugénika) problémái", 323.
58. Ibid., 324.
59. Ibid., 694.
60. Ibid., 324–5.
61. Ibid., 170.
62. Ibid., 327.
63. Ibid.
64. Ibid., 332.
65. Ibid., 333–6.
66. Ibid., 336.
67. Pók, "The Social Function of Sociology", 279.
68. Gluck, "Politics versus Culture", 137.
69. Jászi's scientistic proclivities were already displayed in his opening article for *Huszadik Század*. See "Tudományos publicisztika", *Huszadik Század* 1, 1 (1900): 2–12.
70. Marius Turda, "Nationalizing Eugenics", in Ash and Surman, eds, *The Nationalization of Scientific Knowledge*, 183–208.
71. Karl Pearson, *The Academic Aspect of the Science of National Eugenics* (London: Dulau, 1911), 4.

3 At a Crossroads

1. The list of lectures was published in "Társulati ügyek", *Huszadik Század* 12, 7 (1911): 100–1.
2. These lectures were part of a much larger educational agenda promoted by the Társadalomtudományi Társaság. In 1906, a Free School of Social Sciences (Társadalomtudományok Szabad Iskolája) was established, which offered public lectures and courses. Eugenic themes were often discussed; for example, Gusztáv Dirner's lecture on 23 November 1911, published as "A nemi betegségek és a család", *Huszadik Század* 13, 3 (1912): 346–57. The *Magyar Társadalomtudományi Egyesület* established its own "free school" in 1909.
3. "Társulati ügyek", 101. Ferenc Bene and Pál Bugát founded the Association of Hungarian Doctors and Naturalists in 1841. It functioned until 1933.
4. This awareness was accompanied by the increased coverage of other eugenic societies, particularly the Eugenics Education Society, in the Hungarian press. See, for example, "Ólommérgezés és eugenika", *Ipari Jogvédelem* 1, 14 (1911): 7.
5. The Kalotaszeg region, situated between Bánffyhunyad (Huedin) and Kolozsvár (Cluj), was just such a place. For two *fin de siècle* ethnographic accounts of this region see János Jankó, *Kalotaszeg magyar népe: néprajzi tanulmány* (Budapest: Athenaeum Társulat, 1892) and Károly Kós, *Régi Kalotaszeg* (Budapest: Athenaeum nyomda, 1911).
6. Antal Herrmann, "A magyar turista-tanító és Erdély", *Erdély* 22, 5 (1913): 73. On the uses of domestic tourism for the wider Hungarian nation-building project, see Alexander Vari, "From Friends of Nature to Tourist-Soldiers. Nation Building and Tourism in Hungary, 1873–1914", in Anne E. Gorsuch

and Diane P. Koenker, eds, *Turizm: The Russian and East European Tourist under Capitalism and Socialism* (Ithaca: Cornell University Press, 2006), 64–81.

7. Lajos Dienes, "Eugenika", *A Társadalmi Muzeum Értesítője* 3, 3 (1911): 196–216, and idem, "Eugenika", *A Társadalmi Muzeum Értesítője* 3, 4 (1911): 321–36.

8. Ibid., 198–200.

9. Ibid., 204.

10. Ibid.

11. Ibid., 205.

12. This was in addition to Pearson's reworking of Galton's law on ancestral heredity, as developed, for example, in his "On the Laws of Inheritance in Man, I.", *Biometrika* 2, 4 (1903): 357–462, and idem, "On the Laws of Inheritance in Man, II.", *Biometrika* 3, 2–3 (1903): 131–90.

13. Dienes, "Eugenika", 210.

14. In addition to Pearson's work, Dienes also mentioned that of Ethel M. Elderton, particularly her *The Relative Strength of Nurture and Nature* (London: Dulau, 1909).

15. Dienes, "Eugenika", 322.

16. Pearson, *The Problem of Practical Eugenics*, 25.

17. Dienes, "Eugenika", 331.

18. Ibid., 332.

19. The British journal *The Lancet*, for example, favourably recommended "the methods employed in Hungary for the care of those who would be termed in England 'pauper children' [...] to all interested in social reform, and in particular to local administrators concerned with the care of children and the feeble-minded". See "The Children of the State in Hungary", *The Lancet* 176, 4553 (1910): 1653.

20. Dienes, "Eugenika", 333.

21. Ibid., 335.

22. Jenő Vámos, "Az alkalmazott eugenika", *Huszadik Század* 12, 12 (1911): 571–7.

23. Ibid., 571.

24. Ibid.

25. Ibid., 572.

26. See Francis Galton, "A Theory of Heredity", *The Journal of the Anthropological Institute of Great Britain and Ireland* 5 (1876): 329–48.

27. Vámos, "Az alkalmazott eugenika", 575.

28. See, for example Géza Kenedi, "Feminismus és biológia", *Magyar Társadalomtudományi Szemle* 2, 3 (1909): 218–34, and idem, *Feminista tanulmányok* (Budapest: Lampel R., 1912).

29. Vámos, "Az alkalmazott eugenika", 575.

30. Ibid.

31. Ibid., 576.

32. Ibid., 577.

33. Ibid.

34. See Weindling, *Health, Race and German Politics*, 147–54.

35. See *Offizieller Katalog der Internationalen Hygiene Ausstellung Dresden Mai bis Oktober 1911* (Berlin: Verlag Rudolf Mosse, 1911). See also "International Hygiene Exhibition, Dresden, 1911. The Opening Ceremonies", *The British Medical Journal* 13, 1 (May 1911): 1132–3, and Klaus Vogel, ed., *Das Deutsche*

Hygiene-Museum Dresden 1911–1990 (Dresden: Stiftung Deutsches Hygiene-Museum, 2003).
36. "International Hygiene Exhibition Dresden 1911", *Journal of Hygiene* 10, 1 (1910): 131–4.
37. Weindling, *Health, Race and German Politics*, 230, and Michael Hau, *The Cult of Health and Beauty in Germany: A Social History, 1890–1930* (University of Chicago Press, 2003), esp. 107–10.
38. Max von Gruber, Ernst Rüdin, eds, *Fortpflanzung, Vererbung, Rassenhygiene*, 2nd edn (Munich: J. F. Lehmanns, 1911).
39. Ibid., 121–2.
40. For the coverage of the 1911 Dresden Exhibition in Hungarian newspapers, specialized journals and books, see the following: "A drezdai nemzetközi hygiene kiállítás magyar pavillonja", *Vasárnapi Ujság* 58, 15 (1911): 290–291; René Berkovits, "A drezdai szociálhigiénai kiállítás néhány tanulsága", *Huszadik Század* 13, 9–10 (1912): 405–13; D. L. [Lajos Dienes], "A fajhigiéne a drezdai kiállításon", *A Társadalmi Muzeum Értesítője* 3, 6 (1911): 694–5; Emil Grósz, "A drezdai nemzetközi hygiene kiállítás", *Budapesti Szemle* 39, 414 (1911): 452–6; "Kongresszusok", *A nő és a társadalom* 5, 9 (1911): 152; and Gusztáv Rigler, *A drezdai higiéne-kiállítás* (Budapest: Franklin, 1911).
41. Teleki's letter to Apáthy (dated 16 January 1914), OSZK, Manuscript Collection, Fond *Apáthy István iratai*, no. 2453, file 180, 15–16. According to Teleki, he was the only Hungarian to attend the meeting. Recalling this meeting with Gruber and other German eugenicists in Dresden in an interview with the newspaper *Az Est* twenty years later, Teleki mentioned, however, that "his friend", Géza Hoffmann, also participated. See "A fajnemesítés és sterilizáció magyar tudós társaságáról beszél gróf Teleki Pál és Nékám professzor", *Az Est* 24, 207 (13 September 1933): 11.
42. Teleki's letter to Apáthy, 16.
43. See the overview provided for Hungarian readers by Grete Meisel-Hess, a celebrated Austrian feminist and author of *Die sexuelle Krise. Eine sozialpsychologische Untersuchung* (1909), in her "Népesedési és erkölcsproblémák", *Huszadik Század* 12, 11 (1911): 459–63.
44. In 1896, the French anarchist and neo-Malthusian advocate Paul Robin created the League for Human Regeneration (Ligue pour la régéneration humaine), which, in 1900, was renamed the Universal Federation of Human Regeneration (or the Federation of neo-Malthusian Leagues). European countries represented by the Federation included England, Holland, Germany, France, Spain, Belgium and Austria (Bohemia). Brazil, Cuba and the USA were also members. For the context of this organization, see William H. Schneider, *Quality and Quantity: The Quest for Biological Regeneration in Twentieth-Century France* (Cambridge University Press, 1990), and Mary Lynn Stewart, *For Health and Beauty: Physical Culture for French Women, 1880s–1930s* (Baltimore: Johns Hopkins University Press, 2001).
45. "IV. Internationaler Kongress für Neumalthusianismus, Dresden 24–27 September 1911". Programme held in the International Institute of Social History, Amsterdam. Collection: NMB Archive, no. 152.
46. In this context, feminism referred to a wide range of topics, including control over women's bodies through the use of contraception, social protection for mothers and children, education as well as the right to vote.

47. See Amy Hackett, "Helene Stöcker: Left-Wing Intellectual and Sex Reformer", in Renate Bridenthal, Atina Grossmann and Marion Kaplan, eds, *When Biology Became Destiny: Women in Weimar and Nazi Germany* (New York: Monthly Review Press, 1984), 109–30, and Christl Wickert, Brigitte Hamburger and Marie Lienau, "Helene Stöcker and the Bund für Mutterschutz (The Society for the Protection of Motherhood)", *Women's Studies International Forum* 5, 6 (1982): 611–18.

48. Max Rosenthal, ed., *Mutterschutz und Sexualreform. Referate und Leitsätze des I. Internationalen Kongress für Mutterschutz und Sexualreform in Dresden 28.–30. September 2011* (Breslau: Verlag von Preuss und Jünger, 1912).

49. The League for the Protection of Mothers (Bund für Mutterschutz) was established in 1905 but renamed the German League for the Protection of Mothers in 1908 (Deutscher Bund für Mutterschutz). See Bernd Nowacki, *Der Bund für Mutterschutz, 1905–1933* (Husum: Matthiesen Verlag, 1983).

50. Rosenthal, ed., *Mutterschutz und Sexualreform*, 25–6.

51. Amélie Neumann's report in ibid., 49–51; Rosika Schwimmer's report also in ibid., 74.

52. Hungarian contributions to the First International Congress of Pedology held in Brussels between 12 and 18 August 1911 also bear mentioning in this context. These included the child psychologists Margit Dosai-Révész and László Nagy; the psychologist, László Nógrády; the aesthetician and philosopher, Károly Pekár; and the psychologist, Pál Ranschburg. See I. Ioteyko, ed., *Premier Congrès International de Pédologie*, 2 vols (Brussels: Librarie Misch et Thron, 1912).

53. "Einladung zum I. Internationalen Kongreß für Mutterschutz und Sexualreform". International Institute of Social History, Amsterdam. Collection: NMB Archive, no. 151.

54. In addition to playing a crucial role in the establishment of the Feminist Association in 1904, Schwimmer was also the editor of the most prominent feminist journals in Hungary: *A nő és a társadalom* (1907–13) and *A nő: feminista folyóirat* (1914–28). For her remarkable life and career, both in Hungary and abroad, see Francisca de Haan, Krassimira Daskalova, Anna Loutfi, eds, *A Biographical Dictionary of Women's Movements and Feminisms: Central, Eastern, and South Eastern Europe, 19th and 20th Centuries* (Budapest: CEU Press, 2006), 484–90.

55. See, for example, Rosika Schwimmer, "Der Stand der Frauenbildung in Ungarn", in Helene Lange and Gertrud Bäumer, eds, *Handbuch der Frauenbewegung*, vol. 3 (Berlin: W. Moeser Buchhandlung, 1902), 191–206, and idem, *Staatlicher Kinderschutz in Ungarn* (Leipzig: Dietrich, 1909).

56. See, for example, Rózsa B.-Schwimmer, "Az anyaság védelme", *A nő és a társadalom* 2, 5 (1908): 73–6.

57. Susan Zimmermann, *Die bessere Hälfte: Frauenbewegung und Fraubestrebungen in Ungarn der Habsburgermonarchie 1848 bis 1918* (Vienna: Promedia, 1999) and Agatha Schwartz, *Shifting Voices: Feminist Thought and Women's Writing in Fin-de-Siècle Austria and Hungary* (Montreal: McGill-Queen's University Press, 2008).

58. One such feminist was Mrs Sándor Szegvári. For instance, see her articles "A nő zsenialitása"; *A nő és a társadalom* 5, 10 (1911): 164–5; and "Anyaság", *A nő és a társadalom* 7, 3 (1913): 48–50.

59. G. Spiller, ed., *Papers on Inter-Racial Problems* (London: P. S. King, 1911). See also Michael D. Biddiss, "The Universal Races Congress of 1911", *Race and Class* 13, 1 (1971): 37–46.

60. Ákos Timon, "Theory of the Holy Crown, or the Development and Significance of the Conception of Public Rights of the Holy Crown in the Constitution", in Spiller, ed., *Papers*, 184–95.

61. Ákos Navratil, "Investment and Loans", in ibid., 208–11.

62. Dénes Nagy, "Az emberfajok első egyetemes kongresszusa", *Huszadik Század* 12, 12 (1911): 345–8. The French edition of Spiller's book was published as *Mémoires sur le contact des races communiqués au 1. Congrès universel des races tenu à l'université de Londres 1911* (London: P. S. King, 1911), and reviewed as "Fajok kongresszusa", *Huszadik Század* 13, 8 (1912): 291–3.

63. "Előadás a fajok harczáról", *Magyar Társadalomtudományi Szemle* 4, 10 (1911): 822–5.

64. Ibid., 825.

65. For detailed coverage, see Edgar Schuster, "The First International Eugenics Congress", *The Eugenics Review* 4, 3 (1912): 223–56.

66. *Problems in Eugenics*, vol. 2 (London: The Eugenics Education Society, 1913), 7.

67. Ibid.

68. Szász's Letter to Apáthy (Dated 8 July 1912). OSZK, Manuscript Collection, Fond *Apáthy István iratai*, no. 2453, file 1037, pp. 257–8.

69. Szász's Letter to Apáthy (Dated 16 August 1912). OSZK, Manuscript Collection, Fond *Apáthy István iratai*, no. 2453, file 1166, pp. 250–1.

70. Zsombor Szász, "Az első nemzetközi fajegészségügyi (eugenikai) congressus", *Magyar Társadalomtudományi Szemle* 5, 8 (1912): 650–7.

71. See Bleeker van Wagenen, "Preliminary Report of the Committee of the Eugenic Section of the American Breeders' Association to Study and to Report on the Best Practical Means for Cutting Off the Defective Germ-Plasm in the Human Population", in *Problems in Eugenics*, vol. 1 (London: The Eugenics Education Society, 1912), 460–79. Tellingly, this Committee was created in 1911.

72. Ibid., 464. See also Szász, "Az első nemzetközi fajegészségügyi (eugenikai) congressus", 655–6.

73. See Benny Kraut, *From Reform Judaism to Ethical Culture: The Religious Evolution of Felix Adler* (Cincinnati, OH: Hebrew Union College Press, 1979).

74. Mazumdar, *Eugenics, Human Genetics and Human Failings*, 17.

75. Tracie Matysik, *Reforming the Moral Subject: Ethics and Sexuality in Central Europe, 1890–1930* (Ithaca: Cornell University Press, 2008), 4.

76. J. W. Slaughter, "Eugenics and Moral Education", in Gustav Spiller, ed., *Papers on Moral Education*, 2nd edn (London: David Nutt, 1909), 381.

77. See Jenő Gergely, *A kereszténysocializmus Magyarországon, 1903–1923* (Budapest: Akadémiai Kiadó, 1977), and Paul Hanebrink, *In Defense of Christian Hungary: Religion, Nationalism, and Antisemitism, 1890–1944* (Ithaca: Cornell University Press, 2006), esp. 35–9.

78. Ottokár Prohászka, "Ethical Co-operation of Home and School", in Spiller, ed., *Papers on Moral Education*, 305.

79. Sarolta Geőcze, "Environment and Moral Development", in ibid., 386.

80. Ibid., 387. Geőcze was not the only congress participant to touch upon the topic of degeneration. None other than Cesare Lombroso presented a paper

on juvenile delinquency. See Cesare Lombroso, "Traitement moral du jeune criminel", in ibid., 216–22.

81. It can be argued that Geőcze's Catholic and nationalist feminism falls within the broad tradition of "relational feminism", one that – according to Karen Offen – "emphasized women's rights *as women* (defined principally by their childbearing and/or nurturing capacities) in relation to men. It insisted on *women's* distinctive contributions in these roles to the broader society and made claims on the commonwealth on the basis of these contributions." See Karen Offen, "Defining Feminism: A Comparative Historical Approach", *Signs* 14, 1 (1988): 136.

82. Giesswein's paper was not published in the proceedings. However, it is noted that he spoke at the Congress on the topic of "Bürgerkunde und Patriotism". In Spiller, ed., *Papers on Moral Education*, 270.

83. Mór Kármán, "Aufgaben der sittlichen Erziehung", in ibid., 23–39.

84. Kemény presented three papers at the congress: "Der Interkonfessionalismus ein zwillingsbruder des Internationalismus" (194–8); "Physische Kultur und Character-Building" (468–76); and "Die Erziehung zum Mut" (482–7). All papers are published in Attie G. Dyserinck, ed., *Mémoires sur l'éducation morale* (The Hague: Martinus Nijhoff, 1912). See also Ferenc Kemény, "Nemzetköziség és felekezetköziség", *Szociálpolitikai Szemle* 2, 12 (1912): 178–80.

85. Kemény, "Physische Kultur und Character-Building", 474.

86. Mór Kármán, "Ethisch-Historischen Gesichtspunkte zur Teorie des Lehrplans"; Sarolta Geőcze, "Sittliche Erziehung und Nazionalen Leben" and "Sittliche und Soziale Bildung in Lehrer-Seminaren", both in Dyserinck, ed., *Mémoires sur l'éducation morale*, 593–600; 984–90 and 1031–6, respectively.

87. Pál Angyal, "Rèforme du caractère vicieux et déliquants", in ibid., 860–6.

88. John Russell, "The Eugenic Appeal in Moral Education", in Dyserinck, ed., *Mémoires sur l'éducation morale*, 570–4.

89. C. W. Saleeby, " Eugenic Education or Education for Parenthood", in ibid., 580–3.

90. "Közegészségügyi értekezlet", *Magyar Társadalomtudományi Szemle* 5, 2 (1912): 152.

91. Ibid., 158.

92. Not to be mistaken for the Marxist philosopher with the same name.

93. Other participants included the engineer Imre Forbáth; Flóra Perczelné-Kozma, a Unitarian women's activist; the legal expert and economist Gyula Mandelló; Dezső Okolicsányi-Kuthy, a pulmonologist; the editor of the journal *Gyógyászat*, Miksa Schächter; the dermatologist Adolf Havas; the architect Lajos Schodits; and the chemist Lajos Ilosvay.

94. See László Felkai, "A kultuszminiszter Lukács György", *Magyar Pedagógia* 102, 1 (2002): 3–9.

95. "Közegészségügyi értekezlet", 168.

96. Ibid., 168–9.

97. "Közegészségügyi értekezlet", *Magyar Társadalomtudományi Szemle* 5, 3 (1912): 237.

98. Ibid., 239.

99. Ibid., 246.

100. Ibid., 247. See also "Közegészségügyi szaktanácskozmány", *Magyar Társadalomtudományi Szemle* 5, 4 (1912): 350.
101. Ibid., 341.
102. See Henrik Pach, *A Társadalmi Múzeum és a közegészségügy fejlesztése* (Budapest: Pesti Lloyd-Társulat, 1909).
103. Menyhért Szántó, ed., *Tájékoztató a Társadalmi Múzeum által Győrött 1913. augusztus 14-től szeptember 10-ig rendezett népegészségügyi kiállításról* (Budapest: Pesti nyomda, 1913), and idem, *Tájékoztató a Társadalmi Múzeum által Magyaróvárott 1914 február 1-től február 15-ig rendezett népegészségügyi kiállításról* (Budapest: Pesti nyomda, 1914).
104. M. Szántó, *The Museum of Social Service in Buda-Pest* (Budapest: Garden City Press, 1914), 2.
105. Ibid., 11.
106. "Az olvasókhoz", *"Darwin"* 2, 18 (1913): 1.
107. A Darwin Circle (*Darwin Kör*) was already established in Nagyvárad in 1910, which organized conferences and public lectures on Darwinism and evolution. See "Krónika", *Szabadgondolat* 3, 1 (1913): 33–5.
108. Sidney and Beatrice Webb, "Szegénység és fajszépség", *Szociálpolitikai Szemle* 2, 12 (1912): 180–2.
109. See also Sidney Webb, "Eugenics and the Poor Law", *The Eugenics Review* 2, 3 (1910): 233–41, which further contains many of the arguments presented in the Hungarian article.
110. Sidney and Beatrice Webb, "Szegénység és fajszépség", 181.
111. Ibid.
112. Emil Torday, "Primitive Eugenics", *The Mendel Journal* 3, 4 (1914): 30–6.
113. Ibid., 31.
114. Ibid., 32. Classical eugenic literature often invokes the ancient Spartans and their brutal method for disposing of deformed offspring on Mount Taygetus. See Plutarch, *Lives* (Theseus and Romulus, Lycurgus and Numa, Solon and Publicola) vol. 1 (London: W. Heinemann, 1959), xv, xvi, 255.
115. For an extensive review of Malthusian theories at the end of the nineteenth century, as well as their impact on Hungarian authors, see Gábor Kovács, *A népesedés elmélete* (Debrecen: Hegedűs és Sándor, 1908).
116. C. V. Drysdale, *Neo-Malthusianism and Eugenics* (London: William Bell, 1912).
117. Ibid., 9.
118. Ibid., 21.
119. Alfred Ploetz, "Neo-Malthusianism and Race Hygiene", in *Problems in Eugenics: Report of Proceedings of the First International Eugenics Congress*, vol. 2 (London: The Eugenics Education Society, 1913), 183–9.
120. See Ildikó Vásáry, "'The Sin of Transdanubia': The One-Child System in Rural Hungary", *Continuity and Change* 4 (1989): 429–68. A good overview is also provided in Béla Bodó, "Progress or National Suicide: The Single-Child Family in Hungarian Political Thought, 1840–1945", *Hungarian Studies Review* 28, 1–2 (2001): 185–208.
121. C. V. Drysdale, "Some Impressions of Hungary. I", *The Malthusian* 27, 7 (1913): 49.
122. Ibid.
123. Ibid., 50, and C. V. Drysdale, "Some Impressions of Hungary. II", *The Malthusian* 27, 8 (1913): 57.

124. For a good overview, see Zimmermann, *Divide, Provide and Rule*, 48–56.
125. Drysdale, "Some Impressions of Hungary. II", 58.
126. Drysdale devoted one chapter to both the Memorandum and the Council's subsequent response in his *The Small Family System: Is It Injurious or Immoral?* (London: A. C. Fifield, 1913), 115–19.
127. Quoted in ibid., 115.
128. Quoted in ibid., 116.
129. Ibid.
130. Quoted in ibid., 117.
131. Quoted in ibid., 118–19.
132. Quoted in ibid., 119.
133. Quoted in ibid.

4 Towards National Eugenics

1. See, for example, István Apáthy, *Über das leitende Element des Nervensystems und seine Lagebeziehungen zu den Zellen bei Wirbeltieren und Wirbellosen* (Leiden: Brill, 1896).
2. See Ambrus Ábrahám, "Stephan von Apáthy, 1863–1922", in Hugo Freund and Alexander Berg, eds, *Geschichte der Mikroskopie*, vol. 1 (Frankfurt: Umschau Verlag, 1963), 65–75; Endre Réti, "Apáthy István", in Réti, ed., *A magyar orvosi*, 217–24; György Kiszely, "Apáthy István (1863–1922)", *Orvosi Hetilap* 129, 40 (1988): 2147–9; István Benedeczky, *Apáthy István: a tudós és a hazafi* (Budapest: Szenczi Molnár Társaság, 1995); and Bálint Markó, "Apáthy István", in Gyöngy Kovács Kiss, ed., *Hivatás és tudomány: Az Erdélyi Múzeum-Egyesület kiemelkedő" személyiségei* (Kolozsvár: Erdélyi Múzeum-Egyesület, 2009), 9–36.
3. Margó's main work *Általános Állattan* (*General Zoology*) was published in 1868. On the impact of his theories of heredity and adaptation on his Hungarian students, see Árpád Szállási, "Margó Tivadar", *Orvosi Hetilap* 117, 14 (1976): 851–854 and Sándor Soós, "The Scientific Reception of Darwin's Work in Nineteenth-Century Hungary", in Engels and Glick, eds, *The Reception of Charles Darwin*, vol. 2, 434–5.
4. The German biologist, Felix Anton Dohrn, founded the Zoological Station of Naples in 1872 as a research institute for zoology and comparative anatomy. See Ernst Florey, "The Zoological Station at Naples and the Neuron: Personalities and Encounters in a Unique Institution", *Biological Bulletin* 168 (supplement, 1985): 137–52, and Irmgard Müller, "The Impact of the Zoological Station in Naples on Developmental Physiology", *International Journal of Developmental Biology* 40, 1 (1996): 103–11.
5. The opening lecture was initially published as István Apáthy, *A fejlődésnek nevezett átalakulásról* (Kolozsvár: Ajtai nyomda, 1904).
6. Nándor Nagy, "Apáthy Istvánról és a Kolozsvári Állattani Intézetről", *Collegium Biologicum* 2 (1998): 91–6.
7. See István Apáthy, *Néhány lap önismeretünk történetéből: élettudományi vázlat* (Budapest: Hornyánszky nyomda, 1900).
8. According to the German physiologist, Albrecht J. Th. Bethe (who met and befriended Apáthy during his stay in Italy), Apáthy's Hungarian nationalism

was overtly anti-Austrian/German. Tellingly, he returned any letter addressed to him at Klausenburg (instead of Kolozsvár) unopened, and refused to travel through Austria. See Florey, "The Zoological Station at Naples", 146. This intense nationalism may also explain Apáthy's powerful dislike of German eugenicists, an aspect to be discussed in the following chapters.

9. See, for example, István Apáthy, *A nemzeti dalról* (Kolozsvár: Újhelyi nyomda, 1906); idem, *Magántulajdon, csere és élet az állatországban* (Pécs: Taizs József, 1908); idem, *Öregség és halál* (Budapest: Hornyánszky, 1909); and idem, *A hosszú életről* (Kolozsvár: Gámán János nyomda, 1909).

10. As argued in István Apáthy, "Széchenyi István és a nemzeti sajátságok az emberi továbbfejlődés szempontjából", *Magyar Társadalomtudományi Szemle* 5, 10 (1912): 771–90.

11. Examples include Jenő Varga, *A magyar faj védelme* (Makó: Kovács Antal, 1901), and Gábor Jánossy, *Közművelődési egyesületeink és a magyar faj (állam) jövője* (Szombathely: Egyházmegyei könyvnyomda, 1904).

12. His lectures were published in the Transylvanian Museum's journal, *Értesítő az 'Erdélyi Múzeum-Egylet' Orvos-Természettudományi Szakosztályából*, between 1890 and 1906.

13. Carl E. Schorske, *Thinking with History: Explorations in the Passage to Modernism* (Princeton University Press, 1998), 61.

14. This was a complete version of Apáthy's contribution to the "Eugenika vita". See István Apáthy, "A faj egészségtana", *Magyar Társadalomtudományi Szemle* 4 (1911): 265–79. The article was also published as an offprint in 1912.

15. Ibid., 265.

16. Ibid., 265.

17. Ibid., 268.

18. Ibid., 270. Apáthy's interpretation of the relationship between eugenics and socialism also resembled Schallmayer's. For the latter's views, see Weiss, *Race Hygiene and National Efficiency*, 86 and 104–6, and Weindling, *Health, Race and German Politics*, 130–1.

19. Apáthy, "A faj egészségtana", 271.

20. Ibid., 278.

21. Ibid., 279.

22. István Apáthy, *A fejlődés törvényei és a társadalom* (Budapest: Magyar Társadalomtudományi Egyesület, 1912), esp. 269–76. See also the review published in *Huszadik Század* 14, 6 (1913): 795–800.

23. Apáthy, *A fejlődés törvényei és a társadalom*, 275.

24. Ibid., 277.

25. See, for example, Pál Balogh, *Magyar faj uralma* (Budapest: Lampel R., 1903), and Mihály Réz, *Magyar fajpolitika* (Budapest: Kilán Frigyes, 1905). For a discussion of these cultural and political trends, see Marius Turda, *The Idea of National Superiority in Central Europe, 1880–1918* (New York: Edwin Mellen Press, 2005), esp. 133–57, and Ildikó Nagy, "Jenő Rákosi and the Hungarian Empire of 30 Million People", in Ferenc Gereben, ed., *Hungarian Minorities and Central Europe: Regionalism, National and Religious Identity* (Piliscsaba: Pázmány Péter Catholic University, 2002), 203–19.

26. In fact, the Hungarian-speaking population increased significantly between 1880 and 1910, on account of both the expansion of the population and the assimilation of other nationalities. This growth was reflected in the 1910

census, which confirmed that more than 50% of the population in Hungary (excluding Croatia-Slavonia) spoke Hungarian. See Dorothy Good, "Some Aspects of Fertility Change in Hungary", *Population Index* 30, 2 (1964): 137–71, and László Katus, "Magyarok, nemzetiségek a népszaporulat tükrében 1850–1918", *História* 4–5 (1982): 18–21.

27. "A magyarországi fajok harczára vonatkozó kérdőpontok", OSZK, Manuscript Collection, Fond *Magyar Társadalomtudományi Egyesület iratai*, no. 2454, vol. 1, file 185, 1–2.

28. Ibid., p. 1.

29. Ibid., p. 2. See also István Apáthy, "Miért nem népszerű ma a magyar nemzet?", *Új Nemzedék* 1, 11 (1914): 1–3, and idem, "Radikálizmus és magyarság", *Új Nemzedék* 3, 46 (1916): 1–4.

30. For Erdélyi Múzeum-Egyesület's activities and relationship to other cultural associations in Transylvania, see Borbála Zsuzsanna Török, "The Ethnic Design of Scholarship: Learned Societies and State Intervention in 19th Century Transylvania", in Victor Karady and Borbála Zsuzsanna Török, eds, *Cultural Dimensions of Elite Formations in Transylvania (1770–1950)* (Cluj-Napoca: Ethnocultural Diversity Resource Center, 2008), 115–37.

31. The Austro-Hungarian Foreign Service was divided into three branches: an internal branch, based at the Ballhausplatz headquarters in Vienna, the diplomatic corps and the consular service. The consular service, to which Hoffmann belonged, had the following ranking: consular attaché, vice-consul, consul, consul general second class and finally consul general first class. See István Diószegi, *Hungarians in the Ballhausplatz: Studies in the Austro-Hungarian Common Foreign Policy* (Budapest: Corvina, 1983); Erwin Matsch, *Der auswärtige Dienst von Österreich-Ungarn 1720–1920* (Vienna Böhlau, 1986); and William D. Godsey, *Aristocratic Redoubt: The Austro-Hungarian Foreign Office on the Eve of the First World War* (West Lafayette: Purdue University Press, 1999). Hoffmann began his career as consular attaché in 1908, was then promoted to vice-consul in 1910 and in 1916 to consul. He became consul general first class in 1919. See *Magyar Országos Levéltár* (hereafter MOL), Budapest, Fond K59-7/d (1921).

32. The Austrian author Heimito von Doderer described the Academy's ritualized student life and quality teaching in his 1951 novel *Die Strudlhofstiege oder Melzer und die Tiefe der Jahre*.

33. MOL, Fond K59-7/d (1921).

34. Hugó Hoffmann was Lieutenant General (*Feldzeugmeister*) in the artillery. For the imperial decree, see MOL, Fond K20-989 (1908).

35. Godsey, *Aristocratic Redoubt*, 165.

36. Hoffmann arrived in New York on 8 November 1908. See *New York Passenger Lists, 1820–1957*, Year 1908; microfilm serial: T715; microfilm roll: T715–1166; line 1; 102. For the Consulate General in New York see Rudolf Agstner, *Austria (-Hungary) and Its Consulates in the United States of America since 1820* (Münster: LIT Verlag, 2012), esp. 242–58.

37. It is estimated that around over one million Hungarian citizens emigrated to the USA between 1901 and 1910. See Leslie Konnyu, *Hungarians in the USA: An Immigration Study* (St Louis: The American Hungarian Review, 1967), 22. See also Julianna Puskás, *From Hungary to the United States, 1880–1914* (Budapest: Akadémiai Kiadó, 1982).

38. Ibid.
39. Godsey, *Aristocratic Redoubt*, 145.
40. Between May and September 1902, the Országos Magyar Gazdasági Egyesület (Hungarian National Economic Association) organized a number of congresses on emigration in the regions most greatly affected (Felvidék, Dunántúl and Délvidék). See Puskás, *From Hungary to the United States*, 96.
41. See, for example, *A kivándorlás: a Magyar Gyáriparosok Országos Szövetsége által tartott országos ankét tárgyalásai* (Budapest: Pesti Lloyd nyomda, 1907); Pál Farkas, *Az amerikai kivándorlás* (Budapest: Singer and Wolfner, 1907); and Andor Löherer, *Az amerikai kivándorlás és visszavándorlás* (Budapest: Pátria, 1908).
42. Bertalan Neményi, *A magyar nép állapota és az amerikai kivándorlás* (Budapest: Athenaeum, 1911).
43. Géza Hoffmann, "Az amerikai magyarság", *Közgazdasági Szemle* 44, 1–2 (1910): 455–85.
44. Géza Hoffmann, *Csonka munkásosztály: az amerikai magyarság* (Budapest: Magyar Közgazdasági Társaság, 1911).
45. A complete list of Hungarian associations, churches (Catholic, Greek-Orthodox and Protestant) and libraries in the USA, together with the names and collections of Hungarian newspapers consulted for Hoffmann's study was provided in the appendix.
46. Ibid., 325.
47. Ibid., 334–5.
48. Ibid., 336.
49. Ibid., 336–7.
50. Franz Boas, *Changes in Bodily Form of Descendants of Immigrants* (Washington: U.S. Government Printing Office, 1910), 5.
51. Clarence C. Gravlee, H. Russell Bernard and William R. Leonard, "Heredity, Environment, and Cranial Form: A Reanalysis of Boas's Immigrant Data", *American Anthropologist* 105, 1 (2003): 125–38.
52. Hoffmann, *Csonka munkásosztály*, 345.
53. B. R., "Magyar kivándorlás Amerikába", *Huszadik Század* 13, 5 (1912): 649–54.
54. See Mark A. Largent, *Breeding Contempt: The History of Coerced Sterilization in the United States* (New Brunswick: Rutgers University Press, 2008), and Paul Lombardo, ed., *A Century of Eugenics in America: From the Indiana Experiment to the Human Genome Era* (Bloomington: Indiana University Press, 2011).
55. See Géza Hoffmann, "Die Einschränkung der Einwanderung in der Vereinigten Staaten von Amerika", *Ungarische Rundschau für historische und soziale Wissenschaften* 1,1 (1912): 104–15, and idem, "A bevándorlás és a munkanélküliség az Egyesült Államokban", in Imre Ferenczi, Géza Hoffmann and Imre Illés, *A munkanélküliség és a munkásvándorlások* (Budapest: Benkő, 1913), 64–84; on the subject of "acculturation", see his two-part study, "Akkulturation unter den Magyaren in Amerika", *Zeitschrift für Sozialwissenschaft* 4, 5 (1913): 309–25, and idem, "Akkulturation unter den Magyaren in Amerika", *Zeitschrift für Sozialwissenschaft* 4, 6 (1913): 393–407.
56. Géza Hoffmann, "Die Regelung der Ehe im Rassenhygienischen Sinne in den Vereinigten Staaten von Nordamerika", *Archiv für Rassen- und Gesellschaftsbiologie* 9, 6 (1912): 730–61.

57. See, for example, Géza Hoffmann "Anregung zur Einführung von Gesundheitszeugnissen in Ungarn", *Archiv für Rassen- und Gesellschaftsbiologie* 10, 5 (1913): 696; and his review of László Nógrády's book, "Die Einkinderehe in der ungarischen Landbevölkerung", *Archiv für Rassen- und Gesellschaftsbiologie* 10, 6 (1913): 813–14.

58. See Paula Lukács (Hoffmann Gézáné), "Az eugenika oktatása az Egyesült Államokban", *Huszadik Század* 14, 4 (1913): 519–20, and idem, "Anyák 'nyugdíja' az északamerikai egyesült államokban", *Huszadik Század* 14, 7–8 (1913): 100–3. For a discussion of eugenics as a "new religion", see Marius Turda, "A New Religion: Eugenics and Racial Scientism in Pre-First World War Hungary", *Totalitarian Movements and Political Religions* 7, 3 (2006): 303–25.

59. Géza von Hoffmann, *Rassenhygiene in den Vereinigten Staaten von Nordamerika* (München: J. F. Lehmans Verlag, 1913).

60. Ibid., ix.

61. Ibid., 2. In his first footnote, Hoffmann acknowledged that in Germany the less precise but naturalized term "Rassenhygiene" was regularly used for eugenics. For the sake of his readership, he similarly adopted this term in this book.

62. Ibid., 2–9.

63. Ibid., 10.

64. Ibid., 11.

65. Ibid.

66. Ibid., 13, and Wagenen, "Preliminary Report of the Committee of the Eugenic Section of the American Breeders' Association", 462.

67. Hoffmann, *Rassenhygiene*, 11.

68. Ibid., 14.

69. Ibid.

70. By the end of the 1910s, the eugenics movement in the USA was widely supported by politicians, social activists, private philanthropists and important members of the medical profession. As Laura L. Lovett noted, "by 1912, the entire annual meeting of the congress was devoted to the 'conservation of human life'". See Laura L. Lovett, *Conceiving the Future: Pronatalism, Reproduction, and the Family in the United States, 1890–1938* (Chapel Hill: University of North Carolina Press, 2007), 123. For broader developments on American eugenics during this period see Wendy Kline, *Building a Better Race: Gender, Sexuality, and Eugenics from the Turn of the Century to the Baby Boom* (Berkeley: University of California Press, 2001).

71. Hoffmann, *Rassenhygiene*, 17.

72. Ibid., 18–20.

73. These organizations included the Race Betterment Foundation in Battle Creek, Michigan (established in 1906), and the Eugenics Records Office, Cold Spring Harbor, New York (founded in 1911). See ibid., 21–4 and 26–31. Other countries in which professional eugenics organizations existed were Germany (International Society for Racial Hygiene), Britain (Eugenics Education Society) and Sweden (Swedish Society for Racial Hygiene). France and Hungary are also mentioned as countries where eugenics societies were soon to be established. See ibid., 24–7.

74. Ibid., 33.

75. Ibid., 34–41.

76. Ibid., 51.
77. Ibid., 64.
78. Ibid., 74–5.
79. Ibid., 75–95.
80. Ibid., 96.
81. Ibid., 110–25.
82. Ibid., 118–19.
83. Ibid., 126–48.
84. Hoffmann's bibliography is thematically organized. The first part deals with various topics, such as "publications dealing with racial hygiene in general", "publications on 'euthanasia'" and a "list of journals and periodicals" concerning eugenics (he also included all relevant journals: *Archiv für Rassen-und Gesellschaftsbiologie, Biometrika, Eugenics Review, Journal of Race Development, Journal of Genetics* and *Politisch-Anthropologische Revue*). The second part of his bibliography includes an alphabetical list of all publications on eugenics. As this indicates, Hoffmann's bibliography was a remarkable achievement in itself. It was not until 1924 that another extensive bibliography of eugenic literature was published in any other language, namely Samuel J. Holmes, *A Bibliography of Eugenics* (Berkeley: University of California Press, 1924).
85. As reflected, for example, in work by the Russian eugenicist, Evgenij A. Šepilevskij, *Osnovy i sredstva rasovoj gigieny* (Yuryev: Tipografia K. Matticeia, 1914). I would like to thank Björn Felder for drawing my attention to this reference.
86. Hoffmann continued to report on eugenic developments in the USA for German journals. See, for example, his "Rassenhygienische Jahresversammlung in den Vereinigten Staaten von Nordamerika", *Archiv für Rassen- und Gesellschaftsbiologie* 10, 6 (1913): 829–30, and idem, "Der nächste Internationale Kongreß für Rassenhygiene soll im Sept. 1915 in Neuyork stattfinden", *Archiv für Rassen- und Gesellschaftsbiologie* 10, 6 (1913): 831.
87. Edgar Schuster, "Von Hoffmann Geza. *Die Rassenhygiene in den Vereinigten Staaten von Nordamerika*", *The Eugenics Review* 5, 3 (1913): 279.
88. Amey Eaton Watson, "Recent Books on Human Heredity", *The Journal of Heredity* 5, 9 (1914): 373.
89. "Notices", *Bulletin of the American Academy of Medicine* 15, 5 (1913): 377.
90. Fritz Lenz, "Hoffmann, Géza: Die Rassenhygiene in den Vereinigten Staaten von Nordamerika", *Archiv für Rassen- und Gesellschaftsbiologie* 10, 1–2 (1913): 252. It would be almost a decade until another book on the history and practices of eugenic sterilization in the USA would be published. See Harry H. Laughlin, *Eugenical Sterilization in the United States* (Chicago: Psychopathic Laboratory of the Municipal Court of Chicago, 1922).
91. Bernard Glueck, "Die Rassenhygiene in den Vereinigten Staaten von Nordamerika. Von Geza von Hoffmann", *Journal of the American Institute of Law and Criminology* 4, 6 (1914): 934.
92. Géza Hoffmann, "Az első állami fajhygieniai hivatal", *Szociálpolitikai Szemle* 3, 11 (1913): 161–2.
93. René Berkovits, "Újabb tanulmányok a szociálbiológia köréből", *Huszadik Század* 14, 12 (1913): 610–21.
94. Emil Oberholzer, "Kastration und Sterilisation von Geisteskranken in der Schweiz", *Juristisch-psychiatrische Grenzfragen* 8, 1–3 (1911): 25–144.

95. Berkovits, "Újabb tanulmányok a szociálbiológia köréből", 617.
96. See, for example, Weindling, *Health, Race and German Politics*, and Kühl, *The Nazi Connection*.
97. See Ödön Tuszkai, "Cultura és hygiene", *Magyar Társadalomtudományi Szemle* 6, 1 (1913): 51–7.
98. Maria Björkman and Sven Widmalm, "Selling Eugenics: The Case of Sweden", *Notes and Records of the Royal Society* 64, 4 (2010): 379–400.
99. Schneider, *Quality and Quantity*, 84–115.
100. Francesco Cassata, *Building the New Man: Eugenics, Racial Sciences and Genetics in Twentieth Century Italy* (Budapest: CEU Press, 2011), 40–2.
101. Karl Herfort and Arthur Brožek, "Die eugenische Zentrale des Ernestinums", *EOS: Vierteljahresschrift für die Erkenntnis und Behandlung jugendlicher Abnormer* 10, 3 (1914): 161–7. See also "Eugenics Research in Bohemia", *The Journal of Heredity* 7, 4 (1916): 157.
102. "Eugenics in Austria", *The Eugenics Review* 5, 4 (1914): 387. References to the Viennese Section for Social Biology and Eugenics are also found in Hoffmann's letter to H. H. Laughlin (dated 27 December 1913). 'C' Box, C-2-1-2. Harry H. Laughlin Papers (hereafter HHLP), Pickler Memorial Library, Truman State University, Kirksville, MO. For the context, see Weindling, "A City Regenerated", 86, and especially 108, n. 22, and idem, *Health, Race and German Politics*, 140.
103. Harry H. Laughlin, *Eugenics Record Office. Report No. 1* (Cold Spring Harbor: Eugenics Record Office, 1913), 24. Laughlin also noted Berkovits's "proposed eugenic society in Nagyvárad" in a letter to Hoffmann sent on 31 October 1913. This letter did not survive either, although it is referred to by Hoffmann in a letter to H. H. Laughlin dated 27 December 1913. See 'C' Box, C-2-1-2 (hereafter HHLP), Pickler Memorial Library, Truman State University, Kirksville, MO.
104. Hoffmann's letter to Apáthy (dated 22 October 1913), OSZK, Manuscript Collection, Fond *Apáthy István iratai*, no. 2453, file 116–17.
105. Ibid.
106. For the Gyermektanulmányi Múzeum's activities, see László Nagy and Károly Ballai, eds, *A Gyermektanulmányi Múzeum szervezete és az anyaggyűjtés szabályai* (Budapest: Hungaria, 1910). The Society for Child Study organized its first national congress and exhibition during March 1913. See Dániel Répay and Károly Ballai, eds, *Az Első Magyar Országos Gyermektanulmányi Kongresszus naplója és a vele kapcsolatos Kiállítás leírása* (Budapest: Fritz Ármin, 1913).
107. Hoffmann's letter to Apáthy, OSZK, Manuscript Collection, Fond *Apáthy István iratai*, no. 2453, File 118–19.
108. As Farkas Heller confirmed in a letter to Apáthy dated 6 November 1913. In ibid., File 109–10.
109. Ibid., File 116–17. In another letter, dated 28 October 1913, Hoffmann expressed an interest in Apáthy's publications.
110. Ibid.
111. See the Papers of the *Budapesti Királyi Orvosegyesület*, SOL, II6/913.
112. See Farkas Heller's letter to Apáthy dated 27 December 1913. In OSZK, Manuscript Collection, Fond *Apáthy István iratai*, no. 2453.
113. István Apáthy, "Fajegészségügy és fajegészségtan", *Magyar Társadalomtudományi Szemle* 7, 1 (1914): 52–65.

114. Ibid., 53. In German, as Apáthy remarked, there were two terms for eugenics: *Rassenhygiene* and *Rassenveredelung*.
115. In ibid., 53–4.
116. Ibid., 54 (emphasis in the original).
117. Ibid.
118. Ibid.
119. Hayden White, *Tropics of Discourse: Essays in Cultural Criticism* (Baltimore: Johns Hopkins University Press, 1978), 134. Apáthy's advocacy for a eugenic vocabulary in Hungary was an integral part of a broader endeavour to create a Hungarian scientific language to replace German and Latin, both extensively used during the eighteenth and nineteenth centuries. See Gábor Palló, "Scientific Nationalism: A Historical Approach to Nature in Late Nineteenth-Century Hungary", in Ash and Surman, eds, *The Nationalization of Scientific Knowledge*, 102–12.
120. Weindling, *Health, Race and German Politics*, 125.
121. Apáthy, "Fajegészségügy és fajegészségtan", 57.
122. In an earlier text, Apáthy expressly argued that socialism was the key to the emergence of a new eugenic morality in Hungary. See his *A socialismus az emberi továbbfejlődés szempontjából* (Budapest: Országos Ismeretterjesztő Társulat, 1913). Here too, Apáthy's views are similar to those expressed by Wilhelm Schallmayer, particularly regarding the latter's political ideal, described by Sheila Weiss as a "meritocracy, whereby the most advanced form of human and economic organization, a form of state socialism, would be created in order to promote a higher level of national efficiency". See Weiss, *Race Hygiene and National Efficiency*, 86. See also Weindling, *Health, Race and German Politics*, 95.
123. Apáthy, "Fajegészségügy és fajegészségtan", 58.
124. Ibid.
125. Ibid., 65.
126. Apáthy's letter to Teleki (dated 14 January 1914), OSZK, Manuscript Collection, Fond *Apáthy István iratai*, no. 2453, Files 65–6.
127. Teleki's letter to Apáthy (dated 16 January 1914), in ibid., File 180.
128. Géza Hoffmann, "Ausschüsse für Rassenhygiene in Ungarn", *Archiv für Rassen und Gesellschaftsbiologie* 10, 6 (1913): 831.
129. This sharply contrasts with what Hoffmann claimed in a letter to Harry H. Laughlin on 26 May 1914. 'See 'C' Box, C-2-1-2 (HHLP).
130. Géza Hoffmann, "Eugenika", *Magyar Társadalomtudományi Szemle* 7, 2 (1914): 91–106.
131. Ibid., 91–5.
132. Ibid., 96.
133. Ibid.
134. Ibid., 99–100.
135. Ibid., 100.
136. Ibid.
137. ibid., 101.
138. Ibid., 105–6.
139. Ibid., 106.
140. Ibid.

141. "A fajegészségügyi (eugenikai) szakosztály megalakulása", *Magyar Társadalomtudományi Szemle* 7, 2 (1914): 165–72.

142. Ibid., 168.

143. Ibid., 169.

144. Apáthy was reacting to what had become, by the end of the nineteenth century, a debate over the "Hungarian racial type". Aspects of this debate are discussed in Tibor Frank, "Anthropology and Politics: Craniology and Racism in the Austro-Hungarian Monarchy", in idem, *Ethnicity, Propaganda, Myth-Making: Studies on Hungarian Connections to Britain and America 1848– 1945* (Budapest: Akadémiai Kiadó, 1999), 15–34; Emese Lafferton, "The Magyar Moustache: The Faces of Hungarian State Formation, 1867–1918", *Studies in the History and Philosophy of Biological and Medical Sciences* 38 (2007), 706–32; Marius Turda, "Entangled Traditions of Race: Physical Anthropology in Hungary and Romania, 1900–1940", *Focaal* 58, 3 (2010): 32–46; and idem, "Race, Politics and Nationalist Darwinism in Hungary, 1880–1918", *Ab Imperio Quarterly* 1 (2007): 139–64.

145. "A fajegészségügyi (eugenikai) szakosztály megalakulása", 169.

146. Ibid.

147. *The Eugenics Education Society: The Annual Report, 1908* (London: The Eugenics Education Society, 1908), 21.

148. Quoted in Sheila F. Weiss, "The Race Hygiene Movement in Germany, 1904–1945", in Mark B. Adams, ed., *The Wellborn Science: Eugenics in Germany, France, Brazil, and Russia* (New York: Oxford University Press, 1990), 23–4.

149. "A fajegészségügyi (eugenikai) szakosztály megalakulása", 170.

150. Ibid. Apáthy's criticism of psychoanalysis did not go unnoticed. The next day (25 January 1914) the Budapest-based Jewish newspaper *Pesti Hírlap* attacked Apáthy for his anti-Semitic comments. Within days, Sigmund Freud himself was aware of Apáthy's remarks. On 29 January 1914, the Hungarian psychoanalyst Sándor Ferenczi wrote to Freud that Apáthy "has put himself at the head of the eugenic movement" and that "from this position has let loose against psychoanalysis – as *a panerotic aberration of the Jewish spirit*". Quoted in Eva Brabant et al., eds, *The Correspondence of Sigmund Freud and Sándor Ferenczi*, vol. 1, 1908–1914 (Cambridge, MA, MIT Press, 1993), 535 (emphasis in the original).

151. "A fajegészségügyi (eugenikai) szakosztály megalakulása", 171.

152. Ibid.

153. Ibid., 172.

154. Teleki's letter to Apáthy (dated 24 March 1914), OSZK, Manuscript Collection, Fond *Apáthy István iratai*, no. 2453, File 63.

155. "A fajegészségügyi bizottság értekezlete", *Magyar Társadalomtudományi Szemle* 7, 4 (1914): 317.

156. Teleki's letter to Apáthy (dated 16 January 1914), OSZK, Manuscript Collection, Fond *Apáthy István iratai*, no. 2453, File 180.

157. Teleki's letter to Apáthy (dated 13 March 1914), ibid., File 64.

158. By 1914, Oszkár Jászi and his circle had moved decidedly towards a "radical" form of counter-politics, culminating in the creation of the National Bourgeois Radical Party (Országos Polgári Radikális Párt) in June 1914. See Litván, *A Twentieth-Century Prophet*, 93–6.

159. In contrast, the Gesellschaft für Rassenhygiene had 20 members (18 Germans and 2 non-Germans), in 1905, while the Eugenics Education Society's council had 24 members in 1909. Correspondingly, the French Eugenics Society had 44 members in 1912, while 16 members formed the Italian Committee of Eugenic Studies in 1913. For particular details on these societies' memberships, see Weindling, *Health, Race and German Politics*, 145; *The Eugenics Education Society: The Annual Report*, 19; Schneider, *Quality and Quantity*, 97; and Cassata, *Building the New Man*, 40.

160. See the Biographical Information section at the end of the book.

161. "A personal network", according to Adams, emerged among individuals sharing an "extended family, old school ties, mutual experience, hobbies, private passions, and shared interests". See Mark B. Adams, "Networks in Action: The Khrushchev Era, the Cold War, and the Transformation of Soviet Science", in Garland E. Allen and Roy M. MacLeod, eds, *Science, History and Social Activism: A Tribute to Everett Mendelsohn* (Dordrecht: Kluwer Academic Publishers, 2001), 261.

162. Nikolai Krementsov, *International Science between the Two World Wars: The Case of Genetics* (London: Routledge, 2005), 6. Also see the insightful discussion in Randall Collins, *The Sociology of Philosophies: A Global Theory of Intellectual Change* (Cambridge, MA: Harvard University Press, 1998).

163. "A fajegészségügyi bizottság értekezlete", 317.

164. Ibid.

165. Ibid., 318.

166. "A választmány jelentése az 1914. évi közgyűléshez", *Magyar Társadalomtudományi Szemle* 7, 7 (1914): 588–9.

167. Géza Hoffmann, "Rassenhygienische Gedanken bei Platon", *Archiv für Rassen- und Gesellschaftsbiologie* 11, 2 (1914): 174–83.

168. Géza Hoffmann, "A fajegészségtan irodalma", *Magyar Társadalomtudományi Szemle* 7, 3 (1914): 221–4. See also his Bibliographie der Rassenhygiene", *Archiv für Rassen- und Gesellschaftsbiologie* 11, 1 (1914): 131–2.

169. Hoffmann continued to publish on American eugenics, particularly in the *Archiv für Rassen- und Gesellschaftsbiologie*. See, for example, Géza Hoffmann, "Die rassenhygienischen Gesetze des Jahres 1913 in den Vereinigten Staaten von Nordamerika", *Archiv für Rassen- und Gesellschaftsbiologie* 11, 1 (1914): 21–32; idem, "Das Sterilisierungsprogram in den Vereinigten Staaten von Nordamerika", *Archiv für Rassen- und Gesellschaftsbiologie* 11, 2 (1914): 184–92; and idem, "Rassenhygiene im Lehrplan der nordamerikanischen Universitäten", *Archiv für Rassen- und Gesellschaftsbiologie* 11, 2 (1914): 281.

170. Géza Hoffmann, "Fajegészségügyi (eugenikai) társaságok és működésük", *Magyar Társadalomtudományi Szemle* 7, 5 (1914): 350–6.

171. "A nagyváradi feministák egylete", *A Nő* 1, 2 (1914): 41.

172. Weindling, *Health, Race and German Politics*, 258.

173. Mrs Szirmay Oszkár, "Feminizmus és anyaság", *A Nő* 1, 9 (1914): 186.

174. Zoltán Szász, *A szerelem* (Budapest: Pallas nyomda, 1913), 7.

175. Ibid., 276–81.

176. Ibid., 281.

177. Margit Kaffka, "Szász Zoltán: *A szerelem*", *Nyugat* 6, 15 (1913): 197–9.

178. M. M. E., "Fajegészségügy, fajjavítás – anyák nélkül", *A Nő* 1, 3 (1914): 54.

179. Ibid.
180. Ibid., 55.
181. Géza Hoffmann, "A fajegészségtan (eugenika) és a nő", *A Nő* 1, 8 (1914): 159–61.
182. Ibid., 159.
183. Ibid.
184. Ibid., 160.
185. Ibid.
186. Lipót Nemes, *A kültelki gyermekek élete és jövője* (Budapest: Hungaria könyvnyomda, 1913).
187. Géza Hoffmann, "Környezet vagy átöröklés?", *Magyar Társadalomtudományi Szemle* 7, 4 (1914): 295–7.
188. OSZK, Manuscript Collection, Fond *Magyar Társadalomtudományi Egyesület iratai*, no. 2454, File 65. The title of his lecture was "Rassenhygienische Erfahrungen in Nordamerika". The *Berliner Gesellschaft für Rassenhygiene* was also known as "Ortsgruppe Berlin". Other "local groups" were established in Munich, Freiburg and Stuttgart.
189. For instance, Hoffmann attended the Society for Racial Hygiene's Congress in Jena (6–7 June 1914) and its annual meetings in Munich (22–3 July 1916 and 29 July 1917, respectively).
190. Generalverwaltung der Kaiser-Wilhelm-Gesellschaft. I. Abt., Rep. 1A, Nr. 1350, Bl. 80f. Archiv der Max-Planck-Gesellschaft, Berlin-Dahlem.
191. Ibid.
192. Weindling, *Health, Race and German Politics*, 240.
193. See, for example, Géza Hoffmann, "Rassenhygiene und Familienforschung. Leitsätze der Deutschen Gesellschaft für Rassenhygiene über die Geburtenfrage", *Monatsblatt der Kais. Kön. Heraldischen Gesellschaft "Adler"* 7, 43 (1914): 373–5; idem, "A születések csökkenő száma és a német fajegészségtani (eugenikai) társaság", *Magyar Társadalomtudományi Szemle* 7, 7 (1914): 560–3; idem, "A népesség csökkenő szaporodása és a fajegészségügy (eugenika)", *Közgazdasági Szemle* 38, 11 (1914): 526–7; and idem, "Eugenics in Germany: Society of Race Hygiene Adopts Resolution Calling for Extensive Program of Positive Measures to Check Decline in Birth-Rate", *The Journal of Heredity* 5, 10 (1914): 435–6.
194. Hoffmann, "Eugenics in Germany", 435.
195. Ibid.
196. Ibid., 436.
197. Ibid.
198. For a list of speakers and topics, see *Die Erhaltung und Mehrung der deutschen Volkskraft: Verhandlungen der 8. Konferenz der Zentralstelle für Volkswohlfahrt vom 26.–28. Oktober 1915* (Berlin: Heymanns, 1916).
199. Fritz Lenz, "Aus der Gesellschaft für Rassenhygiene", *Archiv für Rassen- und Gesellschaftsbiologie* 11, 4 (1915): 561.
200. Géza Hoffmann, "Race Hygiene in Germany", *The Journal of Heredity* 7, 1 (1916): 32.
201. On Britain, see Soloway, *Demography and Degeneration* and Deborah Dvork, *War is Good for Babies & Other Young Children: A History of the Infant and Child Welfare Movement* (London: Tavistock Publications, 1987); on Germany, see Weindling, *Health, Race and German Politics*, and Annette F. Timm, *The Politics*

of Fertility in Twentieth-Century Berlin (Cambridge University Press, 2010). Finally, on France, see Schneider, *Quality and Quantity*, and Joshua Cole, *The Power of Large Numbers: Population, Politics, and Gender in Nineteenth-Century France* (Ithaca: Cornell University Press, 2000).
202. Béla Tomka, "Social Integration in 20th Century Europe: Evidence from Hungarian Family Development", *Journal of Social History* 35, 2 (2001): 327–48.
203. Soloway, *Demography and Degeneration*, 59.

5   Health Anxieties and War

1. The pervasiveness of eugenic themes in American films has been addressed in Martin Pernick's *The Black Stork: Eugenics and the Death of "Defective" Babies in American Medicine and Motion Pictures since 1915* (New York: Oxford University Press, 1996).
2. *Wien im Krieg* was written and directed by Fritz Freisler and Heinz Hanus. For a discussion of its immediate context and cultural impact, see Sema Colpan, "Geschlechterrollen im Film im Ersten Weltkrieg am Beispiel von 'Wien im Krieg' (1916)" (dissertation (Mag. Phil.), University of Vienna, 2009). For a description of this scene, see Maureen Healy, *Vienna and the Fall of the Habsburg Empire: Total War and Everyday Life in World War I* (Cambridge University Press, 2007), 266.
3. For a general discussion, see Roger Chickering and Stig Förster, eds, *Great War, Total War: Combat and Mobilization on the Western Front, 1914–1918* (Cambridge University Press, 2000).
4. József Galántai, *Hungary in the First World War* (Budapest: Akadémiai Kiadó, 1989), 93–5. See also Béla K. Király, Peter Pastor and Ivan Sanders, eds, *Essays on World War I: Total War and Peacemaking. A Case Study of Trianon* (New York: Social Science Monographs, 1984).
5. Wilhelm Schweisheimer, "Bevölkerungsbiologische Bilanz des Krieges 1914/19", *Archiv für Rassen- und Gesellschaftsbiologie* 13, 2/4 (1920): 176–93.
6. See Clemens Pirquet, "Einleitung", in idem, ed., *Volksgesundheit im Krieg*, vol. 1 (Vienna: Hölder Pichler Tempsky, 1926), 4. For more data, particularly with respect to Hungary, see Jenő Kollarits, "Beiträge zur Biologie des Krieges mit besonderer Berücksichtigung des Ungartums", *Archiv für Rassen- und Gesellschaftsbiologie* 25, 1 (1931): 19–41, and Theodore Szél, "The Genetic Effects of the War in Hungary", in *A Decade of Progress in Eugenics. Scientific Papers of the Third International Congress of Eugenics Held at the American Museum of Natural History, New York, August 21–23, 1932* (Baltimore: Williams & Wilkins, 1934), 249–54.
7. See József Melly, "A nemi betegségek elterjedettsége, különös tekintettel a székesfővárosra", in Gábor Doros and József Melly, *A nemi betegségek kérdése Budapesten*, vol. 1 (Budapest: Székesfőváros Házinyomdája, 1930), 396–489. See also József Guszman, "Zur Frage der Bekämpfung der venerischen Krankheiten", in Wilhelm Manninger, Karl M. John and Josef Parassin, eds, *Erstes Jahrbuch des Kriegsspitals der Geldinstitute in Budapest. Beiträge zur Kriegsheilkunde* (Berlin: Julius Springer, 1917), 691–704.
8. Zsigmond Somogyi, "A háború és a fertőző betegségek", *Természettudományi Közlöny* 46, 18–19 (1914): 652–7; Tamás Marschalkó, "Háború és venereás

bántalmak", *Orvosi Hetilap* 59, 26 (1915): 347–50; idem, "Háború és venereás bántalmak", *Orvosi Hetilap* 59, 27 (1915): 365–9; idem, "Háború és venereás bántalmak", *Orvosi Hetilap* 59, 28 (1915): 381–3; Dezső Hahn, *A fertőző nemibetegségek és a háború* (Budapest: Népszava könyvkereskedés, 1916); and Béla Entz, "Küzdelem a fertőző betegségek ellen a háborúban", *Természettudományi Közlöny* 48, 15–16 (1916): 489–512. See also Menyhért Szántó, *Küzdelem a népbetegségek ellen* (Budapest: Társadalmi Múzeum, 1916), and Béla Schmidt, "A fertőző nemibetegségek és a háború", *Huszadik Század* 17, 3–4 (1916): 286–8.

9. See Ernő Deutsch, *Az orvos szociális munkája a háború alatt* (Budapest: Radó nyomda, 1915), and László Nagy, *A háború és a gyermek lelke. Adatok a gyermek értelmi, érzelmi és erkölcsi fejlődéséhez* (Budapest: Eggenberger, 1915).

10. Géza Farkas, "A hadsereg táplálása háborúban", *Természettudományi Közlöny* 46, 20–1 (1914): 673–83 and Károly Schaffer, "A háború és az idegrendszer", *Budapesti Szemle* 43, 459 (1915): 396–407.

11. Lajos Száhlender, "A háborúban használható fojtó, mérges és könnyezést fakasztó gázokról", *Természettudományi Közlöny* 48, 3–4 (1916): 120–1.

12. Paul J. Weindling, *Epidemics and Genocide in Eastern Europe, 1890–1945* (Oxford University Press, 2000), 75.

13. Richard Pearson Strong et al., *Typhus Fever with Particular Reference to the Serbian Epidemic* (Cambridge, MA: Harvard University Press, 1920).

14. Gábor Kiss, "Megfigyelőállomások és sebesültszállítmányt kísérő osztagok tevékenysége az első világháborúban", *Orvostörténeti Közlemények* 49, 3–4 (2004): 69–83.

15. Béla Johan, "Über Schutzimpfstoffe", in Wilhelm Manninger, Karl M. John and Josef Parassin, eds, *Erstes Jahrbuch des Kriegsspitals der Geldinstitute in Budapest. Beiträge zur Kriegsheilkunde* (Berlin: Verlag von Julius Springer, 1917), 567–80.

16. Gábor Kiss, "Orvosok a m. kir. honvédségben (1868–1918)", *Orvostörténeti Közlemények* 50, 1–4 (2005): 135–47, and idem, "Honvéd, valamint császári és királyi egészségügyi intézmények az első világháború idején", *Orvostörténeti Közlemények* 51, 3–4 (2006): 191–204.

17. See Wilhelm Raschofsky, "Militärärztliche Organization und Leistungen der Epidemiespitäler der österreichisch-ungarischen Armee", in Pirquet, ed., *Volksgesundheit im Krieg*, vol. 1, 122–32.

18. Menyhért Szántó and Ernő Tomor, eds, *Had- és Népegészségügyi kiállítás katalógusa* (Budapest: Hornyánszky, 1915).

19. Two societies were particularly active: the National Association for the Protection against Venereal Diseases (Venereás Betegségek Elleni Országos Védő Egyesület) and the National Committee for the Fight against Tuberculosis (Tuberkulózis Elleni Küzdelem Országos Bizottsága).

20. György Lukács, "Előszó", in Szántó and Tomor, eds, *Had- és Népegészségügyi kiállítás*, 5. On the German exhibition, see Stefan Goebel, "Exhibitions", in Jay Winter and Jean-Louise Robert, eds, *Capital Cities at War: Paris, London, Berlin 1914–1919*, vol. 2 (Cambridge University Press, 2007), 147–8, and Deborah Cohen, *The War Come Home: Disabled Veterans in Britain and Germany, 1914–1939* (Berkeley: University of California Press, 2001), 64. Other exhibitions devoted to similar themes were the German War Exhibition, organized in a number of cities including Berlin, Leipzig, Flensburg, Hamburg, Frankfurt

and Hannover (1916–17), the Vienna War Exhibition in 1916 and the Trieste War Exhibition in 1917. See Britta Lange, *Einen Krieg ausstellen: die "Deutsche Kriegsausstellung" 1916 in Berlin* (Berlin: Verbrecher, 2003), and Healy, *Vienna and the Fall of the Habsburg Empire*, 87–121.

21. Aladár Bálint, "Hadegészségügyi kiállítás", *Nyugat* 8, 8 (1915): 460–1.
22. Ernő Tomor, "Had- és népegészségügyi kiállítás", *Orvosi Hetilap* 58, 19 (1915): 264.
23. Ibid.
24. Ibid.
25. Szántó and Tomor, eds, *Had- és Népegészségügyi kiállítás*, 197–234.
26. Tomor, "Had- és Népegészségügyi kiállítás", 265.
27. Ernő Tomor, *A socialis egészségtan biológiai alapjai* (Budapest: Singer és Wolfner, 1915).
28. For a discussion of Grotjahn's interpretation of social hygiene, see S. Milton Rabson, "Alfred Grotjahn, Founder of Social Hygiene", *Bulletin of the New York Academy of Medicine* 12, 2 (1936): 43–58, and Weindling, *Health, Race and German Politics*, 220–6.
29. Tomor, *A socialis egészségtan*, 14.
30. Ibid., 15–16.
31. Ibid., 74.
32. Ibid., 75.
33. Ibid., 76.
34. Also in 1915 Általános Közjótékonysági Egyesület initiated the publication of a book dealing with the (social, economic, moral, eugenic and medical) impact of the war on Hungarian society. Contributors included János Bókay (on child protection), Ernő Deutsch (on disability), Mihály Lenhossék (on eugenics) and Lajos Nékám (on venereal diseases). See Márton Lányi, ed., *A háború és a jövő* (Budapest: Grill Károly, 1916).
35. "Első Magyarországi Közjóléti Kongresszus Előkészítő Bizottsága". OSZK, Manuscript Collection, Fond *Magyar Társadalomtudományi Egyesület iratai*, no. 2453, File 51.
36. On this eugenic narrative, see Wilhelm Schallmayer, "Eugenik, ihre Grundlagen und ihre Beziehungen zur kulturellen Hebung der Frau", *Archiv für Frauenkunde und Eugenetik* 1, 3 (1914): 271–91.
37. The language of sacrifice and rebirth characterized all countries involved in war, and clearly not only the medical profession. See, for example, Robert Wohl, *The Generation of 1914* (Cambridge, MA: Harvard University Press, 1979); George Mosse, *Fallen Soldiers: Reshaping the Memory of the World Wars* (New York: Oxford University Press, 1991); and Modris Eksteins, *Rites of Spring: The Great War and the Birth of the Modern Age* (Boston: Houghton Mifflin, 1989).
38. On this point, see Jane Lewis, *The Politics of Motherhood: Child and Maternal Welfare in England, 1900–1939* (London: Croom Helm, 1980).
39. See Árpád Szállási, "Bársony János professzor pályafutása, a tévedések anatómiája", *Orvosi Hetilap* 144, 28 (2003): 1402.
40. János Bársony, "Eugenika a háború után", *Orvosi Hetilap* 59, 34 (1915): 451–4, and idem, "Eugenetik nach dem Kriege", *Archiv für Frauenkunde und Eugenetik* 2, 2 (1915): 267–75. In the following, I will use the enlarged, German version of the article.
41. Bársony, "Eugenetik", 267.

42. Ibid.
43. Ibid., 268.
44. Ibid.
45. Ibid., 269.
46. Ibid., 270.
47. Ibid., 272.
48. Ibid., 273–4.
49. Ibid., 275.
50. See László Kiss, "Egészség és politika – az egészségügyi prevenció Magyarországon a 20. század első felében", *Korall* 17 (2004): 109–10.
51. See the letter from the Stefánia Association to Alajos Kovács (dated 19 July 1915). Központi Statisztikai Hivatal (Central Statistical Office, hereafter KSH), VB 1282/1.
52. See Tara Zahra, *Kidnapped Souls: National Indifference and the Battle for Children in the Bohemian Lands, 1900–1948* (Ithaca: Cornell University Press, 2008), esp. 79–80.
53. The official opening was, however, in 1909. The Chairman of the KAVH Board of Directors was the Berlin physician and eugenicist, Carl von Behr-Pinnow. See Stacey Freeman, "Constructing the Pediatric Nurse: Eugenics and the Gendering of Infant Hygiene in Early Twentieth Century Berlin", *Dynamis* 19 (1999): 353–78, and Sigrid Stöckel, *Säuglingsfürsorge zwischen sozialer Hygiene und Eugenik* (Berlin: Walter de Gruyter, 1996), 253–60. See also Weindling, *Health, Race and German Politics*, 206–9, and Edward Ross Dickinson, *The Politics of German Child Welfare from the Empire to the Federal Republic* (Cambridge, MA: Harvard University Press, 1996), 53–4.
54. Freeman, "Constructing the Pediatric Nurse", 360.
55. Weindling, *Health, Race and German Politics*, 287–8.
56. H. J. Gerstenberger, "The Methods of the System Employed in Caring for Institutional Infants Abroad – More Especially in Germany and Austria-Hungary", in *American Association for Study and Prevention of Infant Mortality. Transactions of the Fifth Annual Meeting 1914* (Baltimore: Franklin, 1915), 142.
57. Gerstenberger identified four principles underpinning institutional care for infants in Hungary and Germany: "(1) to keep mother and baby together; (2) to give the infant breast milk; (3) to place the infant in a properly chosen and supervised family-home, and (4) to use institutions – unless mother and child are together and the baby at the breast and well – only as temporary stopping and observation places." In ibid., 148.
58. See, for example, Pál Ruffy, "Az állami gyermekvédelem", *A Gyermekvédelem Lapja* 9, 2 (1914): 339–94; idem, *Állami gyermekvédelem és a hadi árvák* (Budapest: Franklin, 1916); and László Zombory, *Gyermekvédelem a háború alatt* (Budapest: Stephaneum nyomda, 1916).
59. It would be remembered that Teleki refused to have the Társadalomtudományi Társaság become a member of the Eugenics Committee. Madzsar was also in a personal relationship with Oszkár Jászi, having married the latter's sister, Alice.
60. Hoffmann positively reviewed Madzsar's work on infant mortality in Budapest and commented on Madzsar's knowledge of the recent trends in eugenics as revealed in his lecture on "The Protection of Future Generations and the War" in the *Archiv für Rassen und Gesellschaftsbiologie* 12, 2 (1917): 228 and 229.

61. It is therefore not true that it was only in Germany, as Peter Weingart argued, "that the small community of race hygienists (as eugenicists called themselves there), seeking status and recognition, formed a coalition with politicians of the conservative and radical right". See Peter Weingart, "German Eugenics between Science and Politics", *Osiris* 5 (1989): 260.

62. See József Madzsar, "Az anya- és csecsemővédelem a háborúban", *Városi Szemle* 7, 10–11 (1914): 726–30.

63. József Madzsar, *Az anya- és csecsemővédelem országos szervezése: A Stefánia Szövetség alapszabályainak tervezete* (Budapest: Székesfőváros házinyomda, 1915), and idem, *Mit akar a Stefánia-Szövetség?* (Budapest: Pfeiffer, 1916).

64. Madzsar, *Mit akar a Stefánia-Szövetség*, 3–4.

65. Ibid., 14.

66. Madzsar, *Az anya- és csecsemővédelem*, 2–3.

67. Ibid., 6.

68. József Madzsar, "A jövő nemzedék védelme és a háború", *Huszadik Század* 17, 1 (1916): 1–22. The lecture, together with comments from Pál Ruffy, Eugénia Mellerné Miskolczy, Sándor Szana, Miklós Berend, Zoltán Rónai, Zsigmond Kunfi and Sándor Giesswein, was also published as a book in 1916.

69. Ibid., "A jövő nemzedék", 1.

70. From 54,491 in October 1914 to 25,589 in October 1915. In ibid., 5.

71. Ibid., 12–13.

72. Ibid., 13.

73. Ibid.

74. Ibid., 16–17.

75. József Madzsar, *A meddő Budapest* (Budapest: Pfeiffer, 1916). See also the review: "A meddő Budapest", *Új Nemzedék* 4, 7 (1917): 105–7, and Sándor Szana, "A meddő Temesvár", *Népjóléti Közlöny* 1, 4 (1917): 53.

76. *Az anya- és csecsemővédelem a képviselőházban* (Budapest: Pfeiffer, 1916).

77. Ibid., 3.

78. Ibid., 4.

79. Ibid., 5.

80. Ibid., 30.

81. "Meghívó". Invitation sent by Jenő Gaál on 16 December 1915. OSZK, Manuscript Collection, Fond *Magyar Társadalomtudományi Egyesület iratai*, no. 2453, File 53.

82. OSZK, Manuscript Collection, Fond *Magyar Társadalomtudományi Egyesület iratai*, no. 2453, File 54.

83. "Körlevél az Egyesületközi Fajegészségügyi Bizottsághoz" (dated 20 January 1916). OSZK, Manuscript Collection, Fond *Magyar Társadalomtudományi Egyesület iratai*, no. 2453, File 19.

84. Ibid.

85. Ibid.

86. Géza Hoffmann, "Eugenics in Hungary", *The Journal of Heredity* 7, 3 (1916): 105.

87. From a growing literature see, for example, Turda, "The Biology of War", 238–64; idem, *Modernism and Eugenics*, 40–63, and the special issue on "Sexual Deviance and Social Control in Late Imperial Eastern Europe" edited by Keely Stauter-Halsted and Nancy M. Wingfield for the *Journal for the History of Sexuality* in 2011.

88. "Küzdelem a kéttős morál ellen", *A Nő* 4, 12 (1917): 198–9.
89. Géza Hoffmann, "Rassenhygienische Anträge im ungarischen Abgeordneten-hause", *Archiv für Rassen- und Gesellschaftsbiologie* 12, 2 (1917): 253.
90. Ibid.
91. Ibid.
92. Professional itineraries often intersected personal and family connections. It is thus worth mentioning that Hoffmann's wife, Paula, was Lukács's sister.
93. Lajos Nékám, *A háború és a nemi betegségek* (Budapest: Franklin, 1915), and idem, "A háború és a nemi betegségek", in Lányi, ed., *A háború és a jövő*, 139–53.
94. Lajos Nékám, *A Budapesti Kir. M. Tud. Egyetem új klinikája bőr- és nemibetegek számára* (Budapest: Universitas, 1915); idem, *A venereás betegségek társadalmi kihatása* (Budapest: Nemzetvédő Szövetség, 1917); and idem, *A nemi betegek kötelező bejelentésének, nyilvántartásának és gyógyításának kérdése* (Budapest: Nemzetvédő Szövetség, 1917). On Nékám, see Ferenc Földvári, "Nékám Lajos (1868–1957)", *Orvosi Hetilap* 98, 13 (1957): 313–14.
95. *A "Nemzetvédő Szövetség a Nemibajok Ellen" alapszabályai* (Budapest: Stephaneum, 1916), 3–4. See also "Küzdelem a nemi bajok ellen", *Szociál-politikai Szemle* 6, 12 (1916): 344–6.
96. Bashford, *Imperial Hygiene*, 180–4.
97. See "Ankét a nemi betegségek leküzdése tárgyában", *A Nő* 4, 5 (1917): 75–8; "Ankét a nemi betegségek elleni küzdelemről", *A Nő* 4, 6 (1917): 94–6; and "Ankét a nemi betegségek elleni küzdelemről", *A Nő* 4, 7 (1917): 114–16.
98. The contributions were collected and published by Lajos Nékám, ed., *A nemibajok leküzdésének irányítása. A Nemzetvédő Szövetség Törvényelőkészítő Szakértekezletének munkálatai*, vol. 1 (Budapest: Nemzetvédő Szövetség, 1918).
99. This was, indeed, a much wider, transnational phenomenon. See Bashford, *Imperial Hygiene*, 166–7.

6   Eugenics Triumphant

1. See Turda, *Modernism and Eugenics*, 40–63.
2. For American and British perspectives, see S. J. Holmes, "The Decadence of Human Heredity", *The Atlantic Monthly* 114, 3 (1914): 302–8, and Leonard Darwin, "Eugenics and the War", *The Eugenics Review* 6, 3 (1914): 195–203, and idem, "Eugenics during and after the War", *The Eugenics Review* 7, 2 (1915): 91–106. See also Dénes Nagy, "Háború és eugenika", *Huszadik Század* 17, 1 (1916): 77–9 (a review of David Star Jordan's 1914 *War's Aftermath*).
3. Ploetz, *Grundlinien einer Rassen-Hygiene*, vol. 1, 147.
4. The lecture was subsequently published as an article by Mihály Lenhossék, "A háború és a létért való küzdelem tétele", *Természettudományi Közlöny* 47, 619–20 (1915): 91–5. Other speakers included the ophthalmologist Emil Grósz, who lectured on medicine and war and the philosopher Bernát Alexander who discussed war as a national educator. See *Háborús előadások a Budapesti K. M. Tudományegyetemen* (Budapest: Franklin, 1915).
5. Mihály Lenhossék, "A fejlődés mibenléte", *Természettudományi Közlöny* 46, 614–15 (1914): 721–40 (part 1), and idem, "A fejlődés mibenléte", *Természettudományi Közlöny* 46, 616 (1914): 761–71 (part 2).

6. Lenhossék, "A háború", 95.
7. Ibid.
8. Mihály Lenhossék, "Háború és eugenika", in Lányi, ed., *A háború és a jövő*, 95–105.
9. Sándor Giesswein, *A háború és a társadalomtudomány* (Budapest: Stephaneum, 1915), and idem, "A háború szociális problémái", in Lányi, ed., *A háború és a jövő*, 35–42.
10. Ottokár Prohászka, *A háború lelke* (Budapest: Élet, 1915).
11. Lóránt Hegedűs, *A magyarság jövője a háború után. Politikai tanulmány* (Budapest: Athenaeum, 1916).
12. Lajos Méhely, *A háború biológiája* (Budapest: Pallas Irodalmi és Nyomdai Részvénytársaság, 1915). For a discussion of Méhely's ideas during this period, see also Gyurgyák, *A zsidókérdés Magyarországon*, 387–90.
13. Méhely, *A háború biológiája*, 5.
14. Ibid.
15. Ibid.
16. Ibid., 12 (emphasis in the original).
17. Ibid., 19.
18. Ibid., 15 (emphasis in the original).
19. Ibid., 19.
20. Ibid., 19.
21. Since 1838 when the poet Ferenc Kölcsey (1790–1838) expressed his romantic vision of *nemzethalál* in his poem 'The Second Song of Zrínyi', this theme has become one of the cornerstones of Hungarian nationalism.
22. Ibid., 24.
23. Ibid.
24. The accusation that Hungarian Jews avoided military service during the First World War was largely unfounded. Yet anti-Jewish pamphlets and sentiments in Hungary increased after the outbreak of the war in 1914. Even progressive journals like *Huszadik Század* engaged with the topic. See, for example, Dezső Szabó, "A magyar zsidóság organikus elhelyezkedése", *Huszadik Század* 15, 3 (1914): 340–7; and the public debate organized under the title "A zsidóság problémája", *Huszadik Század* 15, 4 (1914): 561–6. See also Oszkár Jászi, ed., *A zsidókérdés Magyarországon. A Huszadik Század körkérdése* (Budapest: A Társadalomtudományi Társaság kiadása, 1917). For reactions in the nationalist press see Győző Concha, "A zsidó-kérdésről", *Új Nemzedék* 4, 32 (1917): 505–7, and idem, "A zsidó-kérdésről", *Új Nemzedék* 4, 33 (1917): 521–4.
25. Within this framework eugenic concern with the racial qualities of the Hungarian nation dovetailed with the emerging racism and anti-Semitism of various Hungarian nationalist physicians, as exemplified by the political career of András Csilléry – future chairman of the Hungarian National Association of Physicians (Magyar Orvosok Nemzeti Egyesülete) and Minister of Public Health in the Friedrich Government (7 August–24 November 1919).
26. Méhely, *A háború biológiája*, 26.
27. Ibid.
28. Ibid.
29. Géza Hoffmann, "Fajegészségtan és eugenika", *Természettudományi Közlöny* 48, 13–14 (1916): 450.

30. Ibid., 452.
31. Ibid. (emphasis in the original).
32. Ibid.
33. Hoffmann will return to this point in a short overview of German racial hygiene written for *The Journal of Heredity* in 1917. "The German conception of race hygiene", he argued, "embraces more than the American idea of pure eugenics, as it is thought that eugenics and other phases of social life are linked together in a way which demands a common consideration of all of our social institutions from the viewpoint of the benefit of future generations." In Géza Hoffmann, "Race Hygiene in Germany", *The Journal of Heredity* 8, 3 (1917): 112.
34. Hoffmann, "Fajegészségtan és eugenika", 453.
35. Ibid.
36. Ibid.
37. Ibid., 454.
38. See Hermann W. Siemens, *Die biologischen Grundlagen der Rassenhygiene und der Bevölkerungspolitik* (Munich: J. F. Lehmanns Verlag, 1917).
39. Géza Hoffmann, *Krieg und Rassenhygiene. Die bevölkerungspolitischen Aufgaben nach dem Kriege* (Munich: J. F. Lehmanns Verlag, 1916). A review of the book was published by László Fenyvessy in *Huszadik Század* 18, 2 (1917): 165–7.
40. Hoffmann, *Krieg und Rassenhygiene*, 7.
41. Ibid., 8.
42. Ibid., 9.
43. Once again Hoffmann specified the difference between eugenics, considered to be "reproductive hygiene", and a "branch of racial hygiene". In ibid.
44. Ibid., 10.
45. Ibid., 11.
46. Ibid., 12.
47. Ibid., 16.
48. Ibid., 17.
49. Ibid., 18.
50. Ibid., 21.
51. Ibid., 21–3.
52. Ibid., 25–6.
53. Ibid., 29.
54. Fritz Lenz, "Hoffmann, Geza. Krieg und Rassenhygiene", *Archiv für Rassen- und Gesellschaftsbiologie* 12, 5–6 (1918): 510–11.
55. Géza Hoffmann, "Háború és fajhygiéne", *A Cél* 7, 5 (1916): 428–37.
56. Géza Hoffmann, "Fajegészségtan és népesedési politika", *Természettudományi Közlöny* 48, 659–60 (1916): 617–21.
57. Ibid., 618.
58. Ibid., 620.
59. Ibid., 620–1 (emphasis in the original).
60. Ibid., 621.
61. See also Géza Hoffmann, "A magyar nép elpóriasodása", *Természettudományi Közlöny* 49, 683–4 (1917): 708–9.
62. Hoffmann, "Fajegészségtan és népesedési politika", 621.
63. Ibid., 620.
64. Ibid., 621.

65. Julius Tandler, "Krieg und Bevölkerung", *Wiener klinische Wochenschrift* 29, 15 (1916): 445. See also Doris Byer, *Rassenhygiene und Wohlfahrtspflege: zur Entstehung eines sozialdemokratischen Machtdispositivs in Österreich bis 1934* (Frankfurt: Campus Verlag, 1988), 75–84.
66. Ibid., 446. For a discussion of Tandler's views on population politics and eugenics, see Gudrun Exner, Josef Kytir and Alexander Pinwinkler, *Bevölkerungswissenschaft in Österreich in der Zwischenkriegszeit (1918–1938): Personen, Institutionen, Diskurse* (Vienna: Böhlau, 2004), 37–8, and Britta McEwen, *Sexual Knowledge: Feeling, Fact, and Social Reform in Vienna, 1900–1934* (New York: Berghahn, 2012), 29–31.
67. Sándor Szana, "Krieg und Bevölkerung", *Wiener Klinische Wochenschrift* 29, 16 (1916): 485–9.
68. Ibid., 485.
69. Ibid.
70. Ibid., 487.
71. Ibid., 488.
72. Sándor Szana, *Irányeszmék a magyar népesedési politikához* (Budapest: Lloyd Társulat, 1916). The text was first published in the journal *Budapesti Orvosi Újság* in six consecutive issues in 1916. See also idem, "A hadiárvák egészségügyi védelme", in Lipót Nemes, ed., *A gyermekmentés útjai* (Budapest: Bethlen Gábor, 1918), 126–38.
73. Szana, *Irányeszmék*, 1.
74. Ibid., 2–4.
75. Ibid., 4–6.
76. Ibid., 7–10.
77. Ibid., 17.
78. Ibid., 17–18.
79. Ibid., 20.
80. Ibid., 21.
81. Ibid., 32–3.
82. Ibid., 33–4.
83. Ibid., 34–5
84. Ibid., 35. See also Sándor Szana, *A társadalmi egészségügy és az orvos* (Budapest: Franklin, 1917).
85. Géza Hoffmann, "Nyelvében él a nemzet", *A Cél* 7, 6 (1916): 221–3. The title is a direct quote from the great Hungarian reformer of the nineteenth century, István Széchenyi.
86. István Apáthy, "Háború és fajegészségügy", *Új Nemzedék* 4, 18–19 (1917): 289–93.
87. Tilkovszky, *Pál Teleki*, 17, and Ablonczy, *Teleki Pál*, 120–5.
88. Pál Petri, *A magyar hadigondozás történetének vázlata* (Budapest: Fritz Ármin, 1917), 5–11. For a description of the *Rokkantügyi Hivatal*'s activities, see Kuno Klebelsberg, *A magyar rokkantügy szervezete* (Budapest: Bíró Miklós könyvnyomdája, 1916).
89. See Ernő Tomor, *Belső telepítés és rokkantügy* (Budapest: Globus, 1915), and Sándor Pályi, ed., *A hadi rokkantak, özvegyek és árvák ügye* (Budapest: Franklin, 1916). In Austria, the Imperial-Royal Austrian Military Widows and Orphans Fund (Kaiserlich-Königlich Österreichischer Militär Witwen

und Waisenfond) was created in 1914. See Healy, *Vienna and the Fall of the Habsburg Empire*, 221–2.

90. Géza Hoffmann, "Eugenics in the Central Empire since 1914", *Social Hygiene* 7 3 (1921): 292.

91. See Béla Molnós-Kovács, *Hadigondozó szociálpolitika* (Budapest: Bethlen Gábor, 1918); Farkas Heller, "Hadigondozásunk fejlődése", *Magyar Társadalomtudományi Szemle* 9, 1 (1918): 6–21; and Aladár Pettkó-Szandtner, *A magyar hadigondozás: visszapillantás és tájékoztató* (Budapest: Pesti nyomda, 1924). For a recent overview, see János Suba, "Az Országos Hadigondozó Hivatal", *Rendvédelem-történeti Füzetek* 18, 21 (2010): 123–39.

92. Pál Teleki, *Szociálpolitika és hadigondozás* (Budapest: Országos Hadigondozó Hivatal, 1918).

93. See "Az Országos Hadigondozó Hivatal népesedéspolitikai és fajegészségügyi tevékenysége", *A Cél* 9, 3 (1918): 441–4.

94. The speech was first included in Antal Papp, ed., *Gróf Teleki Pál. Országgyűlési beszédei*, vol. 1 (1917–38) (Budapest: Stádium, 1941), 7–24. Balázs Ablonczy republished it under the title "Szociálpolitika és fajegészségügy" in his collection *Teleki Pál. Válogatott politikai írások és beszédek* (Budapest: Osiris, 2000), 27–48. Hereafter reference will be made to this edition.

95. Teleki, "Szociálpolitika és fajegészségügy", 28.

96. Ibid., 29.

97. Ibid., 44.

98. Ibid.

99. Ibid., 46.

100. Ibid., 47.

101. Ibid.

102. Ibid., 48.

103. Pál Teleki, "A rokkantak szaktanácsadói", *Népjóléti Közlöny* 2, 1 (1918): 1.

104. Pál Teleki, "Körlevél az eugenikáról", *Szociálpolitikai Szemle* 6, 9–10 (1917): 169–71.

105. Ibid., 169.

106. Ibid.

107. Ibid (emphasis in the original).

108. Ibid.

109. Ibid., 170.

110. Ibid.

111. Ibid.

112. Ibid., 171 (emphasis in the original).

113. Hoffmann, "Eugenics in the Central Empire", 291.

114. Ibid., 292.

115. Ibid.

116. Ibid.

117. Ibid.

118. Ibid., 294. See also "A gyermektelenek és kevésgyermekűek adótöbblete és a sokgyermekű család adókedvezménye", *Nemzetvédelem* 1, 1–2 (1918): 85–9, and Géza Hoffmann, "A sokgyermekű család közvetlen gazdasági támogatása fajegészségügyi szempontból", *Közgazdasági Szemle* 42, 12 (1918): 610–19.

119. Géza Hoffmann, "Rassenhygiene in Ungarn", *Archiv für Rassen und Gesellschaftsbiologie* 13, 1 (1918): 62–3.
120. István Bárczy, "A Népjóléti Központ megalakulására", *Népjóléti Közlöny* 1, 1 (1917): 1.
121. See "Az anya- és csecsemővédő szakosztály alakuló ülése", *Népjóléti Közlöny* 1, 9 (1917): 98–100, and "A népegészségügyi szakosztály alkoholizmus elleni alosztályának ülése", *Népjóléti Közlöny* 2, 1 (1918): 2–3.
122. "Az Anya- és Csecsemővédő Központi Intézet", *Népjóléti Közlöny* 1, 9 (1917): 105.
123. The result was an almost 600-page volume edited by Béla Fenyvessy and József Madzsar under the title *A Népegészségi Országos Nagygyűlés munkálatai* (Budapest: Eggenberger, 1918).
124. Ibid., 40–52.
125. Ibid., 52. A Ministry of Welfare (Népjóléti Minisztérium) was, in fact, created in August 1917 and given to Tivadar Batthyány. See Marcel Kadosa, "Népjóléti Minisztérium", *Szociálpolitikai Szemle* 7, 8 (1917): 129–31. For an overview, see Iván Bognár, "A Népjóléti Minisztérium és a Népjóléti Népbiztosság szervezete, 1917–1919", *Levéltári Közlemények* 37, 2 (1966): 293–343.
126. Fenyvessy and Madzsar, eds, *A Népegészségi*, 54–71.
127. Ibid., 72.
128. Ibid.
129. Ibid., 76–7. This Ministry eventually became the Ministerium für soziale Fürsorge, established in December 1917. See "Errichtung des Ministeriums für soziale Fürsorge", in Elizabeth Kovács, ed., *Ungergang oder Rettung der Donaumonarchie? Politische Dokumente zu Kaiser und König Karl I. (IV)* (Vienna: Böhlau Verlag, 2004), 273–8.
130. Fenyvessy and Madzsar, eds, *A Népegészségi*, 86.
131. Ibid.
132. Ibid., 87–92.
133. Ibid., 92.
134. As illustrated in his 1916 lecture to the *Berliner Gesellschaft für Rassenhygiene*, also published as Géza Hoffmann, "Rassenhygiene und Fortpflanzungshygiene (Eugenik)", *Öffentliche Gesundheitspflege* 12, 1 (1917): 1–11. See also idem, "Über die Begriffe Rassenhygiene und Fortpflanzungshygiene (Eugenik)", *Münchener Medizinische Wochenschrift* 44, 4 (1917): 110–11, and "Über die Begriffe Rassenhygiene und Fortpflanzungshygiene (Eugenik)", *Archiv für Soziale Hygiene und Demographie* 12, 1–2 (1917): 49–55.
135. Fenyvessy and Madzsar, eds, *A Népegészségi*, 93.
136. Ibid., 95.
137. Ibid.
138. Ibid., 99.
139. Ibid., 102.
140. Ibid., 103.
141. Ibid.
142. Ibid., 104.
143. Ibid.
144. Ibid., 105–24. See also Vilmos Tauffer, *Az anya- és csecsemővédelem szervezete* (Budapest: Pfeiffer, 1918).

145. Fenyvessy and Madzsar, eds, *A Népegészségi*, 152–72.
146. Ibid., 190–233.
147. Ibid., 238–92.
148. Ibid., 293–320.
149. Ibid., 321–78.
150. Ibid., 379–437.
151. Ibid., 438–86.
152. Ibid., 508–52.
153. Ibid., 492–508.
154. See also the report on the congress published by Mrs Sándor Szegvári, "Fajpusztulás", *A Nő* 4, 11 (1917): 178–9.
155. See "Népegészségügyi országos nagygyűlés", *Magyar Társadalomtudományi Szemle* 8, 1 (1917): 50–6; "A népegészségügyi nagygyűlés", *Népjóléti Közlöny* 1, 16 (1917): 167; "A népegészségügyi nagygyűlés", *Népjóléti Közlöny* 1, 17 (1917): 171–2; and, "A népegészségügyi országos nagygyűlés munkálatai", *Nemzetvédelem* 1, 3 (1918): 133.
156. Teleki's letter to Apáthy, OSZK, Manuscript Collection, Fond *Apáthy István iratai*, no. 2453, File 44, and Teleki's letter to Alajos Kovács. KSH, VB 1282/2. Both letters are dated 4 November 1917.
157. Hoffmann's letter to Apáthy (dated 17 November 1917), OSZK, Manuscript Collection, Fond *Apáthy István iratai*, no. 2453, File 45, 1–3.
158. Ibid., 1.
159. Ibid., 2.
160. The aims of the Society were published in *Mitteilungen der Deutschen Gesellschaft für Bevölkerungspolitik* 1, 1 (1916): 1–4.
161. See Gudrun Exner, "Die 'Österreichische Gesellschaft für Bevölkerungspolitik' (1917–1938) – eine Vereinigung mit sozialpolitischen Zielsetzungen im Wien der 20er und 30er Jahre", *Demographische Informationen* (2001): 93–107, and idem, "Eugenisches Gedankengut im bevölkerungswissen-schaftlichen und bevölkerungspolitischen Diskurs in Österreich in der Zwischenkriegszeit", in Baader, Hofer and Mayer, eds, *Eugenik in Österreich*, 184–207.
162. Hoffmann, "Eugenics in the Central Empire", 291.
163. "A Magyar Fajegészségtani és Népesedéspolitikai Társaság alapszabály-tervezete", OSZK, Manuscript Collection, Fond *Apáthy István iratai*, no. 2453, Files 46–7, 1.
164. Ibid.
165. Ibid.
166. Hoffmann, "Rassenhygiene in Ungarn", 64, and "Az Országos Hadigondozó Hivatal népesedéspolitikai és fajegészségügyi tevékenysége", *Nemzetvédelem* 1, 1–2 (1918): 58.
167. "A Magyar Fajegészségtani és Népesedéspolitikai Társaság alapszabály-tervezete", OSZK, Manuscript Collection, Fond *Apáthy István iratai*, no. 2453, Files 46–7, 2.
168. Ibid. To understand how much these sums meant one can compare it with the monthly wage of a clerk at the time, which was between 100 and 150 koronas.
169. Ibid.
170. Ibid., 3–4.

171. On 16 January 1918, for example, Lajos Holló, an MP for the United Party of Independence, claimed in Parliament that "Before the war 765,000 children a year were born in Hungary. In the first year of the war, 1914, the number of births was reduced by 18,000; in 1915 only 481,000 children were born – that is, 284,000 less than in time of peace. In 1916, the number of births was 333,000 – that is a reduction of 432,000. In 1917, the births amounted to 328,000 – that is, the reduction was 438,000. Therefore our losses (in Hungary alone) behind the front line reached the number of 1,172,866 individuals. Deaths. – Whereas in time of peace, the infant mortality, for a period of seven years, was 34 per cent, in 1915, the proportion increased to 48 per cent, and in 1916 to 50 per cent." See "Lowered Birth Rate in Germany and Hungary", *The Journal of Heredity* 9, 6 (1918): 281.

172. See "Magyar Fajegészségtani és Népesedéspolitikai Társaság alakulása", *Szociálpolitikai Szemle* 7, 11–12 (1917): 212–14; "Magyar Fajegészségtani és Népesedéspolitikai Társaság", *Közgazdasági Szemle* 41, 12 (1917): 880; "Magyar Fajegészségtani és Népesedéspolitikai Társaság alakulása", *"Darwin"* 7, 1 (1918): 5–7; and, "Magyar fajegészségtani és népességpolitikai társaság alakulása", *A Cél* 9, 1–2 (1918): 103–6.

173. Hoffmann, "Rassenhygiene in Ungarn", 58.

174. Teleki's letter to Apáthy (dated 27 November 1917) and the text of the lecture in OSZK, Manuscript Collection, Fond *Apáthy István iratai*, no. 2453, File 40–3.

175. Hoffmann, "Eugenics in the Central Empire", 292–3.

176. Ibid., 293.

177. "A Fajegészségtani és népesedéspolitikai irodalom néhány olvasásra ajánlható terméke", OSZK, Manuscript Collection, Fond *Apáthy István iratai*, no. 2454, File 39.

178. Hoffmann, "Rassenhygiene in Ungarn", 59.

179. The street – situated now in the 2nd District of Budapest – was named after Ferenc Heltai, Lord Mayor of Budapest (1912), following his death in 1913. The name of the street was changed in 1920 (Heltai was Jewish) to Károly Keleti, the founder and first director of the Központi Statisztikai Hivatal (Central Statistical Office). I want to thank László András Magyar for sharing some of this information with me. The building still exists today.

180. Teleki's letter to members of the Society. OSZK, Manuscript Collection, Fond *Apáthy István iratai*, no. 2454, File 38, 2. The lecture was published under the title *Egészséges magyar családnak soha magva ne szakadjon* (Budapest: A Társadalmi Múzeum kiadványai, 1918).

181. Hoffmann, "Rassenhygiene in Ungarn", 59.

182. Ibid.

183. Ibid.

184. Teleki's letter to members of the Society. OSZK, Manuscript Collection, Fond *Apáthy István iratai*, no. 2454, File 38, 1.

185. Dezső Laky's letter to the members of the Society (dated 21 January 1918). OSZK, Manuscript Collection, Fond *Apáthy István iratai*, no. 2454, File 36, 1–2.

186. Hoffmann, "Rassenhygiene in Ungarn", 59.

187. Dezső Laky's letter to the members of the Society (dated 15 February 1918). OSZK, Manuscript Collection, Fond *Apáthy István iratai*, no. 2454, File 33,

1–2, and "A Magyar Fajegészségtani és Népesedéspolitikai Társaság etnográfiai szakosztálya", *Ethnographia* 29, 1–4 (1918): 166.
188. Hoffmann's letter to Apáthy (dated 22 January 1918). OSZK, Manuscript Collection, Fond *Apáthy István iratai*, no. 2454, File 37.
189. Hoffmann's letter to Apáthy (dated 9 April 1918). OSZK, Manuscript Collection, Fond *Apáthy István iratai*, no. 2454, File 32, 1–2.
190. Hoffmann, "Rassenhygiene in Ungarn", 56.

7   The Fall of the Race

1. József Mailáth, "A magyar faj regeneratiója és a népbetegségek", *Magyar Társadalomtudományi Szemle* 9, 1 (1918): 1–5.
2. One example is provided by Ernő Tomor, *Neubegründung der Bevölkerungspolitik* (Würzburg: Kurt Kabitzch, 1918). See also idem, "Die Grundirrtümer der heutigen Rassenhygiene", *Würzburger Abhandlungen aus dem Gesamtgebiet der praktischen Medizin* 20, 4–5 (1920): 67–89.
3. István Apáthy, "A fajegészségtan köre és feladatai", part I, *Természettudományi Közlöny* 50, 689–90 (1918): 6–21, and "A fajegészségtan köre és feladatai", part II, *Természettudományi Közlöny* 50, 691–2 (1918), 81–101.
4. Apáthy, "A fajegészségtan köre és feladatai", part I, 7.
5. Ibid.
6. Ibid., 9 (emphasis in the original).
7. Ibid.
8. Apáthy, "A fajegészségtan köre és feladatai", part II, 82–3, 85–6 and 98–9.
9. Ibid., 86.
10. Ibid., 100–1.
11. Hoffmann's letter to Apáthy (dated 16 April 1918), OSZK, Manuscript Collection, Fond *Apáthy István iratai*, no. 2454, Files 30–1, 1–3.
12. Ibid., 1.
13. Ibid., 2.
14. Ibid., 3.
15. Ibid.
16. Hoffmann, "Rassenhygiene in Ungarn", 55.
17. Ibid., 56–7.
18. Ibid., 66.
19. Ibid., 67.
20. On Lenhossék's career, see István Krompecher, "Lenhossék Mihály", in Endre Réti, ed., *A magyar orvosi*, 171–80.
21. Mihály Lenhossék, "A népfajok és az eugenika", *Természettudományi Közlöny* 50, 695–6 (1918): 214–16.
22. Ibid., 214.
23. Ibid.
24. Ibid., 230.
25. Lenhossék, "A népfajok", 241.
26. Ibid.
27. As acknowledged by Hoffmann's letter to Apáthy (dated 25 April 1918), OSZK, Manuscript Collection, Fond *Apáthy István iratai*, no. 2453, File 19, 1–2.

28. See, for example, "Nemzetünk jövőjéért", *Népjóléti Közlöny* 2, 5 (1918): 43, and "Gyakorlati fajegészségügy", *Népjóléti Közlöny* 2, 10 (1918): 88.

29. "A kivándorlás, bevándorlás és visszavándorlás szabályozása fajegészségügyi szempontból", and "A belső telepítés alapelvei fajegészségügyi szempontból". OSZK, *Apáthy István iratai*, Fond 2454, vol. 2, Files 10–13, 1–5, and Files 14–18, 1–5, respectively. See also "A kivándorlás, bevándorlás és visszavándorlás szabályozása fajegészségügyi szempontból", *A Cél* 9, no. 9 (1918): 562–6.

30. For a recent analysis of how this selection process of the immigrant body operated, see Jay Dolmage, "Disabled upon Arrival: The Rhetorical Construction of Disability and Race at Ellis Island", *Cultural Critique* 77 (2011): 24–69.

31. "A kivándorlás, bevándorlás és visszavándorlás szabályozása fajegészségügyi szempontból", 1.

32. Ibid., 2–5.

33. "A belső telepítés alapelvei fajegészségügyi szempontból", 1.

34. Ibid.

35. Ibid., 3.

36. Ibid., 4–5.

37. After 1918, recolonization or resettlement (*hazatelepítés*) became one of the main tenets of Hungarian discourse on eugenics, population policy and biopolitics. See Chris Davis, "Restocking the Ethnic Homeland: Ideological and Strategic Motives behind Hungary's 'Hazatelepítés' Schemes during WWII (and the Unintended Consequences)", *Regio: Minorities, Politics, Society* 1, 1 (2007): 155–74.

38. As outlined in "Az Országos Hadigondozó Hivatal népesedéspolitikai és fajegészségügyi tevékenysége", *Népjóléti Közlöny* 2, 9 (1918): 78–9.

39. There is an additional aspect that is worth mentioning here: these eugenic schemes of resettlement provide a perfect example of the continuity of eugenic narratives of social and biological improvement in Hungary. It took another 20 years and another world war for these resettlement schemes, together with a broad set of eugenic discourses on the quality and quantity of the population, to be introduced in Hungary. By then Teleki was more than just the president of a eugenic society; he was Hungary's prime minister and his vision of a eugenically regenerated Hungarian nation was stronger than ever.

40. The name of the journal proposed by Hoffmann in 1914 was *Eugenika: Fajegészségügyi Szemle* (see Chapter 4).

41. "A 'Nemzetvédelem'", *Nemzetvédelem* 1, 1–2 (1918): 1.

42. Ibid., 2–3.

43. Géza Hoffmann, "A fajegészségtan, eugenika és népesedéspolitika rendszere", *Nemzetvédelem* 1, 1–2 (1918): 4–13.

44. Lajos Nékám, "Néhány sürgős teendőnk a fajegészségügy terén", *Nemzetvédelem* 1, 1–2 (1918): 14–23. See also "Nékám tanár a nemi bajokról", *Népjóléti Közlöny* 2, 6 (1918): 56.

45. Pál Teleki, "Die Sicherung der Lebenslage Kriegsbeschädigter", *Nemzetvédelem* 1, 1–2 (1918): 24–33.

46. Dezső Laky, "A Magyar Fajegészségtani és Népesedéspolitikai Társaság munkaköre", *Nemzetvédelem* 1, 1–2 (1918): 34–40.

47. Max Christian, "Sozialpolitik und Bevölkerungspolitik", *Nemzetvédelem* 1, 1–2 (1918): 41–7, and Wilhelm Schallmayer, "Hygiene der Erbverfassung und Hygiene der Erscheinungsbildes", in ibid., 1, 3 (1918): 112–22.

48. Ulrich Patz, "Egészségtan és fajegészségtan", *Nemzetvédelem* 1, 1–2 (1918): 64–5.

49. See also "Nemzetvédelem", *Népjóléti Közlöny* 2, 16 (1918): 147–8.

50. Dezső Buday, "Törvényjavaslat a házasulók kötelező orvosi vizsgálatáról", *Nemzetvédelem* 1, 1–2 (1918): 47–53. The article was also published in *Új Nemzedék* 5, 33 (1918): 9–13.

51. Zoltán Dalmady, "A testnevelés fajegészségügyi jelentősége", *Nemzetvédelem* 1, 1–2 (1918): 53–5.

52. Gyula Donáth, "Az alkohol fajegészségügyi és népesedéspolitikai szempontból", *Nemzetvédelem* 1, 3 (1918): 98–111.

53. "Az Országos Hadigondozó Hivatal népesedéspolitikai és fajegészségügyi tevékenysége", *Nemzetvédelem* 1, 1–2 (1918): 55–60.

54. "A telepítés alapelvei fajegészségügyi szempontból", *Nemzetvédelem* 1, 1–2 (1918): 60–4, and "A kivándorlás, bevándorlás és visszavándorlás szabályozása fajegészségügyi szempontból", *Nemzetvédelem* 1, 3 (1918): 126–32.

55. Géza Hoffmann, "A fajegészségtan irodalmának néhány terméke", *Nemzetvédelem* 1, 1–2 (1918): 66–73.

56. "Mitteilungen der Deutschen Gesellschaft für Bevölkerungspolitik", *Nemzetvédelem* 1, 1–2 (1918): 73–5; "A svéd házassági törvény fajegészségügyi követelményei", ibid., 90–1; and "Mit tesznek az amerikaiak a fajegészségügy terén", ibid., 91–3.

57. "Fajegészségügy és népesedéspolitika a parlamentben", *Nemzetvédelem* 1, 1–2 (1918): 75–9.

58. Ibid., 79–82.

59. "A belügyminiszter két népesedéspolitikai körrendelete", *Nemzetvédelem* 1, 1–2 (1918): 82–5.

60. See, for example, Tibor Vadnay, "Népesedési politika", in idem, *A magyar jövő: közgazdaság- és szociálpolitikai tanulmány*, 2nd edn (Budapest: Athenaeum, 1918), 155–76.

61. "A Hajdúdorogi gör. kath. egyházmegye fajegészségügyi körrendelete a lelkészhez", *Nemzetvédelem* 1, 1–2 (1918): 89–90.

62. Géza Hoffmann, "Milyen fajegészségtani tanulmányokra lenne nálunk szükség?", *Nemzetvédelem* 1, 3 (1918): 132–3.

63. Géza Hoffmann, "Künstliche Unfruchtbarkeit nach den Erfahrungen in den Vereinigten Staaten von Nordamerika", in S. Placzek, ed., *Künstliche Fehlgeburt und künstliche Unfruchtbarkeit (Ihre Indikationen, Technik und Rechtslage)* (Leipzig: Verlag von Georg Thieme, 1918), 413–35.

64. In a letter to the Austrian anthropologist Rudolf Pöch, dated 17 January 1918, Alfred Ploetz pointed out that he might miss the former's lecture on 24 January, "because several board members of the Hungarian Society for Racial Hygiene want to come to Munich to agree on a common plan of action for our Societies of Racial Hygiene". *Rudolf Pöch Sammlung*, Naturhistorisches Museum Wien. I want to thank Maria Teschler-Nicola for allowing me to use this letter.

65. Weindling, *Health, Race and German Politics*, 303.

66. *Deutsch-Österreichische Tagung für Volkswohlfahrt am 12. und 13. März 1916* (Wien: Franz Deuticke, 1916).

67. *Konstituierende Generalversammlung der Ungarischen Waffenbrüderlichen Vereinigung in Budapest am 11. Juni 1916* (Budapest: Eigenverlag der Ungarischen Waffenbrüderlichen Vereinigung, 1916).

68. Byer, *Rassenhygiene und Wohlfahrtspflege*, 84–6.

69. Martin Kirchner and Curt Adam, eds, *Der Wiederaufbau der Volkskraft nach dem Kriege: Sitzungsbericht über die gemeinsame Tagung der ärztlichen Abteilungen der Waffenbrüderlichen Vereinigungen Österreichs, Ungarns und Deutschlands in Berlin, 23. bis 26. Januar 1918* (Jena: G. Fischer, 1918), xx–xxiv.

70. O. Krohne, "Bevölkerungspolitische Probleme und Ziele", in ibid., 87–94, and J. Tandler, "Bevölkerungspolitische Probleme und Ziele", in ibid., 108–10.

71. W. Tauffer, "Die Säuglingssterblichkeit und ihre Bekämpfung in Ungarn", in ibid., 114–19.

72. G. Dollinger, "Schutz und Kräftigung der Jugend", in ibid., 181–5.

73. I. Dóczi, "Zum Wiederaufbau der Volkskraft", in ibid., 298–301.

74. E. Jendrassik, "Verhütung und Bekämpfung der übertragbaren Krankheiten", in ibid., 305–16.

75. Hoffmann, "Eugenics in the Central Empires", 288.

76. "Aus anderen Organizationen", *Mitteilungen der Deutschen Gesellschaft für Bevölkerungspolitik* 1, 3 (1918): 29.

77. See "Vermischte Nachrichten", *Wiener Klinische Wochenschrift* 31, 35 (1918): 977; *Jelentés a német, osztrák, török és magyar bajtársi szövetségek orvosi szakosztályainak és a bolgár kiküldötteknek Budapesten 1918 szeptember hó 21–23-án tartott együttes üléséről* (Budapest: Franklin, 1918); and *Gemeinsame Tagung der ärztliche Abteilung der waffenbrüderlichen Vereinigungen Deutschlands, Österreich, der Türkei und Ungarns am 21.–22. September 1918 in Budapest* (Budapest: Franklin, 1918).

78. *Jelentés a német, osztrák, török és magyar bajtársi szövetségek*, vii–viii.

79. It was widely and wrongly assumed by scholars working on German and Central European eugenics – myself included – that this conference did not take place. See Weindling, *Health, Race and German Politics*, 303, and Turda, "The First Debates on Eugenics in Hungary", 214.

80. Pál Teleki, "Begrüssung", in *Jelentés a német, osztrák, török és magyar bajtársi szövetségek*, 339–44.

81. Ibid., 321–2, 330; and *Gemeinsame Tagung der ärztliche Abteilung*, 5.

82. "Magyar-német-osztrák fajegészségtani nagygyűlés", *Népjóléti Közlöny* 2, 17 (1918): 153.

83. "Vermischte Nachrichten", *Wiener Klinische Wochenschrift* 31, 37 (1918): 1022, and "Ankündigungen", *Mitteilungen der Österreichischen Gesellschaft für Bevölkerungspolitik* 1, 1 (1918): 38–9.

84. Géza Hoffmann, "Fajegészségtan és erkölcs", *Magyar Kultúra* 6, 12 (1918): 257–62. See also Arnold Marosi, "Az átöröklés és az ember", *Katholikus Szemle* 33, 3 (1919): 222–32, and idem, "Az átöröklés és az ember", *Katholikus szemle* 33, 4 (1919): 263–81.

85. Kovács, *Liberal Professions and Illiberal Politics*, 41.

86. In the 1930s, Podach published two acclaimed books on Friedrich Nietzsche: *Gestalten um Nietzsche* (1932) and *Der kranke Nietzsche* (1937).

87. F[rigyes] E[rich] P[odach], "Állami eugenika", *Huszadik Század* 19, 11 (1918): 265–6.

88. Ibid.

89. Ibid., 266.

90. Ibid.

91. See Gusztáv Gratz, *A forradalmak kora. Magyarország története, 1918–1920* (Budapest: Magyar Szemle Társaság, 1935).

92. For the importance afforded to public health by the Országos Polgári Radikális Párt, see Zsigmond Kende, "Közegészségügyi feladatok", *Szabadgondolat* 4, 6 (1914): 164–8.

93. Endre Kárpáti, "Madzsar József egészségpolitikai tevékenysége a magyarországi polgári demokratikus forradalom és a Tanácsköztársaság idején", *Orvostörténeti Közlemények* 28, 5 (1963): 60. See also Sándor Székely, "Madzsar József", *Orvosi Hetilap* 117, 8 (1976): 478–81.

94. On Detre and Péterfi, see Dénes Karasszon, "Emlékezés dr. Detre Lászlóra (1874–1939), az antigén névadójára", *Orvosi Hetilap* 131, 20 (1990): 1089–90, and Tibor Donáth, "Dr Péterfi Tibor (1883–1953)", *Orvosi Hetilap* 124, 50 (1983): 3061–2.

95. Rudolfné Dósa, Ervinné Liptai and Mihály Ruff, *A Magyar Tanácsköztársaság egészségügyi politikája* (Budapest: Medicina, 1959), 41.

96. Zoltán Pártos, "Az egészségügy szocializálása", *Szabadgondolat* 9, 2 (1919): 44–8, and Sándor Flamm, "Az egészségügy szocializálása", *Szabadgondolat* 9, 4 (1919): 91–4.

97. For example, the medical section of the Galilei Circle organized lectures on eugenics in February 1919. See A "Galilei Kör vezetőségének jelentése a Kör januári működéséről", *Szabadgondolat* 9, 3 (1919): 71–3.

98. Kárpáti, ed., *Madzsar József válogatott írásai*, 180.

99. Dósa, Liptai and Ruff, *A Magyar Tanácsköztársaság*, 41.

100. See József Madzsar, *Szociálhigiénés békekészülődés* (Budapest: Franklin, 1918).

101. "New Eugenics Society in Hungary", *The Journal of Heredity* 11, 1 (1920): 41. See also Paul Popenoe and Roswell Hill Johnson, *Applied Eugenics* (New York: Macmillan, 1920), 155.

102. This rhetoric was often used in nationalist – and increasingly racist and anti-Semitic – publications like *A Cél* and *Új Nemzedék*. For this trend, see Sándor Kiss, *Fajunk védelméről* (Budapest: Hornyánszky, 1918), and idem, *Zsidó fajiság, magyar fajiság* (Budapest: Hornyánszky, 1918).

103. "Fajegészségügyi és népesedéspolitikai pályázatok", *Huszadik Század* 20, 1–2 (1919): 119–20.

104. Ibid., 119 (emphasis in the original).

105. Ibid.

106. "Gyakorlati fajegészségügyi intézkedések", *Nemzetvédelem* 2, 1 (1919): 145–54. See also Géza Hoffmann, "Gyakorlati fajegészségügy", *"Darwin"* 7, 12 (1918): 140–2.

107. See, for example, Géza Hoffmann, "Miért van szükség sok gyermek születésére?", *Huszadik Század* 20, 4 (1919): 209–16; idem, "A fajegészségtan gondolatköréből", *A Cél* 10, 1 (1919): 12–16; and idem, "A születések differenciális arányszáma", *"Darwin"* 8, 1 (1919): 5–6.

108. Hoffmann, "Eugenics in the Central Empire", 293.

109. See Loren R. Graham, "Science and Values: The Eugenics Movement in Germany and Russia in the 1920s", *American Historical Review* 82, 5 (1977): 1135–64; Pat Simpson, "Bolshevism and 'Sexual Revolution': Visualizing New Soviet Woman as the Eugenic Ideal", in Fae Brauer and Anthea Callen, eds, *Corpus Delecti: Art, Sex and Eugenics* (London: Aldershot, 2008), 209–38;

and Nikolai Krementsov, "From 'Beastly Philosophy' to Medical Genetics: Eugenics in Russia and the Soviet Union", *Annals of Science* 68, 1 (2011): 61–92.

110. Lynne Attwood, *The New Soviet Man and Woman: Sex-Role Socialization in the USSR* (Bloomington: Indiana University Press, 1990).

111. Loren R. Graham, *Science and Philosophy in the Soviet Union* (London: Allen Lane, 1973); Nikolai Krementsov, *Stalinist Science* (Princeton University Press, 1997).

112. Zsófia Dénes, *A nő a kommunista társadalomban* (Budapest: Közoktatásügyi Népbiztosság, 1919), 17. For a similar glorification of motherhood in the emerging Soviet discourse, see Alexandra Kollontai, *Selected Writings* (New York: W. W. Norton, 1980).

113. See István Simonovits, "A Magyar Tanácsköztársaság szociálpolitikája", *Népegészségügy* 39, 3–4 (1958): 53–62; Zsófia Benke, "A Tanácsköztársaság egészségügyi igazgatásának történetéhez", *Levéltári Szemle* 19, 1 (1969): 57–74. For how these health measures worked in the other cities of Hungary, for example in Debrecen and Nagyvárad, see Zoltánné Mervó, "A két forradalom egészségügyi és szociálpolitikai tevékenysége megyénkben, 1918–1919", *Hajdú-Bihar Megyei Levéltár Évkönyve* 5 (1978): 173–90.

114. Kovács, *Liberal Professions and Illiberal Politics*, 42.

115. Frank Eckelt, "The Internal Policies of the Hungarian Soviet Republic", in Iván Völgyes, ed., *Hungary in Revolution, 1918–1919* (Lincoln: University of Nebraska Press, 1971), 61.

116. See Gábor Barta, "Magyarország közegészségügye", *Egészség* 33, 1–2 (1919): 13–23. See also "Medical Practice in Bolshevist Hungary", *The Lancet* 195, 5029 (1920): 168.

117. Dósa, Liptai and Ruff, *A Magyar Tanácsköztársaság*, 103–6.

118. Kárpáti, "Madzsar József egészségpolitikai", 69. See also Katalin Petrák and György Milei, eds, *A Magyar Tanácsköztársaság szociálpolitikája: Válogatott rendeletek, dokumentumok, cikkek* (Budapest: Gondolat, 1959), and Sándor Kovacsics, "Emlékezés Liebermann Leó professzorra", *Népegészségügy* 40, 2 (1959): 51–5.

119. See Jenő Pongrácz, ed., *A Forradalmi Kormányzótanács és népbiztosok rendeletei*, vol. 1 (Budapest: Franklin, 1919). For a detailed discussion see Ferenc Kemény, "Az egészségügy a Tanácsköztársaságban", *Egészség* 33, 5–6 (1919): 121–44. See also Dósa, Liptai and Ruff, *A Magyar Tanácsköztársaság*, 73–102.

120. "Kommunista hygiéne", *Egészség* 33, 3–4 (1919): 73.

121. Pongrácz, ed., *A Forradalmi Kormányzótanács és népbiztosok rendeletei*, vol. 2, 40–1.

122. Ibid., 42–7. See also Eckelt, "The Internal Policies of the Hungarian Soviet Republic", 65.

123. Dósa, Liptai and Ruff, *A Magyar Tanácsköztársaság*, 88.

124. Lajos Bartucz, "A társadalmi embertanról", *Természettudományi Közlöny* 51, 723 (1919): 281.

125. Hoffmann, "Eugenics in the Central Empires", 294.

126. Dezső Szabó, *Az elsodort falu: regény* (Budapest: Táltos, 1919).

127. Paul A. Hanebrink, "Transnational Culture War: Christianity, Nation and the Judeo-Bolshevik Myth in Hungary, 1890–1920", *The Journal of Modern History* 80, 1 (2008): 55–80.

128. Géza Hoffmann, *A fajegészségtan és népesedéspolitikáról* (Budapest: Lampel R., 1920); Sarolta Geőcze, *A nemzeti megújhodás erkölcsi feltételeiről* (Budapest: Kiadja a Szabad Lyceum, 1920); and Gyula Donáth, "A világháborúban legyőzött népek fajbiológiai jövője", *Természettudományi Közlöny* 52, 751–4 (1920): 353–61.

Conclusions

1. See Oszkár Jászi, *Magyarország jövője és a Dunai Egyesült Államok* (Budapest: Az Új Magyarország, 1918).
2. A list of the Hungarian propaganda materials can be found in *A Catalogue of Paris Peace Conference Delegation Propaganda in the Hoover War Library* (Stanford University Press, 1926), 47–8. In terms of the role played by eugenics and eugenicists in preparing these materials, it is worth mentioning that one pamphlet, dealing with Hungary's economic unity (Documents 4–5), was written by László Buday, also a member of the Society for Racial Hygiene and Population Policy. The president of the Society, Pál Teleki, similarly prepared Hungary's ethnographic map.
3. This pamphlet was a reprint of the article with the same title published in *Nemzetvédelem* 2, 4 (1919): 162–4. It appeared as Document no. 7 in the list of propaganda materials.
4. See Albert Apponyi et al., *Justice for Hungary: Review and Criticism of the Effects of the Treaty of Trianon* (London: Longmans, 1928).
5. "The Consequences of the Division of Hungary", 162 (emphasis in the original).
6. Ibid., 163 (emphasis in the original).
7. Ibid. (emphasis in the original).
8. Ibid.
9. Ibid.
10. Ibid., 163–4.
11. Ibid., 164.
12. Ibid.
13. In the obituary published in the *Archiv für Rassen- und Gesellschaftsbiologie* on the occasion of his death in 1921, Hoffmann was described as "the most enthusiastic, energetic and industrious supporter of the cause of racial hygiene". See "Notizen", *Archiv für Rassen- und Gesellschaftsbiologie* 14, 1 (1922): 99.

Epilogue

1. "Correspondence", *The Eugenics Review* 10, 4 (1919): 223.

Bibliography

Secondary sources that I have used in this book are given in the footnotes. Due to length constrains, only primary sources are included in this bibliography.

Archives

The Eugenics Society, Wellcome Library, London
Galton Papers, University College London
International Institute of Social History, Amsterdam
Központi Statisztikai Hivatal, Budapest
Magyar Nemzeti Múzeum Történeti Fényképtár, Budapest
Magyar Országos Levéltár, Budapest
Magyar Természettudományi Múzeum, Budapest
Max Plack Society, Berlin
The National Archives, Kew Gardens, London
The National Archives, Prague
Naturhistorisches Museum, Vienna
Országos Széchényi Könyvtár, Budapest
Pickler Memorial Library, Truman State University, Kirksville, Missouri
The Rockefeller Archive, Sleepy Hollow, New York
Semmelweis Orvostörténeti Múzeum, Könyvtár és Levéltár, Budapest

Abbreviations

HHLP	Harry H. Laughlin Papers
KSH	Központi Statisztikai Hivatal
MOL	Magyar Országos Levéltár
OSZK	Országos Széchényi Könyvtár
SOM	Semmelweis Orvostörténeti Múzeum, Könyvtár és Levéltár

Journals

A Cél
A Nő
A Nő és a Társadalom
A Társadalmi Muzeum Értesítője
American Journal of Sociology
Archiv für Anthropologie
Archiv für Frauenkunde und Eugenetik
Archiv für Rassen- und Gesellschaftsbiologie
Athenaeum
Az alkoholizmus ellen

Az Est
Az Orvos
Biometrika
British Medical Journal
Budapesti Orvosi Újság
Budapesti Szemle
Bulletin of the New York Academy of Medicine
"Darwin"
Deutsche Medizinischer Wochenschrift
Economic Journal, The
Egészség
Egészségügyi Munkás
EOS: Vierteljahresschrift für die Erkenntnis und Behandlung jugendlicher Abnormer
Erdély
Ethnographia
Eugenics Review, The
Fajegészségügy
Gyógyászat
Huszadik Század
Ipari Jogvédelem
Jogállam
Journal of Heredity
Journal of the American Institute of Law and Criminology
Journal of the Anthropological Institute of Great Britain and Ireland
Juristisch-psychiatrische Grenzfragen
Katholikus Szemle
Klinikai Füzetek
Közgazdasági Szemle
Lancet, The
Magyar Kultúra
Magyar Orvosok Lapja
Magyar Társadalomtudományi Szemle
Malthusian, The
Mendel Journal, The
Mitteilungen der Deutschen Gesellschaft für Bevölkerungspolitik
Mitteilungen der Österreichischen Gesellschaft für Bevölkerungspolitik
Monatsblatt der Kais. Kön. Heraldischen Gesellschaft "Adler"
Münchener Medizinischer Wochenschrift
Nemzetvédelem
Népegészségügy
Népjóléti Közlöny
Nyugat
Öffentliche Gesundheitspflege
Orvosi Hetilap
Social Hygiene
Szabadgondolat
Szociális Egészségügy
Szociálpolitikai Szemle
Természettudományi Közlöny

Új Nemzedék
Ungarische Rundschau für Historische und Soziale Wissenschaften
Unsere Gesundheit
Városi Szemle
Vasárnapi Ujság
Volksgesundheit
Vörös Újság
Wiener Klinische Wochenschrift
Würzburger Abhandlungen aus dem Gesamtgebiet der praktischen Medizin
Zeitschrift für Sexualwissenschaft
Zeitschrift für Sozialwissenschaft

Books

A kivándorlás: a Magyar Gyáriparosok Országos Szövetsége által tartott országos ankét tárgyalásai (Budapest: Pesti Lloyd nyomda, 1907).
A "Nemzetvédő Szövetség a Nemibajok Ellen" alapszabályai (Budapest: Stephaneum, 1916).
A társadalmi fejlődés iránya (Budapest: Politzer Zsigmond, 1904).
Alexander, Bernát, and Mihály Lenhossék, eds. *Az ember testi és lelki élete, egyéni és faji sajátságai*, 2 vols (Budapest: Athenaeum, 1905, 1907).
Apáthy, István. *Über das leitende Element des Nervensystems und seine Lagebeziehungen zu den Zellen bei Wirbeltieren und Wirbellosen* (Leiden: Brill, 1896).
———. *Néhány lap önismeretünk történetéből: élettudományi vázlat* (Budapest: Hornyánszky nyomda, 1900).
———. *A fejlődésnek nevezett átalakulásról* (Kolozsvár: Ajtai nyomda, 1904).
———. *A nemzeti dalról* (Kolozsvár: Újhelyi nyomda, 1906).
———. *Magántulajdon, csere és élet az állatországban* (Pécs: Taizs József, 1908).
———. *Öregség és halál* (Budapest: Hornyánszky, 1909).
———. *A hosszú életről* (Kolozsvár: Gámán János nyomda, 1909).
———. *A Magyar Társadalomtudományi Egyesület legelső teendői* (Pécs: Taizs József, 1908).
———. *A Darwinismus birálata és a társadalomtan* (Budapest: Pesti könyvnyomda, 1910).
———. *A fejlődés törvényei és a társadalom* (Kolozsvár: Ajtai K. Albert, 1912).
———. *A socialismus az emberi továbbfejlődés szempontjából* (Budapest: Országos Ismeretterjesztő Társulat, 1913).
Apponyi, Albert, and János Sándor. *Az anya- és csecsemővédelem a képviselőházban* (Budapest: Pfeiffer, 1916).
Apponyi Albert et al. *Justice for Hungary: Review and Criticism of the Effects of the Treaty of Trianon* (London: Longmans, 1928).
Az anya- és csecsemővédelem a képviselőházban (Budapest: Pfeiffer, 1916).
Baernreither, Joseph-Maria, ed. *Schriften des 2. Österreischischen Kinderschutzkongresses in Salzburg, 1913* (Wien: Perles, 1913).
Balás, Károly. *A népesedés* (Budapest: Politzer, 1905).
———. *A család és a magyarság* (Budapest: Grill Károly, 1908).
Balogh, Pál. *A magyar faj uralma* (Budapest: Lampel R., 1903).
Bársony, János. *A gyermekágyi lázról* (Budapest: Pesti Lloyd nyomda, 1912).

Bartucz, Lajos. *Arad megye népének anthropológiai vázlata* (Arad: Réthy Lipót, 1912).

Bibliography of Eugenics and Related Subjects, Compiled by the Bureau, A (The Bureau of Analysis and Social Welfare, Bulletin no. 3, The Capitol. Albany, NY: 1913).

Boas, Franz. *Changes in Bodily Form of Descendants of Immigrants* (Washington: U.S. Government Printing Office, 1910).

Bosnyák, Zoltán, and L. Edelsheim-Gyulai, eds *Le droit de l'enfant abandonné et le système Hongrois de protection de l'enfance* (Budapest: Athenaeum, 1909).

Buday, Dezső. *A házasság társadalmi védelme* (Budapest: Politzer Zsigmond, 1902).

———. *Az egyke* (Budapest: Deutsch Zsigmond, 1909).

———. *Az emberiség táplálkozásának szocziológiai vonatkozásai* (Budapest: Pesti Lloyd nyomda, 1916).

———. *A társadalmi ideál* (Budapest: Eggenberger, 1917).

Catalogue of Paris Peace Conference Delegation Propaganda in the Hoover War Library, A (Stanford University Press, 1926).

Chyzer, Béla. *A gyermekmunka Magyarországon* (Budapest: Az Országos Gyermekvédő Liga kiadása, 1909).

Compte-Rendu. XVIe Congrès International de Médecine, Budapest August–September 1909 (Budapest: Franklin, 1910).

Czirbusz, Géza. *Nemzetek alakulása. Anthropo-geographiai szempontból* (Nagybecskerek: Pleitz Pál, 1910).

Dénes, Zsófia. *A nő a kommunista társadalomban* (Budapest: Közoktatásügyi Népbiztosság, 1919).

Deutsch, Ernő. *Az orvos szociális munkája a háború alatt* (Budapest: Radó nyomda, 1915).

Deutsch-Österreichische Tagung für Volkswohlfahrt am 12. und 13. März 1916 (Wien: Franz Deuticke, 1916).

Donáth, Gyula. *Der Arzt und die Alkoholfrage* (Vienna: Moritz Perles, 1907).

Drysdale, C. V. *Neo-Malthusianism and Eugenics* (London: William Bell, 1912).

———. *The Small Family System. Is It Injurious or Immoral?* (London: A. C. Fifield, 1913).

Elderton, Ethel M. *The Relative Strength of Nurture and Nature* (London: Dulau, 1909).

Engel, Zsigmond. *Grundfragen des Kinderschutzes* (Dresden: Bohmert, 1911).

Epstein, László, ed. *Első Országos Elmeorvosi Értekezlet Munkálatai* (Budapest: Pallas, 1901).

Die Erhaltung und Mehrung der deutschen Volkskraft: Verhandlungen der 8. Konferenz der Zentralstelle für Volkswohlfahrt vom 26.–28. Oktober 1915 (Berlin: Heymanns, 1916).

Erstes Jahrbuch des Kriegsspitals der Geldinstitute in Budapest. Beiträge zur Kriegsheilkunde (Berlin: Julius Springer, 1917).

Eugenics Education Society: The Annual Report, 1908, The (London: Eugenics Education Society, 1908).

Fenyvessy, Béla, and József Madzsar, eds. *A Népegészségi Országos Nagygyűlés munkálatai* (Budapest: Eggenberger, 1918).

Farkas, Pál. *Az amerikai kivándorlás* (Budapest: Singer and Wolfner, 1907).

Frank, Ödön, ed. *A millenniumi közegészségi és orvosügyi kongresszus tárgyalásai* (Budapest: Franklin, 1897).

Fülöp, Zsigmond. *Mi az élet?* (Budapest: Európa, 1909).

Galton, Francis. *Inquiries into Human Faculty and Its Development* (London: Macmillan, 1883).

———. *Memories of My Life*, 2nd edn (London: Methuen, 1908).

———. *Essays in Eugenics* (London: Eugenics Education Society, 1909).

Gemeinsame Tagung der ärztliche Abteilung der waffenbrüderlichen Vereinigungen Deutschlands, Österreich, der Türkei und Ungarns am 21.–22. September 1918 in Budapest (Budapest: Franklin, 1918).

Geőcze, Sarolta. *A nemzeti megújhodás erkölcsi feltételeiről* (Budapest: Kiadja a Szabad Lyceum, 1920).

Gerlóczy, Zsigmond, ed. *Jelentés az 1894. szeptember hó 1-től 9-ig Budapesten tartott VIII-ik nemzetközi közegészségi és demografiai congressusról és annak tudományos munkálatairól*, vol. 7 (Budapest: Pesti könyvnyomda, 1896).

Giesswein, Sándor. *A háború és a társadalomtudomány* (Budapest: Stephaneum, 1915).

———. *Egyén és társadalom* (Budapest: Magyar Tudományos Akadémia, 1915).

Gruber, Max von, and Ernst Rüdin, eds. *Fortpflanzung, Vererbung, Rassenhygiene*, 2nd edn (Munich: J. F. Lehmanns, 1911).

Gratz, Gusztáv. *A forradalmak kora. Magyarország története, 1918–1920* (Budapest: Magyar Szemle Társaság, 1935).

Grósz, Emil, ed. *XVIe Congrès International de Medicine. Compte-Rendu. Volume général* (Budapest: Franklin, 1910).

Háborús előadások a Budapesti K. M. Tudományegyetemen (Budapest: Franklin, 1915).

Hahn, Dezső. *A fertőző nemibetegségek és a háború* (Budapest: Népszava könyvkereskedés, 1916).

Hajós, Lajos. *Az idegélet egészségtana* (Budapest: Athenaeum, 1910).

Harkányi, Ede. *A holnap asszonyai* (Budapest: Politzer Zsigmond, 1905).

———. *Babonák ellen* (Budapest: Grill Károly, 1907).

Hegedűs, Lóránt. *A magyarság jövője a háború után. Politikai tanulmány* (Budapest: Athenaeum, 1916).

Hoffmann, Géza. *Csonka munkásosztály. Az amerikai magyarság* (Budapest: Pesti könyvnyomda, 1911).

———. *Die Rassenhygiene in den Vereinigten Staaten von Nordamerika* (München: J. F. Lehmann, 1913).

———. *Fajegészségtan és eugenika* (Budapest: Pesti könyvnyomda, 1916).

———. *Krieg und Rassenhygiene. Die bevölkerungspolitischen aufgaben nach dem kriege* (Munchen: J. F. Lehnmann, 1916).

———. *Nyelvében él a nemzet* (Budapest: Hornyánszky, 1916).

———. *Egészséges magyar családnak soha magva ne szakadjon* (Budapest: Társadalmi Múzeum kiadványai, 1918).

———. *Gyakorlati fajegészségügyi intézkedések* (Pozsony: Angermayer Károly, 1918).

———. *A fajegészségtan és népesedéspolitikáról* (Budapest: Lampel R., 1920).

Holmes, Samuel J. *A Bibliography of Eugenics* (Berkeley: University of California Press, 1924).

L'Hygiène Publique en Hongrie (Budapest: Wodianer, 1909).

Illyefalvi-Vitéz, Géza. *Születési és termékenységi statisztika* (Budapest: Grill, 1906).

Ioteyko, I, ed. *Premier Congrès International de Pédologie*, 2 vols (Brussels: Libraire Misch et Thron, 1912).

Jankó, János. *Kalotaszeg magyar népe: néprajzi tanulmány* (Budapest: Athenaeum Társulat, 1892).

Jánossy, Gábor. *Közmûvelõdési egyesületeink és a magyar faj (állam) jövõje* (Szombathely: Egyházmegyei könyvnyomda, 1904).

Jászi, Oszkár, ed. *A zsidókérdés Magyarországon. A Huszadik Század körkérdése* (Budapest: A Társadalomtudományi Társaság kiadása, 1917).

———. *Magyarország jövõje és a Dunai Egyesült Államok* (Budapest: Az Új Magyarország, 1918).

Jekelfalussy, József. *The Millennium of Hungary and Its People* (Budapest: Pesti könyvnyomda, 1897).

Jelentés a német, osztrák, török és magyar bajtársi szövetségek orvosi szakosztályainak és a bolgár kiküldötteknek Budapesten 1918 szeptember hó 21–23-án tartott együttes ülésérõl (Budapest: Franklin, 1918).

Jelentés a Társadalmi Muzeum berendezésérõl és annak elsõ évi munkásságáról (Budapest: A Társadalmi Múzeum kiadása, 1903).

Kende, Mór. *Die Entartung des Menschengeschlechts, ihre Ursachen und die Mittel zu ihrer Bekämpfung* (Halle: Carl Marhold, 1901).

Kenedi, Géza. *Szocziológiai nyomozások*, 2 vols (Budapest: Franklin, 1910).

———. *Feminista tanulmányok* (Budapest: Lampel R., 1912).

Kirchner, Martin, and Curt Adam, eds. *Der Wiederaufbau der Volkskraft nach dem Kriege: Sitzungsbericht über die gemeinsame Tagung der ärztlichen Abteilungen der Waffenbrüderlichen Vereinigungen Österreichs, Ungarns und Deutschlands in Berlin, 23. bis 26. Januar 1918* (Jena: G. Fischer, 1918).

Kiss, Sándor. *Fajunk védelmérõl* (Budapest: Hornyánszky, 1918).

———. *Zsidó fajiság, magyar fajiság* (Budapest: Hornyánszky, 1918).

Klebelsberg, Kuno. *A magyar rokkantügy szervezete* (Budapest: Bíró Miklós könyvnyomdája, 1916).

Kollontai, Alexandra. *Selected Writings* (New York: W. W. Norton, 1980).

Konstituierende Generalversammlung der Ungarischen Waffenbrüderlichen Vereinigung in Budapest am 11. Juni 1916 (Budapest: Eigenverlag der Ungarischen Waffenbrüderlichen Vereinigung, 1916).

Kós, Károly. *Régi Kalotaszeg* (Budapest: Athenaeum nyomda, 1911).

Kovács, Gábor. *A népesedés elmélete* (Debrecen: Hegedûs és Sándor, 1908).

Kozáry, Gyula. *Az átöröklés problémája* (Budapest: Athenaeum, 1894).

———. *Átöröklés és nemzeti nevelés. Természetbölcseleti és neveléstani tanulmány* (Budapest: Athenaeum, 1905).

Lányi, Márton, ed. *A háború és a jövõ* (Budapest: Grill Károly, 1916).

Laughlin, Harry H. *Eugenical Sterilization in the United States* (Chicago: Psychopathic Laboratory of the Municipal Court of Chicago, 1922).

Leitgedanken der Referate. Nemzetközi Alkoholizmus Elleni X. Kongresszus. Budapest, 1905. Szept. 11–16 (Budapest: Hausdruckerei der Haupt- u. Residenzstadt, 1905).

Lenhossék, Mihály. *Az anthropológiáról és teendõinkrõl az anthropológia terén* (Budapest: Franklin, 1915).

———. *Az ember helye a természetben*, 2nd edn (Budapest: Franklin, 1921).

Löherer, Andor. *Az amerikai kivándorlás és visszavándorlás* (Budapest: Pátria, 1908).

Madzsar, József. *Darwinizmus és lamarckizmus* (Budapest: Deutsch Márkus, 1909).

———. *Az anya- és csecsemõvédelem országos szervezése* (Budapest: Stefánia Szövetség, 1915).

————. *A jövő nemzedék védelme és a háború* (Budapest: Politzer, 1916).

————. *A meddő Budapest* (Budapest: Pfeiffer, 1916).

————. *Mit akar a Stefánia-Szövetség?* (Budapest: Pfeiffer, 1916).

————. *Az ember származása és a származástan vázlata* (Budapest: Új Magyarország, 1918).

————. *Szociálhigiénés békekészülődés* (Budapest: Franklin, 1918).

Méhely, Lajos. *Az élettudomány bibliája* (Budapest: Pesti Lloyd, 1909).

————. *A háború biológiája* (Budapest: Pallas, 1915).

Méray-Horváth, Károly. *Die Physiologie unserer Weltgeschichte und der kommende Tag: Die Grundlagen der Sociologie* (Budapest: S. Politzer, 1902).

Molnós-Kovács, Béla. *Hadigondozó szociálpolitika* (Budapest: Bethlen Gábor, 1918).

Nagy, László. *A háború és a gyermek lelke. Adatok a gyermek értelmi, érzelmi és erkölcsi fejlődéséhez* (Budapest: Eggenberger, 1915).

Nagy, László, and Károly Ballai, eds. *A Gyermektanulmányi Múzeum szervezete és az anyaggyűjtés szabályai* (Budapest: Hungaria, 1910).

Nékám, Lajos. *A Budapesti Kir. M. Tud. Egyetem új klinikája bőr és nemibetegek számára* (Budapest: Universitas, 1915).

————. *A háború és a nemi betegségek* (Budapest: Franklin, 1915).

————. *A nemi betegek kötelező bejelentésének, nyilvántartásának és gyógyításának kérdése* (Budapest: Nemzetvédő Szövetség, 1917).

————. *A venereás betegségek társadalmi kihatása* (Budapest: Nemzetvédő Szövetség, 1917).

Nékám, Lajos, ed. *A nemibajok leküzdésének irányítása*, vol. 1 (Budapest: Nemzetvédő Szövetség, 1918).

Neményi, Bertalan. *A magyar nép állapota és az amerikai kivándorlás* (Budapest: Athenaeum, 1911).

Nemes, Lipót. *A kültelki gyermekek élete és jövője* (Budapest: Hungaria, 1913).

Nemes, Lipót, ed. *A gyermekmentés útjai* (Budapest: Bethlen Gábor, 1918).

Nordau, Max. *Entartung* (Berlin: G. Dunkers, 1892–93).

Offizieller Katalog der Internationalen Hygiene Ausstellung Dresden Mai bis Oktober 1911 (Berlin: Verlag Rudolf Mosse, 1911).

Pach, Henrik. *Magyar munkásegészségügy* (Budapest: Benkő Gyula, 1907).

————. *A Társadalmi Múzeum és a közegészségügy fejlesztése* (Budapest: Pesti Lloyd-Társulat, 1909).

Pályi, Sándor, ed. *A hadi rokkantak, özvegyek és árvák ügye* (Budapest: Franklin, 1916).

Pándy, Kálmán. *Gondoskodás az elmebetegekről más államokban és nálunk* (Gyula: Vértesi Arnold, 1905).

————. *Die Irrenfürsorge in Europa. Eine vergleichende Studie* (Berlin: Georg Reiner, 1908).

Papp, Antal, ed. *Gróf Teleki Pál. Országgyűlési beszédei*, vol. 1 (1917–38) (Budapest: Stádium, 1941).

Pearson, Karl. *The Groundwork of Eugenics* (London: Dulau, 1909).

————. *The Scope and Importance to the State of the Science of National Eugenics* (London: Dulau, 1909).

————. *The Academic Aspect of the Science of National Eugenics* (London: Dulau, 1911).

Pearson, Karl, ed. *The Life, Letters and Labours of Francis Galton*, 3 vols (Cambridge University Press, 1914, 1924, 1930).

Petri, Pál. *A magyar hadigondozás történetének vázlata* (Budapest: Fritz Ármin, 1917).

Pettkó-Szandtner, Aladár. *A magyar hadigondozás: visszapillantás és tájékoztató* (Budapest: Pesti nyomda, 1924).

Pirquet, Clemens, ed. *Volksgesundheit im Krieg*, vol. 1 (Vienna: Hölder Pichler Tempsky, 1926).

Ploetz, Alfred. *Grundlinien einer Rassen-Hygiene. vol. 1. Die Tüchtigkeit unserer Rasse und der Schutz der Schwachen* (Berlin: S. Fischer, 1895).

Plutarch, *Lives* (Theseus and Romulus, Lycurgus and Numa, Solon and Publicola) vol. 1 (London: W. Heinemann, 1959).

Pongrácz, Jenő, ed. *A Forradalmi Kormányzótanács és népbiztosok rendeletei*, 5. vols (Budapest: Franklin, 1919).

Popenoe, Paul, and Roswell Hill Johnson. *Applied Eugenics* (New York: Macmillan, 1920).

Problems in Eugenics: Papers Communicated to the First International Eugenics Congress Held at the University of London, July 24th to 30th, 1912, 2 vols (London: Eugenics Education Society, 1912, 1913).

Proceedings of the First National Conference on Race Betterment (Battle Creek, MI: Gage, 1914).

Prohászka, Ottokár. *A háború lelke* (Budapest: Élet, 1915).

Répay, Dániel, and Károly Ballai, eds. *Az első magyar gyermektanulmányi kongresszus naplója és a vele kapcsolatos kiállítás leírása* (Budapest: Fritz Ármin, 1913).

Réz, Mihály. *A magyar fajpolitika* (Budapest: Kilán Frigyes, 1905).

Rigler, Gusztáv. *A drezdai higiéne-kiállítás* (Budapest: Franklin, 1911).

Roper, Allen G. *Ancient Eugenics* (London: B. H. Blackwell, 1913).

Rosenthal, Max., ed. *Mutterschutz und Sexualreform. Referate und Leitsätze des I. Internationalen Kongress für Mutterschutz und Sexualreform in Dresden 28.–30. September 2011* (Breslau: Preuss & Jünger, 1912).

Ruffy, Pál. *Állami gyermekvédelem és a hadiárvák* (Budapest: Franklin, 1916).

Saleeby, Caleb W. *The Methods of Race-Regeneration* (London: Cassell, 1911).

Scherer, István, ed. *Nemzetközi Gyermekvédő Kongresszus Naplója* (Budapest: Pesti könyvnyomda, 1900).

Schreiber, Emil. *A prostitúció* (Budapest: Pátria, 1917).

Schubert, Paul, ed. *Bericht über den I. Internationalen Kongress für Schulhygiene, Nürnberg 4.–9. April 1904* (Nürnberg: J. L. Schrag, 1904).

Schwimmer, Rosika. *Staatlicher Kinderschutz in Ungarn* (Leipzig: Dietrich, 1909).

Siegmund, Heinrich. *Zur sächsischen Rassenhygiene* (Hermannstadt: Peter Drotleff, 1901).

Siemens, Hermann W. *Die biologischen Grundlagen der Rassenhygiene und der Bevölkerungspolitik* (Munich: J. F. Lehmanns Verlag, 1917).

Somló, Bódog. *Jogbölcseleti előadásai*, 2 vols (Kolozsvár: Sonnenfeld, 1906).

———. *Politika és szociológia. Méray rendszere és prognózisai* (Budapest: Deutsch Zsigmond, 1906).

Sommer, Robert, ed. *Bericht über den II. Kurz mit Kongreß für Familienforschung, Vererbungs- und Regenerationslehre in Gießen vom 9. bis 13 April 1912* (Halle: Carl Marhold, 1912).

Spiller, G., ed. *Papers on Moral Education*, 2nd edn (London: David Nutt, 1909).

———. *Papers on Inter-Racial Problems* (London: P. S. King, 1911).

Stein, Fülöp. *Az alkoholkérdés mai állásáról* (Budapest: Posner Károly, 1910).

Strong, Richard Pearson et al. *Typhus Fever with Particular Reference to the Serbian Epidemic* (Cambridge, MA: Harvard University Press, 1920).

Szabó, Dezső. *Az elsodort falu: regény* (Budapest: Táltos, 1919).

Szana, Sándor. *Az állami gyermekvédelem fejlesztéséről* (Temesvár: Uhrmann Henrik, 1903).

———. *A züllött. Gyermek socialhygiéniájának magyar rendszere* (Budapest: Pallas, 1910).

———. *Irányeszmék a magyar népesedési politikához* (Budapest: Lloyd Társulat, 1916).

———. *A társadalmi egészségügy és az orvos* (Budapest: Franklin, 1917).

Szántó, Manó. *A fakultatív sterilitás kérdéséről* (Budapest: Nagel Ottó, 1905).

Szántó, Menyhért, ed. *Tájékoztató a Társadalmi Múzeum által Győrött 1913. augusztus 14-től szeptember 10-ig rendezett népegészségügyi kiállításról* (Budapest: Társadalmi Múzeum, 1913).

———. *The Museum of Social Service in Buda-Pest* (Budapest: Garden City Press, 1914).

———. *Tájékoztató a Társadalmi Múzeum által Magyaróvárott 1914 február 1-től február 15-ig rendezett népegészségügyi kiállításról* (Budapest: Társadalmi Múzeum, 1914).

———. *Küzdelem a népbetegségek ellen* (Budapest: Társadalmi Múzeum, 1916).

Szántó, Menyhért, and Ernő Tomor, eds. *Had- és Népegészségügyi kiállítás katalógusa* (Budapest: Hornyánszky, 1915).

Szász, Zoltán. *A szerelem* (Budapest: Pallas nyomda, 1913).

Šepilevskij, Evgenij A. *Osnovy i sredstva rasovoj gigieny* (Yuryev: Tipografia K. Matticeia, 1914).

Tauffer, Vilmos. *Az anya- és csecsemővédelem szervezete* (Budapest: Pfeiffer, 1918).

Teleki, Pál. *Szociálpolitika és hadigondozás* (Budapest: Országos Hadigondozó Hivatal, 1918).

———. *The Evolution of Hungary and Its Place in European History* (New York: Macmillan, 1923).

Theilhaber, Felix A. *Das Sterile Berlin. Eine volkswirtschaftliche Studie* (Berlin: Eugen Marquardt, 1913).

Tomor, Ernő. *A socialis egészségtan biológiai alapjai* (Budapest: Singer és Wolfner, 1915).

———. *Belső telepítés és rokkantügy* (Budapest: Globus, 1915).

———. *Neubegründung der Bevölkerungspolitik* (Würzburg: Curt Kabitzsch, 1918).

Torday, Ferenc. *Das staatliche Kinderschutzwesen in Ungarn* (Langensalza: H. Beyer, 1908).

Vadnay, Tibor. *A magyar jövő: közgazdaság- és szociálpolitikai tanulmány*, 2nd edn (Budapest: Athenaeum, 1918).

Vámbéry, Rusztem. *A házasság védelme a büntetőjogban* (Budapest: Politzer, 1901).

Varga, Jenő. *A magyar faj védelme* (Makó: Kovács Antal, 1901).

Woltmann, Ludwig. *Politische Anthropologie. Eine Untersuchung über den Einfluss der Descendenztheorie auf die Lehre von der politische Entwicklung der Völker* (Leipzig: Thüringische Verlags-anstalt, 1903).

Zombory, László. *Gyermekvédelem a háború alatt* (Budapest: Stephaneum nyomda, 1916).

Chapters in Collected Volumes

Angyal, Pál. "Rèforme du caractère des enfants vicieux et déliquants". In Attie G. Dyserinck, ed., *Mémoires sur l'èducation morale* (The Hague: Martinus Nijhoff, 1912), 860–6.

Dalmady, Zoltán. "Betegség és egészség". In Bernát Alexander and Mihály Lenhossék, eds, *Az ember testi és lelki élete, egyéni és faji sajátságai*, vol. 2 (Budapest: Athenaeum, 1907), 461–536.

Dóczi, Imre. "Zum Wiederaufbau der Volkskraft". In Martin Kirchner and Curt Adam, eds, *Der Wiederaufbau der Volkskraft nach dem Kriege: Sitzungsbericht über die gemeinsame Tagung der ärztlichen Abteilungen der Waffenbrüderlichen Vereinigungen Österreichs, Ungarns und Deutschlands in Berlin, 23. bis 26. Januar 1918* (Jena: Gustav Fischer, 1918), 298–301.

Dollinger, Julius. "Schutz und Kräftigung der Jugend". In Martin Kirchner and Curt Adam, eds, *Der Wiederaufbau der Volkskraft nach dem Kriege: Sitzungsbericht über die gemeinsame Tagung der ärztlichen Abteilungen der Waffenbrüderlichen Vereinigungen Österreichs, Ungarns und Deutschlands in Berlin, 23. bis 26. Januar 1918* (Jena: Gustav Fischer, 1918), 181–5.

Donáth, Gyula. "Der physische Rückgang der Bevölkerung in den modernen Culturstaaten, mit besonderer Rücksicht auf Oesterreich-Ungarn". In Zsigmond Gerlóczy, ed., *Jelentés az 1894. szeptember hó 1-től 9-ig Budapesten tartott VIII-ik nemzetközi közegészségi és demografiai congressusról és annak tudományos munkálatairól*, vol. 7 (Budapest: Pesti könyvnyomda, 1896), 605–17.

Gaál, Ernő. "Pleasure of the Mind as a Factor in Education". In Gustav Spiller, ed., *Papers on Moral Education*, 2nd edn (London: David Nutt, 1909), 104–5.

Galton, Francis. "Probability, The Foundation of Eugenics". In Francis Galton, *Essays in Eugenics* (London: Eugenics Education Society, 1909), 73–99.

Geőcze, Sarolta. "Environment and Moral Development". In Gustav Spiller, ed., *Papers on Moral Education*, 2nd edn (London: David Nutt, 1909), 386–9.

———. "Sittliche Erziehung und Nationalen Leben". In Attie G. Dyserinck, ed., *Mémoires sur l'èducation morale* (The Hague: Martinus Nijhoff, 1912), 984–90.

———. "Sittliche und Soziale Bildung in Lehrer-Seminaren". In Attie G. Dyserinck, ed., *Mémoires sur l'èducation morale* (The Hague: Martinus Nijhoff, 1912), 1031–6.

Gerstenberger, H. J. "The Methods of the System Employed in Caring for Institutional Infants Abroad – More Especially in Germany and Austria Hungary". In *American Association for Study and Prevention of Infant Mortality. Transactions of the Fifth Annual Meeting 1914* (Baltimore: Franklin, 1915), 139–50.

Giesswein, Sándor. "A háború szociális problémái". In Márton Lányi, ed., *A háború és a jövő* (Budapest: Grill Károly, 1916), 35–42.

Gruber, Max. "Vererbung, Auslese und Hygiene". In Emil Grósz, ed., *XVIe Congrès International de Medicine. Compte-Rendu. Volume général* (Budapest: Franklin, 1910), 228–52.

Guszman, Jószef. "Zur Frage der Bekämpfung der venerischen Krankheiten". In Wilhelm Manninger, Karl M. John and Josef Parassin, eds, *Erstes Jahrbuch des Kriegsspitals der Geldinstitute in Budapest. Beiträge zur Kriegsheilkunde* (Berlin: Julius Springer, 1917), 691–704.

Hoffmann, Géza. "A bevándorlás és a munkanélküliség az Egyesült Államokban". In Imre Ferenczi, Géza Hoffmann and Imre Illés, *A munkanélküliség és a munkásvándorlások* (Budapest: Benkő, 1913), 64–84.

———. "Künstliche Unfruchtbarkeit nach den Erfahrungen in den Vereinigten Staaten von Nordamerika". In S. Placzek, ed., *Künstliche Fehlgeburt und künstliche Unfruchtbarkeit (Ihre Indikationen, Technik und Rechtslage)* (Leipzig: Georg Thieme, 1918), 413–35.

Jendrassik, Ernő. "Az átöröklődő idegbajok (Elsődleges degeneratiók)". In Árpád Bókay, Károly Kétli and Frigyes Korányi, eds, *A belgyógyászat kézikönyve*, vol. 6 (Budapest: Dobrowsky, 1900), 994–1018.

———. "Verhütung und Bekämpfung der übertragbaren Krankheiten". In Martin Kirchner and Curt Adam, eds, *Der Wiederaufbau der Volkskraft nach dem Kriege: Sitzungsbericht über die gemeinsame Tagung der ärztlichen Abteilungen der Waffenbrüderlichen Vereinigungen Österreichs, Ungarns und Deutschlands in Berlin, 23. bis 26. Januar 1918* (Jena: Gustav Fischer, 1918), 315–16.

Johan, Béla. "Über Schutzimpfstoffe". In Wilhelm Manninger, Karl M. John and Josef Parassin, eds, *Erstes Jahrbuch des Kriegsspitals der Geldinstitute in Budapest. Beiträge zur Kriegsheilkunde* (Berlin: Julius Springer, 1917), 567–80.

Kármán, Mór. "Aufgaben der sittlichen Erziehung". In Gustav Spiller, ed., *Papers on Moral Education*, 2nd edn (London: David Nutt, 1909), 23–9.

———. "Ethisch-Historischen Gesichtpunkte zur Teorie des Lehrplans". In Attie G. Dyserinck, ed., *Mémoires sur l'èducation morale* (The Hague: Martinus Nijhoff, 1912), 593–600.

Karsai, Alexander. "La Ligue Nationale de la Protection de l'Enfance". In Zoltán Bosnyák and L. Ederlsheim-Gyulai, eds, *Le droit de l'enfant abandonné et le système Hongrois de protection de l'enfance* (Budapest: Athenaeum, 1909), 303–26.

Kemény, Ferenc. "Die Erziehung zum Mut". In Attie G. Dyserinck, ed., *Mémoires sur l'èducation morale* (The Hague: Martinus Nijhoff, 1912), 482–7.

———. "Der Interkonfessionalismus ein zwillingsbruder des Internationalismus". In Attie G. Dyserinck, ed., *Mémoires sur l'èducation morale* (The Hague: Martinus Nijhoff, 1912), 194–8.

———. "Physische Kultur und Character-Building". In Attie G. Dyserinck, ed., *Mémoires sur l'èducation morale* (The Hague: Martinus Nijhoff, 1912), 468–76.

Krohne, O. "Bevölkerungspolitische Probleme und Ziele". In Martin Kirchner and Curt Adam, eds, *Der Wiederaufbau der Volkskraft nach dem Kriege: Sitzungsbericht über die gemeinsame Tagung der ärztlichen Abteilungen der Waffenbrüderlichen Vereinigungen Österreichs, Ungarns und Deutschlands in Berlin, 23. bis 26. Januar 1918* (Jena: Gustav Fischer, 1918), 87–94.

Lenhossék, Mihály. "Háború és eugenika". In Márton Lányi, ed., *A háború és a jövő* (Budapest: Grill Károly, 1916), 95–105.

Lombroso, Cesare. "Traitement moral du jeune criminel". In Gustav Spiller, ed., *Papers on Moral Education*, 2nd edn (London: David Nutt, 1909), 216–22.

Lukács, György. "Előszó". In Menyhért Szántó and Ernő Tomor, eds, *Had- és Népegészségügyi kiállítás katalógusa* (Budapest: Hornyánszky, 1915), 5–6.

Melly, József. "A nemi betegségek elterjedettsége, különös tekintettel a székesfővárosra". In Gábor Doros and József Melly, *A nemi betegségek kérdése Budapesten*, vol. 1 (Budapest: Székesfőváros Házinyomdája, 1930), 396–489.

Navratil, Ákos. "Investment and Loans". In G. Spiller, ed., *Papers on Inter-Racial Problems* (London: P. S. King & Son, 1911), 208–11.

Nékám, Lajos. "A háború és a nemi betegségek". In Márton Lányi, ed., *A háború és a jövő* (Budapest: Grill Károly, 1916), 139–53.

Ploetz, Alfred. "Die Begriffe Rasse und Gesellschaft und die davon abgeleiteten Disziplinen". In *Verhandlungen des Ersten Deutschen Soziologentages vom 9.–22. Oktober 1910 in Frankfurt a. M* (Tübingen: J. C. B. Mohr, 1911), 111–47.

———. "Neo-Malthusianism and Race Hygiene". In *Problems in Eugenics: Report of Proceedings of the First International Eugenics Congress*, vol. 2 (London: Eugenics Education Society, 1913), 183–9.

Prohászka, Ottokár. "Ethical Co-operation of Home and School". In Gustav Spiller, ed., *Papers on Moral Education*, 2nd edn (London: David Nutt, 1909), 302–5.

Raschofsky, Wilhelm. "Militärärztliche Organization und Leistungen der Epidemiespitäler der österreichisch-ungarischen Armee". In Clemens Pirquet, ed., *Volksgesundheit im Krieg*, vol. 1 (Vienna: Hölder Pichler Tempsky, 1926), 122–32.

Russell, John. "The Eugenic Appeal in Moral Education". In Attie G. Dyserinck, ed., *Mémoires sur l'éducation morale* (The Hague: Martinus Nijhoff, 1912), 570–4.

Saleeby, C. W. "Eugenic Education or Education for Parenthood". In Attie G. Dyserinck, ed., *Mémoires sur l'èducation morale* (The Hague: Martinus Nijhoff, 1912), 580–3.

Schneller, István. "Die Centrale Pädagogisch-Didaktische Bedeutung der Geschichte in der Mittelschule". In Gustav Spiller, ed., *Papers on Moral Education*, 2nd edn (London: David Nutt, 1909), 148–52.

Schwimmer, Rosika. "Der Stand der Frauenbildung in Ungarn". In Helene Lange und Gertrud Bäumer, eds, *Handbuch der Frauenbewegung*, vol. 3 (Berlin: W. Moeser Buchhandlung, 1902), 191–206.

Slaughter, J. W. "Eugenics and Moral Education". In Gustav Spiller, ed., *Papers on Moral Education*, 2nd edn (London: David Nutt, 1909), 380–3.

Szana, Sándor. "Fürsoge für in öffentliche Versorgung gelangende Säuglinge". In *Bericht über den XIV. Internationalen Kongress für Hygiene und Demographie. Berlin. 23–29 September 1907* (Berlin: August Hirschwald, 1908), 439–52.

———. "A hadiárvák egészségügyi védelme". In Lipót Nemes, ed., *A gyermekmentés útjai* (Budapest: Bethlen Gábor, 1918), 126–38.

Szél, Tivadar. "The Genetic Effects of the War in Hungary". In *A Decade of Progress in Eugenics. Scientific Papers of the Third International Congress of Eugenics held at the American Museum of Natural History, New York, August 21–23, 1932* (Baltimore: Williams & Wilkins, 1934), 249–54.

Tandler, Julius. "Bevölkerungspolitische Probleme und Ziele". In Martin Kirchner and Curt Adam, eds, *Der Wiederaufbau der Volkskraft nach dem Kriege: Sitzungsbericht über die gemeinsame Tagung der ärztlichen Abteilungen der Waffenbrüderlichen Vereinigungen Österreichs, Ungarns und Deutschlands in Berlin, 23. bis 26. Januar 1918* (Jena: Gustav Fischer, 1918), 108–10.

Tauffer, Vilmos. "Die Säuglingssterblichkeit und ihre Bekämpfung in Ungarn". In Martin Kirchner and Curt Adam, eds, *Der Wiederaufbau der Volkskraft nach dem Kriege: Sitzungsbericht über die gemeinsame Tagung der ärztlichen Abteilungen der Waffenbrüderlichen Vereinigungen Österreichs, Ungarns und Deutschlands in Berlin, 23. bis 26. Januar 1918* (Jena: Gustav Fischer, 1918), 114–19.

Teleki, Pál. "Begrüssung". In *Jelentés a német, osztrák, török és magyar bajtársi szövetségek orvosi szakosztályainak és a bolgár kiküldötteknek Budapesten 1918 szeptember hó 21–23-án tartott együttes üléséről* (Budapest: Franklin, 1918), 339–44.

Timon, Ákos. "Theory of the Holy Crown, or the Development and Significance of the Conception of Public Rights of the Holy Crown in the Constitution". In G. Spiller, ed., *Papers on Inter-Racial Problems* (London: P. S. King & Son, 1911), 184–95.

Wagenen, Bleeker van. "Preliminary Report of the Committee of the Eugenic Section of the American Breeders' Association to Study and to Report on the Best Practical Means for Cutting Off the Defective Germ-Plasm in the Human Population". In *Problems in Eugenics*, vol. 1 (London: Eugenics Education Society, 1912), 460–79.

Articles

Angyal, Pál. "A castratio és sterilitatis procuratio", *Jogállam* 1, 5 (1902): 403–9.

Anonymous. "Spencer Herbert levele a Huszadik Századhoz", *Huszadik Század* 1, 1 (1900): 1.

———. "Tudományos publicisztika", *Huszadik Század* 1, 1 (1900): 2–12.

———. "Felolvasó ülés: Kende Mór, A degenerációról", *Egészség* 15, 1 (1901): 14–15.

———. "Magyarország elmebetegügye", *Vasárnapi Ujság* 51, 44 (1904): 745–7.

———. "Az olvasóhoz!" *Fajegészségügy* 1, 1 (1906): 1.

———. "Antialkoholos mozgalom", *Fajegészségügy* 2, 2 (1907): 1–2.

———. "Baleset elleni védelem", *Fajegészségügy* 2, 3 (1907): 1–2.

———. "Munkásvédelem", *Fajegészségügy* 2, 2 (1907): 3–4.

———. "A Magyar Néprajzi Társaság új alapszabályai", *Ethnographia* 21, 6 (1910): 379–82.

———. "The Children of the State in Hungary", *The Lancet* 176, 4553 (1910): 1635.

———. "International Hygiene Exhibition Dresden 1911", *Journal of Hygiene* 10, 1 (1910): 131–4.

———. "A drezdai nemzetközi hygiene kiállítás magyar pavillonja", *Vasárnapi Ujság* 58, 15 (1911): 290–1.

———. "A fajnemesítés (eugénika) problémái", *Huszadik Század* 12, 6 (1911): 694–709.

———. "A fajnemesítés (eugénika) problémái", *Huszadik Század* 12, 7 (1911): 29–44.

———. "A fajnemesítés (eugénika) problémái", *Huszadik Század* 12, 8–9 (1911): 157–70.

———. "A fajnemesítés (eugénika) problémái", *Huszadik Század* 12, 10 (1911): 322–36.

———. "Előadás a fajok harczáról", *Magyar Társadalomtudományi Szemle* 4, 10 (1911): 822–5.

———. "International Hygiene Exhibition, Dresden, 1911. The Opening Ceremonies", *The British Medical Journal* 13, 1 (1911): 1132–3.

———. "Kongresszusok", *A Nő és a Társadalom* 5, 9 (1911): 152.

———. "Ólommérgezés és eugenika", *Ipari Jogvédelem* 1, 14 (1911): 7.

———. "Társulati ügyek", *Huszadik Század* 12, 7 (1911): 100–1.

———. "Eszmecsere a közegészségügy szociális vonatkozásairól", *Magyar Társadalomtudományi Szemle* 5, 1 (1912): 78–9.

———. "Fajok kongresszusa" *Huszadik Század* 13, 8 (1912): 291–3.

———. "Közegészségügyi értekezlet", *Magyar Társadalomtudományi Szemle* 5, 2 (1912): 151–71.

———. "Közegészségügyi értekezlet", *Magyar Társadalomtudományi Szemle* 5, 3 (1912): 236–55.

———. "Közegészségügy és orvoskérdés", *Magyar Társadalomtudományi Szemle* 5, 1 (1912): 57–61.

———. "Közegészségügyi szaktanácskozmány", *Magyar Társadalomtudományi Szemle* 5, 4 (1912): 340–64.

———. "Közegészségügyi szaktanácskozmány", *Magyar Társadalomtudományi Szemle* 5, 6 (1912): 520.

———. "Az olvasókhoz", "*Darwin*" 2, 18 (1913): 1.

———. "Kronika", *Szabadgondolat* 3, 1 (1913): 33–5.

———. "Notices", *Bulletin of the American Academy of Medicine* 15, 5 (1913): 377.

———. "Tiltakozás az eugenika ellen", *Szociálpolitikai Szemle* 3, 12 (1913): 181.

———. "A fajegészségügyi bizottság értekezlete", *Magyar Társadalomtudományi Szemle* 7, 4 (1914): 317–18.

———. "A fajegészségügyi (eugenikai) szakosztály megalakulása", *Magyar Társadalomtudományi Szemle* 7, 2 (1914): 165–72.

———. "A nagyváradi feministák egylete", *A Nő* 1, 2 (1914): 41.

———. "A zsidóság problémája", *Huszadik Század* 15, 4 (1914): 561–6.

———. "Eugenics in Austria", *The Eugenics Review* 5, 4 (1914): 387.

———. "A 'Stefánia Szövetség' megalakulása alkalmából", *Orvosi Hetilap* 59, 28 (1915): 383–4.

———. "Eugenics Research in Bohemia", *Journal of Heredity* 7, 4 (1916): 157.

———. "Eugenika a háború után", *Szociálpolitikai Szemle* 6, 3 (1916): 98–102.

———. "Küzdelem a kéttős morál ellen", *A Nő* 4, 12 (1916): 198.

———. "Küzdelem a nemi bajok ellen", *Szociálpolitikai Szemle* 6, 12 (1916): 344–6.

———. "A meddő Budapest", *Új Nemzedék* 4, 7 (1917): 105–7.

———. "A népegészségügyi nagygyűlés", *Népjóléti Közlöny* 1, 16 (1917): 167.

———. "A népegészségügyi nagygyűlés", *Népjóléti Közlöny* 1, 17 (1917): 171–2.

———. "Ankét a nemi betegségek elleni küzdelemről", *A Nő* 4, 6 (1917): 94–6.

———. "Ankét a nemi betegségek elleni küzdelemről", *A Nő* 4, 7 (1917): 114–16.

———. "Ankét a nemi betegségek leküzdése tárgyában", *A Nő* 4, 5 (1917): 75–8.

———. "Az anya- és csecsemővédő szakosztály alakuló ülése", *Népjóléti Közlöny* 1, 9 (1917): 98–100.

———. "Magyar Fajegészségtani és Népesedéspolitikai Társaság", *Közgazdasági Szemle* 41, 12 (1917): 880.

———. "Magyar Fajegészségtani és Népesedéspolitikai Társaság alakulása", *Szociálpolitikai Szemle* 7, 11–12 (1917): 212–14.

———. "Népegészségügyi országos nagygyűlés", *Magyar Társadalomtudományi Szemle* 8, 1 (1917): 50–6.

———. "A belügyminiszter két népesedéspolitikai körrendelete", *Nemzetvédelem* 1, 1–2 (1918): 82–5.

———. "A fajegészségtani és népesedéspolitikai irodalom néhány olvasásra ajánlható terméke", *Szociálpolitikai Szemle* 8, 1–2 (1918): 24.

————. "A gyermektelenek és kevésgyermeküek adótöbblete és a sokgyermekű család adókedvezménye", *Nemzetvédelem* 1, 1–2 (1918): 85–9.

————. "A Hajdúdorogi gör. kath. egyházmegye fajegészségügyi körrendelete a lelkészhez", *Nemzetvédelem* 1, 1–2 (1918): 89–90.

————. "A kivándorlás, bevándorlás és visszavándorlás szabályozása fajegészségügyi, szempontból", *A Cél* 9, 9 (1918): 562–6.

————. "A kivándorlás, bevándorlás és visszavándorlás szabályozása fajegészségügyi szempontból", *Nemzetvédelem* 1, 3 (1918): 126–32.

————. "A Magyar Fajegészségtani és Népesedéspolitikai Társaság etnográfiai szakosztálya", *Ethnographia* 29, 1–4 (1918): 166.

————. "A 'Nemzetvédelem'", *Nemzetvédelem* 1, 1–2 (1918): 1.

————. "A népegészségügyi országos nagygyűlés munkálatai", *Nemzetvédelem* 1, 3 (1918): 133.

————. "A népegészségügyi szakosztály alkoholizmus elleni alosztályának ülése", *Népjóléti Közlöny* 2, 1 (1918): 2–3.

————. "A svéd házassági törvény fajegészségügyi követelményei", *Nemzetvédelem* 1, 1–2 (1918): 90–1.

————. "A telepítés alapelvei fajegészségügyi szempontból", *Nemzetvédelem* 1, 1–2 (1918): 60–4.

————. "Ankündigungen", *Mitteilungen der Österreichischen Gesellschaft für Bevölkerungspolitik* 1, 1 (1918): 38–9.

————. "Aus anderen Organizationen", *Mitteilungen der Deutschen Gesellschaft für Bevölkerungspolitik* 1, 3 (1918): 29.

————. "Az Országos Hadigondozó Hivatal népesedéspolitikai és fajegészségügyi tevékenysége", *A Cél* 9, 7 (1918): 441–4.

————. "Az Országos Hadigondozó Hivatal népesedéspolitikai és fajegészségügyi tevékenysége", *Nemzetvédelem* 1, 1–2 (1918): 55–60.

————. "Az Országos Hadigondozó Hivatal népesedéspolitikai és fajegészségügyi tevékenysége", *Népjóléti Közlöny* 2, 9 (1918): 78–9.

————. "Decline of Birth Rate in Hungary", *Journal of Heredity* 9, 2 (1918): 85–6.

————. "Fajegészségügy és népesedéspolitika a parlamentben", *Nemzetvédelem* 1, 1–2 (1918): 75–82.

————. "Gyakorlati fajegészségügy", *A Cél* 9, 9 (1918): 566–70.

————. "Gyakorlati fajegészségügy", *Közgazdasági Szemle* 42, 2 (1918): 288.

————. "Gyakorlati fajegészségügy", *Népjóléti Közlöny* 2, 10 (1918): 88.

————. "Lowered Birth Rate in Germany and Hungary", *Journal of Heredity* 9, 6 (1918): 281.

————. "Magyar Fajegészségtani és Népesedéspolitikai Társaság alakulása", *"Darwin"* 7, 1 (1918): 5–7.

————. "Magyar Fajegészségtani és Népességpolitikai Társaság alakulása", *A Cél* 9, 1–2 (1918): 103–6.

————. "Magyar-német-osztrák fajegészségtani nagygyűles", *Népjóléti Közlöny* 2, 17 (1918): 153.

————. "Mit tesznek az amerikaiak a fajegészségügy terén", *Nemzetvédelem* 1, 1–2 (1918): 91–3.

————. "Mitteilungen der Deutschen Gesellschaft für Bevölkerungspolitik", *Nemzetvédelem* 1, 1–2 (1918): 73–5.

————. "Nemzetünk jövőjéért", *Népjóléti Közlöny* 2, 5 (1918): 43.

————. "Nemzetvédelem", *Népjóléti Közlöny* 2, 16 (1918): 147–8.

———. "Vermischte Nachrichten", *Wiener Klinische Wochenschrift* 31, 35 (1918): 977.

———. "Vermischte Nachrichten", *Wiener Klinische Wochenschrift* 31, 37 (1918): 1022.

———. "A nagyvárosi lakáskérdés a fajegészségügy szempontjából", *Nemzetvédelem* 2, 1 (1919): 173–5.

———. "The Consequences of the Division of Hungary from the Standpoint of Eugenics", *Nemzetvédelem* 2, 1 (1919): 162–4.

———. "Fajegészségügyi és népesedéspolitikai pályázatok", *Huszadik Század* 20, 1–2 (1919): 119–20.

———. "Galilei Kör vezetőségének jelentése a Kör januári működéséről", *Szabadgondolat* 9, 3 (1919): 71–3.

———. "Gyakorlati fajegészségügyi intézkedések", *Nemzetvédelem* 2, 1 (1919): 145–54.

———. "Kommunista hygiéne", *Egészség* 33, 3–4 (1919): 71–3.

———. "The Request from the Hungarian Society for Eugenics", *The Eugenics Review* 10, 4 (1919): 223.

———. "Medical Practice in Bolshevist Hungary", *The Lancet* 195, 5029 (1920): 168.

———. "New Eugenics Society in Hungary", *Journal of Heredity* 11, 1 (1920): 41.

———. "Notizen", *Archiv für Rassen- und Gesellschaftsbiologie* 14, 1 (1922): 99.

———. "A fajnemesítés és sterilizáció magyar tudós társaságáról beszél gróf Teleki Pál és Nékám professzor", *Az Est* 24, 207 (13 September 1933): 11.

Apáthy, István. "A nemzetalkotó különbözésről általános élettani szempontból", *Magyar Társadalomtudományi Szemle* 1, 7 (1908): 597–615.

———. "A darwinismus bírálata és a társadalomtan", *Magyar Társadalomtudományi Szemle* 2, 4 (1909): 309–39.

———. "A faj egészségtana", *Magyar Társadalomtudományi Szemle* 4, 4 (1911): 265–79.

———. "Széchenyi István és a nemzeti sajátságok az emberi továbbfejlődés szempontjából", *Magyar Társadalomtudományi Szemle* 5, 10 (1912): 771–90.

———. "Fajegészségügy és fajegészségtan", *Magyar Társadalomtudományi Szemle* 7, 1 (1914): 52–65.

———. "Miért nem népszerü ma a magyar nemzet?" *Új Nemzedék* 1, 11 (1914): 1–3.

———. "Radikálizmus és magyarság", *Új Nemzedék* 3, 46 (1916): 1–4.

———. "Háború és fajegészségügy", *Új Nemzedék* 4, 18–19 (1917): 289–93.

———. "A fajegészségtan köre és feladatai", *Természettudományi Közlöny* 50, 689 (1918): 6–21.

———. "A fajegészségtan köre és feladatai", *Természettudományi Közlöny* 50, 692 (1918) : 81–101.

Balás, Károly. "A neomalthusianismusról", *Magyar Társadalomtudományi Szemle* 1, 6 (1908): 500–32.

———. "A népesedés és a család", *Magyar Társadalomtudományi Szemle* 2, 2 (1909): 166–70.

———. "A népesedés és a szociális kérdés", *Magyar Társadalomtudományi Szemle* 2, 2 (1909): 121–49.

———. "Társadalmi és nemzeti assimilatio", *Magyar Társadalomtudományi Szemle* 3, 1 (1910) : 20–40.

Bálint, Aladár. "Hadegészségügyi kiállítás", *Nyugat* 8, 8 (1915): 460.

Bárczy, István. "A Népjóléti Központ megalakulására", *Népjóléti Közlöny* 1, 1 (1917): 1.

Bársony, János. "Eugenika a háború után", *Orvosi Hetilap* 59, 34 (1915): 451–4.

———. "Eugenetik nach dem Kriege", *Archiv für Frauenkunde und Eugenetik* 2, 2 (1915): 267–75.

Barta, Gábor. "Magyarország közegészségügye", *Egészség* 33, 1–2 (1919): 13–23.

Bartók, György. "Evolutio és sociologia", *Magyar Társadalomtudományi Szemle* 3, 1 (1910) : 1–19.

Bartók, Imre. "A betegségekre való hajlamosságról", *"Darwin"* 7, 12 (1918): 140–2.

Bartucz, Lajos. "Die körpergröße der heutigen Magyaren", *Archiv für Anthropologie* 15, 1 (1917): 44–59.

———. "A társadalmi embertanról", *Természettudományi Közlöny* 51, 723 (1919): 273–81.

Berkovits, René. "A drezdai szociálhigiénei kiállítás néhány tanulsága", *Huszadik Század* 13, 9–10 (1912): 405–13.

———. "Újabb tanulmányok a szociálbiológia köréből", *Huszadik Század* 14, 12 (1913): 610–21.

Bókay, János. "A 'Stefánia-Szövetség' megalakulása alkalmából", *Orvosi Hetilap* 59, 28 (1915): 383–4.

B. R. "Magyar kivándorlás Amerikába", *Huszadik Század* 13, 5 (1912): 649–54.

———. "Protection de l'enfance en Hongrie", *L'Enfant* 21, 204 (1912): 185–6.

Bresztovszky, Ede. "A szociális biztosítás és a faji egészségügy", *Huszadik Század* 18, 4 (1917): 379–82.

Buday, Dezső. "Törvényjavaslat a házasulók kötelező orvosi vizsgálatáról", *Nemzetvédelem* 1, 1–2 (1918): 47–53.

———. "Törvényjavaslat a házasulók kötelező orvosi vizsgálatáról", *Új Nemzedék* 5, 33 (1918): 9–13.

Buro, Péter. "Társadalmi és faji egészségtan", *Egészség* 21, 7 (1907): 195–9.

C. C. "Az országos hadigondozó hivatal népesedéspolitikai és fajegészségügyi tevékenysége", *A Cél* 9, 7 (1918): 441–4.

Christian, Max. "Sozialpolitik und Bevölkerungspolitik", *Nemzetvédelem* 1, 1–2 (1918): 41–7.

Concha, Győző. "A zsidó-kérdésről", *Új Nemzedék* 4, 32 (1917): 505–7.

———. "A zsidó-kérdésről", *Új Nemzedék* 4, 33 (1917): 521–4.

Dalmady, Zoltán. "A testnevelés fajegészségügyi jelentősége", *Nemzetvédelem* 1, 1–2 (1918): 53–5.

Darwin, Leonard. "Eugenics and the War", *The Eugenics Review* 6, 3 (1914): 195–203.

———. "Eugenics during and after the War", *The Eugenics Review* 7, 2 (1915): 91–106.

Dienes, Lajos. "Biometrika", *Huszadik Század* 11, 1–2 (1910): 50–1.

———. "A fajhigiéne a drezdai kiállításon", *A Társadalmi Muzeum Értesítője* 3, 6 (1911): 694–5.

———. "A fajnemesítés biometrikai alapjai", *Huszadik Század* 12, 3 (1911): 291–307.

———. "Eugenika", *A Társadalmi Muzeum Értesítője* 3, 3 (1911): 196–216.

———. "Eugenika", *A Társadalmi Muzeum Értesítője* 3, 4 (1911): 321–36.

Dirner, Gusztáv. "A nemi betegségek és a család", *Huszadik Század* 13, 3 (1912): 346–57.

Donáth, Gyula. "Adatok az idegrendszer hadsérüléses megbetegedéseihez", *Orvosi Hetilap* 59, 23 (1915): 311–13.

———. "Az alkohol fajegészségügyi és népesedéspolitikai szempontból", *Nemzetvédelem* 1, 3 (1918): 98–111.

———. "A világháborúban legyőzött népek fajbiológiai jövője", *Természettudományi Közlöny* 52, 751–4 (1920): 353–61.

Drysdale, C. V. "Some Impressions of Hungary", *The Malthusian* 37, 7 (1913): 49–50.

———. "Some Impressions of Hungary", *The Malthusian* 37, 8 (1913): 57–9.

Entz, Béla. "Küzdelem a fertőző betegségek ellen a háborúban", *Természettudományi Közlöny* 48, 655–6 (1916): 489–512.

Farkas, Géza. "A hadsereg táplálása háborúban", *Természettudományi Közlöny* 46, 612–13 (1914): 673–83.

Fenyvessy, László. "Háború és eugenika", *Huszadik Század* 18, 2 (1917): 165–7.

Flamm, Sándor. "Az egészségügy szocializálása", *Szabadgondolat* 9, 4 (1919): 91–4.

Fülöp, Zsigmond. "Bizonyítékok a szerzett sajátságok örökölhetőségére", *Huszadik Század* 8, 5 (1907): 484–6.

———. "A biogenetikai alaptörvény a modern biológiában", *Huszadik Század* 8, 12 (1907): 730–3.

———. "Eugenika", *Huszadik Század* 11, 9 (1910): 161–76.

———. "Átöröklés és kiválasztás a népek életében", *Huszadik Század* 12, 2 (1911): 239–43.

———. "Az eugenetika követelései és korunk társadalmi viszonyai", *Huszadik Század* 12, 3 (1911): 308–19.

———. "Francis Galton", *Huszadik Század* 12, 3 (1911): 335–6.

———. "Fajgyűlölet és való küzdelem", *Huszadik Század* 16, 1 (1915): 62–4.

———. "Van-e háborús ösztön?" *"Darwin"* 7, 20 (1918): 232–4.

Galton, Francis. "Hereditary Improvement", *Fraser's Magazine* 7, 37 (1873): 116–30.

———. "A Theory of Heredity", *The Journal of the Anthropological Institute of Great Britain and Ireland* 5 (1876): 329–48.

———. "Results Derived from the Natality Table of Kőrösi by Employing the Method of Contours or Isogens", *Nature* 49 (1894): 570–1.

———. "Eugenics: Its Definitions, Scope and Aims", *The American Journal of Sociology* 10, 1 (1904): 1–6.

———. "Eugenics: Its Definition, Scope and Aims", *Sociological Papers* 1 (1904): 45–50.

———. "Entwürfe zu einer Fortpflanzungs-Hygiene", *Archiv für Rassen- und Gesellschaftsbiologie* 2, 5–6 (1905): 812–29.

———. "Eugenics as a Factor in Religion", *Sociological Papers* 2 (1905): 52–3.

———. "Mr Galton's Reply", *Sociological Papers* 2 (1905): 49–51.

———. "Restrictions in Marriage", *Sociological Papers* 2 (1905): 3–13.

———. "Studies in National Eugenics", *Sociological Papers* 2 (1905): 14–17.

———. "A valószínűség, mint az eugenetika alapja", *Huszadik Század* 8, 12 (1907): 1013–29.

Gáll, Géza. "Háború és tuberculosis", *Orvosi Hetilap* 59, 8 (1915): 105–6.

Glueck, Bernard. "Die Rassenhygiene in den Vereinigten Staaten von Nordamerika", *Journal of the American Institute of Law and Criminology* 4, 6 (1914): 934–5.

Goldberger, Márk. "A lelkirokkantak sterilisatiója", *Klinikai Füzetek* 22 (1912): 126–33, 141–8.

Goldscheid, Rudolf. "Kultúrperspektívák", *Huszadik Század* 14, 9 (1913): 177–99.

Grósz, Emil. "A drezdai nemzetközi hygiene kiállítás", *Budapesti Szemle* 39, 414 (1911): 452–6.

Grotjahn, A. "Die Eugenik als Hygiene der Fortpflanzung", *Archiv für Frauenkunde und Eugenetik* 1, 1 (1914): 15–18.

Gruber, Max. "Führt die Hygiene zur Entartung der Rasse?" *Münchener Medizinischer Wochenschrift* 50 (1903): 1713–18, 1781–5.

Hahn, Dezső. "A fertőző nemibetegségek és a háború", *Huszadik Század* 17, 8 (1916): 1–19.

Hajós, Lajos. "Az egészség társadalmi védelme", *Huszadik Század* 2, 8 (1901): 81–90.

———. "Az egészség társadalmi védelme", *Huszadik Század* 2, 9 (1901): 185–92.

———. "Az egészség társadalmi védelme", *Huszadik Század* 2, 10 (1901): 286–92.

———. "Az egészség társadalmi védelme", *Huszadik Század* 2, 11 (1901): 370–5.

Heller, Farkas. "A Magyar Társadalomtudományi Egyesület emlékirata közegészségügyi viszonyaink javítása tárgyában", *Magyar Társadalomtudományi Szemle* 5, 6 (1912): 437–54.

———. "Hadigondozásunk fejlődése", *Magyar Társadalomtudományi Szemle* 9, 1 (1918): 6–21.

Heller, Erik and Farkas Heller. "Gyermekvédelem", *Magyar Társadalomtudományi Szemle* 7, 2 (1914): 149–64.

Heltai, István. "A magyar nemesség kialakulásáról", *Nemzetvédelem* 2, 1 (1919): 159–62.

Herfort, Karl and Arthur Brožek. "Die eugenische Zentrale des Ernestinums", *EOS: Vierteljahresschrift für die Erkenntnis und Behandlung jugendlicher Abnormer* 10, 3 (1914): 161–7.

Herrmann, Antal. "A magyar turista-tanító és Erdély", *Erdély* 22, 5 (1913): 73–4.

Hoffmann, Géza. "Az amerikai magyarság", *Közgazdasági Szemle* 44, 1–2 (1910): 455–85.

———. "Die Einschränkung der Einwanderung in den Vereinigten Staaten von America", *Ungarische Rundschau für Historische und Soziale Wissenschaften* 1, 1 (1912): 104–15.

———. "Die Regelung der Ehe im Rassenhygienischen Sinne in den Vereinigten Staaten von Nordamerika", *Archiv für Rassen- und Gesellschaftsbiologie* 9, 6 (1912): 730–61.

———. "Akkulturation unter den Magyaren in Amerika", *Zeitschrift für Sozialwissenschaft* 4, 5 (1913): 309–25.

———. "Akkulturation unter den Magyaren in Amerika", *Zeitschrift für Sozialwissenschaft* 4, 6 (1913): 393–407.

———. "Anregung zur Einführung von Gesundheitszeugnissen in Ungarn", *Archiv für Rassen- und Gesellschaftsbiologie* 10, 5 (1913): 696.

———. "Ausschüsse für Rassenhygiene in Ungarn", *Archiv für Rassen- und Gesellschaftsbiologie* 10, 6 (1913): 830–1.

———. "Az első állami fajhygieniai hivatal", *Szociálpolitikai Szemle* 3, 11 (1913): 161–2.

———. "Die Einkinderehe in der ungarischen Landbevölkerung", *Archiv für Rassen- und Gesellschaftsbiologie* 10, 6 (1913): 813–14.

———. "Der nächste Internationale Kongreß für Rassenhygiene soll im Sept. 1915 in Neuyork stattfinden", *Archiv für Rassen- und Gesellschaftsbiologie* 10, 6 (1913): 831.

———. "Rassenhygienische Jahresversammlung in den Vereinigten Staaten von Nordamerika", *Archiv für Rassen- und Gesellschaftsbiologie* 10, 6 (1913): 829–30.

———. "A fajegészségtan (eugenika) és a nő", *A Nő* 1, 8 (1914): 159–61.

———. "A fajegészségügyi (eugenikai) társaságok és működésük", *Magyar Társadalomtudományi Szemle* 7, 5 (1914): 350–6.

———. "A fajegészségtan irodalma", *Magyar Társadalomtudományi Szemle* 7, 3 (1914): 221–4.

———. "A népesség csökkenő szaporodása és a fajegészségügy (eugenika)", *Közgazdasági Szemle* 38, 11 (1914): 526–7.

———. "A születések csökkenő száma és a német fajegészségtani (eugenikai) társaság", *Magyar Társadalomtudományi Szemle* 7, 7 (1914): 560–3.

———. "Bibliographie der Rassenhygiene", *Archiv für Rassen- und Gesellschaftsbiologie* 11, 1 (1914): 131–2.

———. "Eugenics in Germany: Society of Race Hygiene Adopts Resolution Calling for Extensive Program of Positive Measures to Check Decline in Birthrate", *Journal of Heredity* 5, 10 (1914): 435–6.

———. "Eugenika", *Magyar Társadalomtudományi Szemle* 7, 2 (1914): 91–106.

———. "Környezet vagy átöröklés?" *Magyar Társadalomtudományi Szemle* 7, 4 (1914): 295–7.

———. "Rassenhygiene im Lehrplan der nordamerikanischen Universitäten", *Archiv für Rassen- und Gesellschaftsbiologie* 11, 2 (1914): 281.

———. "Rassenhygiene und Familienforschung. Leitsätze der Deutschen Gesellschaft für Rassenhygiene über die Geburtenfrage", *Monatsblatt der Kais. Kön. Heraldischen Gesellschaft 'Adler'* 7, 43 (1914): 373–5.

———. "Rassenhygienische Gedanken bei Platon", *Archiv für Rassen- und Gesellschaftsbiologie* 11, 2 (1914): 174–83.

———. "Die rassenhygienischen Gesetze des Jahres 1913 in den Vereinigten Staaten von Nordamerika", *Archiv für Rassen- und Gesellschaftsbiologie* 11, 1 (1914): 21–32.

———. "Das Sterilisierungs program in den Vereinigten Staaten von Nordamerika", *Archiv für Rassen- und Gesellschaftsbiologie* 11, 2 (1914): 184–92.

———. "Eugenics in Hungary", *Journal of Heredity* 7, 3 (1916): 105.

———. "Fajegészségtan és eugenika", *Természettudományi Közlöny* 48, 653–4 (1916): 450–4.

———. "Fajegészségtan és népesedési politika", *Természettudományi Közlöny* 48, 659–60 (1916): 617–21.

———. "Háború és fajhygiéne", *A Cél* 7, 5 (1916) : 428–37.

———. "Nyelvében él a nemzet", *A Cél* 7, 6 (1916): 221–3.

———. "Race Hygiene in Germany", *Journal of Heredity* 7, 1 (1916): 32.

———. "A faji öntudat ébredése a német ifjúságban", *A Cél* 8, 6 (1917): 345–50.

———. "A magyar nép elpóriasodása", *Természettudományi Közlöny* 49, 683–4 (1917): 708–9.

———. "A német fajegészségtani társaság működése a háború kitörése óta", *Magyar Társadalomtudományi Szemle* 8, 1 (1917): 57–60.

———. "Drohende Verflachung und Einseitigkeit rassenhygienischer Bestrebungen in Deutschland", *Archiv für Rassen- und Gesellschaftsbiologie* 12, 3–4 (1917): 343–8.

———. "Geburtenbeschränkung in Nordamerika", *Archiv für Rassen- und Gesellschaftsbiologie* 12, 3–4 (1917): 398–400.

———. "Neuere unfruchtbarmachungen Minderwertiger in den Vereinigten Staaten von Nordamerika", *Öffentliche Gesundheitspflege* 12, 6 (1917): 312–16.

———. "Race Hygiene in Germany", *Journal of Heredity* 8, 3 (1917): 112.

———. "Rassenhygiene in America", *Archiv für Rassen- und Gesellschaftsbiologie* 12, 2 (1917): 249–53.

———. "Rassenhygiene und Fortpflanzungshygiene (Eugenik)", *Öffentliche Gesundheitspflege* 12, 1 (1917): 1–11.

———. "Rassenhygienische Anträge im ungarischen Abgeordnetenhause", *Archiv für Rassen- und Gesellschaftsbiologie* 12, 2 (1917): 253.

———. "Rassenhygienische Unfruchtbarmachung", *Archiv für Rassen- und Gesellschaftsbiologie* 12, 3–4 (1917): 400–1.

———. "Über die Begriffe Rassenhygiene und Fortpflanzungshygiene (Eugenik)", *Münchener Medizinische Wochenschrift* 44, 4 (1917): 110–11.

———. "Über die Begriffe Rassenhygiene und Fortpflanzungshygiene (Eugenik)", *Archiv für Soziale Hygiene und Demographie* 12, 1–2 (1917): 49–55.

———. "A fajegészségtan, eugenika és népesedéspolitika rendszere", *Nemzetvédelem* 1, 1–2 (1918): 4–13.

———. "A fajegészségtan irodalmának néhány terméke", *Nemzetvédelem* 1, 1–2 (1918): 66–73.

———. "A sokgyermekű család közvetlen gazdasági támogatása fajegészségügyi szempontból", *Közgazdasági Szemle* 42, 12 (1918): 610–19.

———. "Fajegészségtan és erkölcs", *Magyar Kultúra* 6, 12 (1918): 257–62.

———. "Gyakorlati fajegészségügy", *"Darwin"* 7, 12 (1918): 140–2.

———. "Milyen fajegészségtani tanulmányokra lenne nálunk szükség?" *Nemzetvédelem* 1, 3 (1918): 132–3.

———. "Rassenhygiene in Ungarn", *Archiv für Rassen- und Gesellschaftsbiologie* 13, 1 (1918): 55–67.

———. "Rassenhygiene und Fortpflanzungshygiene (Eugenik)", *Zeitschrift für Sexualwissenschaft* 4 (1918): 44–6.

———. "A fajegészségtan gondolatköréből", *A Cél* 10, 1 (1919): 12–16.

———. "A születések differenciális arányszáma", *"Darwin"* 8, 1 (1919): 5–6.

———. "Miért van szükség sok gyermek születésére?" *Huszadik Század* 20, 4 (1919): 209–16.

———. "Eugenics in the Central Empire since 1914", *Social Hygiene* 7, 3 (1921): 285–96.

Hollós, István. "Az öröklésnek az elmebetegségek fellépésére való jelentősége", *Budapesti Orvosi Újság*, 2, 20 (1904): 420–3.

Holmes, S. J. "The Decadence of Human Heredity", *The Atlantic Monthly* 114, 3 (1914): 302–8.

Horváth, Lajos. "Az átöröklés törvényei", *Magyar Társadalomtudományi Szemle* 4, 1 (1911): 57–67.

Illés, József. "Magyar házassági vagyonjog és önálló nemzeti fejlődés", *Huszadik Század* 1, 11 (1900): 381–6.

Jankulov, Boriszláv. "A talaj és a faj viszonya a történethez", *Magyar Társadalomtudományi Szemle* 5, 4 (1912): 270–309.

———. "A talaj és a faj viszonya a történethez", *Magyar Társadalomtudományi Szemle* 5, 7 (1912): 533–51.

———. "A faj jelentősége a történelemben", *Magyar Társadalomtudományi Szemle* 6, 8 (1913): 577–611.

Jászi, Oszkár. "Lamarckisták és Darwinisták", *Huszadik Század* 1, 8 (1900): 153–5.

———. "Tíz év", *Huszadik Század* 11, 1–2 (1910): 1–10.

Jendrassik, Ernő. "Mi az oka annak, hogy több fiú születik, mint leány? És más öröklési problémákról", *Magyar Orvosi Archivum* 12 (1911): 331–43.

Kadosa, Marcel. "Népjóléti Minisztérium", *Szociálpolitikai Szemle* 7, 8 (1917): 129–31.

Kaffka, Margit. "Szász Zoltán: *A szerelem*", *Nyugat* 6, 15 (1913): 197–9.

Kammerer, Pál. "A szerzett tulajdonságok átöröklése és annak szociológiai jelentősége", *Huszadik Század* 14, 3 (1913): 305–24.

———. "Ernst Haeckel", *Huszadik Század* 15, 2 (1914): 137–50.

———. "Nacionalizmus és biológia", *Huszadik Század* 15, 5 (1914): 625–31.

Kende, Zsigmond. "Közegészségügyi feladatok", *Szabadgondolat* 4, 6 (1914): 164–8.

Kenedi, Géza. "Feminismus és biológia", *Magyar Társadalomtudományi Szemle* 2, 3 (1909): 218–34.

Kemény, Ferenc. "Nemzetköziség és felekezetköziség", *Szociálpolitikai Szemle* 2, 12 (1912): 178–80.

———. "Az egészségügy a Tanácsköztársaságban", *Egészség* 33, 5–6 (1919): 121–44.

Kollarits, Jenő. "Beiträge zur Biologie des Krieges mit besonderer Berücksichtigung des Ungartums", *Archiv für Rassen- und Gesellschaftsbiologie* 25, 1 (1931): 19–41.

Kovács, Alajos. "A magyarság és a nemzetiségek erőviszonyai", *Magyar Társadalomtudományi Szemle* 6, 3 (1913): 175–98.

———. "Az egyke elterjedése", *Magyar Társadalomtudományi Szemle* 6, 6 (1913): 419–34.

Kozáry, Gyula. "Az Eugenics kérdése (Lapok az Eugenics kérdésenek történetéből)", part 1, *Athenaeum* 15, 2 (1906): 242–8.

———. "Az Eugenics kérdése (Lapok az Eugenics kérdésének történetéből)", part 2, *Athenaeum* 15, 3 (1906): 351–7.

———. "Az Eugenics kérdése (Lapok az Eugenics kérdésének történetéből)", part 3, *Athenaeum* 15, 4 (1906): 458–64.

———. "Az Eugenics kérdése (Lapok az Eugenics kérdésének történetéből)", part 4, *Athenaeum* 16, 1 (1907): 59–69.

Laky, Dezső. "A Magyar Fajegészségtani és Népesedéspolitikai Társaság munkaköre", *Nemzetvédelem* 1, 1–2 (1918): 34–40.

Lantos, Emil. "Az anyaság és a csecsemő védelme", *Budapesti Orvosi Újság* 21, 2 (1908): 19–21.

Lengyel, Sándor. "Eugenika és sterilizáció", *Magyar Kultúra* 7, 42 (1934): 289–95.

Lenhossék, Mihály. "A fejlődés mibenléte" (part 1), *Természettudományi Közlöny* 46, 614–15 (1914): 721–740.

———. "A fejlődés mibenléte" (part 2), *Természettudományi Közlöny* 46, 616 (1914): 761–71.

———. "A háború és a létért való küzdelem tétele", *Természettudományi Közlöny* 47, 619–20 (1915): 91–5.

———. "A magyarság anthropológiai vizsgálata", *Természettudományi Közlöny* 47, 639–40 (1915): 757–83.

———. "Über Anthropologie im allgemeine and über die Aufgaben Ungarns auf dem Gebiete der Anthropologie", *Archiv für Anthropologie* 15, 2 (1917): 142–54.

————. "A népfajok és az eugenika", *Természettudományi Közlöny* 50, 695–6 (1918): 213–41.

————. "Európa lakosságának eredete és fajbeli összetétele", *Természettudományi Közlöny* 50, 697–8 (1918): 269–93.

————. "Az emberi test alkotásának néhány törvényszerűsége", *Természettudományi közlöny* 44, 799–802 (1922): 321–34.

Lenz, Fritz. "Hoffmann, Géza. Die Rassenhygiene in den Vereinigten Staaten von Nordamerika", *Archiv für Rassen- und Gesellschaftsbiologie* 10, 1–2 (1913): 249–52.

————. "Aus der Gesellschaft für Rassenhygiene", *Archiv für Rassen- und Gesellschaftsbiologie* 11, 4 (1915): 561–2.

Lukács, József. "A jövendő nemzedék védelme", *Szociálpolitikai Szemle* 5, 7–9 (1915): 193–6.

Lukács Paula (Hoffmann Gézáné). "Az eugenika oktatása az Egyesült Államokban", *Huszadik Század* 14, 4 (1913): 519–20.

————. "Anyák 'nyugdíja' az északamerikai egyesült államokban", *Huszadik Század* 14, 7–8 (1913): 100–3.

M. M. E. "Fajegészségügy, fajjavítás – anyák nélkül", *A Nő* 1, 3 (1914): 54–55.

Madzsar, József. "A szaporodás higiénéje", *Huszadik Század* 7, 4 (1906): 336–7.

————. "Gyakorlati eugenika", *Huszadik Század* 11, 1–2 (1910): 115–17.

————. "Fajromlás és fajnemesítés", *Huszadik Század* 12, 2 (1911): 145–60.

————. "Az anya- és csecsemővédelem a háborúban", *Városi Szemle* 7, 10–11 (1914): 726–30.

————. "A jövő nemzedék védelme és a háború", *Huszadik Század* 17, 1 (1916): 1–22.

Mailáth, József. "A magyar faj regeneratiója és a népbetegségek", *Magyar Társadalomtudományi Szemle* 9, 1 (1918): 1–5.

Manninger, Vilmos. "A háború hatása a szellemi munkára", *Természettudományi Közlöny* 49, 681–2 (1917): 635–9.

Marosi, Arnold. "Az átöröklés és az ember", *Katholikus Szemle* 33, 3 (1919): 222–32.

————. "Az átöröklés és az ember", *Katholikus szemle* 33, 4 (1919): 263–81.

Marschalkó, Tamás. "Háború és venereás bántalmak", *Orvosi Hetilap* 59, 26 (1915): 347–50.

————. "Háború és venereás bántalmak", *Orvosi Hetilap* 59, 27 (1915): 365–9.

————. "Háború és venereás bántalmak", *Orvosi Hetilap* 59, 28 (1915): 381–3.

Mattyasovszky, Miklós. "A népesedés és a család", *Magyar Társadalomtudományi Szemle* 2, 1 (1909): 65–72.

Méhely, Lajos. "A háború biológiája", *Természettudományi Közlöny* 48, 617–618 (1915): 3–28.

Meisel-Hess, Grete. "Népesedési és erkölcsproblémák", *Huszadik Század* 12, 11 (1911): 459–63.

Nagy, Dénes. "Az emberfajok első egyetemes kongresszusa", *Huszadik Század* 12, 10 (1911): 345–8.

————. "Háború és eugenika", *Huszadik Század* 17, 1 (1916): 77–9.

Nékám, Lajos. "Néhány sürgős teendőnk a fajegészségügy terén", *Nemzetvédelem* 1, 1–2 (1918): 14–23.

Oberholzer, Emil. "Kastration und Sterilisation von Geisteskranken in der Schweiz", *Juristisch-psychiatrische Grenzfragen* 8, 1–3 (1911): 25–144.

Obonyu, Endre. "Közegészségügyi és orvoskérdés", *Magyar Társadalomtudományi Szemle* 5, 1 (1912): 57–61.

P[odach]. E[rich]. F[rigyes]. "Állami eugenika", *Huszadik Század* 19, 11 (1918): 265–6.

Paizs, Lajos. "A nőkérdés és a nőnevelés", *Magyar Társadalomtudományi Szemle* 1, 8 (1908): 751–68.

Pártos, Zoltán. "Az egészségügy szocializálása", *Szabadgondolat* 9, 2 (1919): 44–8.

Patz, Ulrich. "Egészségtan és fajegészségtan", *Nemzetvédelem* 1, 4 (1919): 64–5.

Pearson, Karl. "On the Laws of Inheritance in Man, I", *Biometrika* 2, 4 (1903): 357–462.

———. "On the Laws of Inheritance in Man, II", *Biometrika* 3, 2–3 (1903): 131–90.

Petri, Pál. "Háborús tapasztalatok és a szocializmus", *Nemzetvédelem* 2, 1 (1918): 154–9.

Péterfi, Tibor. "A szociogenezis", *Huszadik Század* 11, 1–2 (1910): 69–73.

Pikler, Julius. "Über die biologischen Funktion des Bewußtseins", *Archiv für Rassen- und Gesellschaftsbiologie* 8, 2 (1911): 227–230.

Ploetz, Alfred. "Die Begriffe Rasse und Gesellschaft und die davon abgeleiteten Disciplinen", *Archiv für Rassen- und Gesellschaftsbiologie* 1, 1 (1904): 1–27.

Rentoul, Robert R. "Proposed Sterilization of Certain Mental Degenerates", *American Journal of Sociology* 12, 3 (1906): 319–27.

Révész, Béla. "Der Einfluss des Alters der Mutter auf die körperhöhe", *Archiv für Anthropologie* 4, 3 (1906): 160–7.

———. "Rassen und Geisteskrankheiten. Ein Beitrag zur Rassenpathologie", *Archiv für Anthropologie* 6, 3 (1907): 180–7.

Rónai, Zoltán. "Cesare Lombroso antroposzociológiája", *Huszadik Század* 10, 12 (1909): 453–63.

Ruffy, Pál. "Az állami gyermekvédelem", *A Gyermekvédelem Lapja* 9, 2 (1914): 339–94.

Schaffer, Károly. "A háború és az idegrendszer", *Budapesti Szemle* 43, 459 (1915): 396–407.

Schallmayer, Wilhelm. "Eugenik, ihre Grundlagen und ihre Beziehungen zur kulturellen Hebung der Frau", *Archiv für Frauenkunde und Eugenetik* 1, 3 (1914): 271–91.

———. "Hygiene der Erbverfassung und Hygiene der Erscheinungsbildes", *Nemzetvédelem* 1, 3 (1918): 112–22.

Schmidt, Béla. "A tüdővész elleni védekezés rationalis útja", *Fajegészségügy* 2, 1 (1907): 1–3.

———. "A fertőző nemibetegségek és a háború", *Huszadik Század* 17, 10–11 (1916): 286–8.

Schuster, Edgar. "The First International Eugenics Congress", *The Eugenics Review* 4, 3 (1912): 223–56.

———. "Von Hoffmann Geza", *Die Rassenhygiene in den Vereinigten Staaten von Nordamerika*", *The Eugenics Review* 5, 3 (1913): 279.

Schweisheimer, Wilhelm. "Bevölkerungsbiologische Bilanz des Krieges 1914/19", *Archiv für Rassen- und Gesellschaftsbiologie* 13, 2/4 (1920): 176–93.

Schwimmer, Rosika. "Az anyaság védelme", *A nő és társadalom* 2, 5 (1908): 73–6.

Sebestyén, Gyula. "Turáni Társaság", *Ethnographia* 21, 6 (1910): 324–6.

Seeck, Otto. "Hellasz és Róma bukásáról", *Nemzetvédelem* 2, 1 (1919): 168–72.

Somló, Bódog. "Kidd Benjámin áldarwinizmusa", *Huszadik Század* 5, 1 (1904): 29–37.

Somogyi, Zsigmond. "A háború és a fertőző betegségek", *Természettudományi Közlöny* 46, 18–19 (1914): 652–7.

Szabó, Dezső. "A magyar zsidóság organikus elhelyezkedése", *Huszadik Század* 15, 3 (1914): 340–7.

Száhlender, Lajos. "A háborúban használható fojtó, mérges és könnyezést fakasztó gázokról", *Természettudományi Közlöny* 48, 643–4 (1916): 120–1.

Szana, Sándor. "Die Pflege kranker Säuglinge in Anstalten", *Wiener Klinische Wochenschrift* 18, 2 (1904): 46–52.

———. "A csecsemővédelem hatása a csecsemőhalálozásra", *Magyar Társadalomtudományi Szemle* 4, 10 (1911): 780–801.

———. "Krieg und Bevölkerung", *Wiener Klinische Wochenschrift* 29, 16 (1916): 485–9.

———. "A meddő Temesvár", *Népjóléti Közlöny* 1, 4 (1917): 53.

Szántó, Menyhért. "The Labour Insurance Law in Hungary", *The Economic Journal* 18, 72 (1908): 631–6.

Szász, Zsombor. "Az első nemzetközi fajegészségügyi (eugenikai) congressus", *Magyar Társadalomtudományi Szemle* 5, 8 (1912): 650–7.

Szegvári, Sándorné. "A nő zsenialitása", *A nő és társadalom* 5, 10 (1911): 164–5.

———. "Anyaság", *A nő és a társadalom* 7, 3 (1913): 48–50.

———. "Fajpusztulás", *A Nő* 4, 11 (1917): 178–9.

Szirmay, Oszkárné. "Feminizmus és anyaság", *A Nő* 1, 9 (1914): 186.

———. "Háború és anyavédelem", *A Nő* 2, 1 (1915): 9.

Tandler, Julius. "Krieg und Bevölkerung", *Wiener Klinische Wochenschrift* 29, 15 (1916): 445–52.

Teleki, Pál. "Az emberi természetről", *Huszadik Század* 5, 9 (1904): 241–3.

———. "Politikai embertan", *Huszadik Század* 5, 7 (1904): 73–5.

———. "Társadalomtudomány biológiai alapon", *Huszadik Század* 5, 4 (1904): 318–23.

———. "Körlevél az eugenikáról", *Szociálpolitikai Szemle* 7, 9–10 (1916): 169–71.

———. "A rokkantak szaktanácsadói", *Népjóléti Közlöny* 2, 1 (1918): 1.

———. "Die Sicherung der Lebenslage kriegsbeschädigter", *Nemzetvédelem* 1, 1–2 (1918): 24–33.

Tomor, Ernő. "Had- és népegészségügyi kiállítás", *Orvosi Hetilap* 58, 19 (1915): 264–5.

———. "Die Grundirrtümer der heutigen Rassenhygiene", *Würzburger Abhandlungen aus dem Gesamtgebiet der praktischen Medizin* 20, 4–5 (1920): 67–89.

Torday, Emil. "Primitive Eugenics", *The Mendel Journal* 4, 3 (1912): 30–6.

Torday, Ferenc. "A jövő nemzedék egészségének biztosításáról", *Egészség* 22, 6 (1908): 166–81.

Tornai, József. "Adatok a háborús haemothorax kórtanához és orvoslásához", *Orvosi Hetilap* 59, 24 (1915): 326–9.

Török, Aurél. "Versuch einer systematischen Charakteristik des Kephalindex", *Archiv für Anthropologie* 4, 3 (1906): 110–29.

Tuszkai, Ödön. "Cultura és hygiene", *Magyar Társadalomtudományi Szemle* 6, 1 (1913): 51–7.

Vacher de Lapouge, Georges. "The Fundamental Laws of Anthropo-sociology", *The Journal of Political Economy* 6, 1 (1897): 54–92.

Vajda, Mihály. "Társadalmi problémák", *Szociálpolitikai Szemle* 4, 3 (1914): 135–46.

Vámos Jenő. "Az alkalmazott eugenika", *Huszadik Század* 12, 12 (1911): 571–7.

Watson, Amey Eaton. "Recent Books on Human Heredity", *The Journal of Heredity* 5, 9 (1914): 373.

Webb, Sidney. "Eugenics and the Poor Law", *The Eugenics Review* 2, 3 (1910): 233–41.

Webb, Sidney, and Beatrice Webb. "Szegénység és fajszépség", *Szociálpolitikai Szemle* 2, 12 (1912): 180–2.

Weber, Max. "On Race and Society", *Social Research* 38, 1 (1978): 30–41.

Zsoldos, Benő. "A szellemi degeneratio társadalmi korlátozása", *Magyar Társadalomtudományi Szemle* 5, 10 (1912): 831–3.

Index

Page numbers in **bold** refer to figures, page numbers in *italic* refer to tables.

Lightning Source UK Ltd.
Milton Keynes UK
UKOW06f0855040816
279868UK00019B/116/P